PUMP USERS HANDBOOK

4th Edition

PUMP USERS HANDBOOK

4th Edition

R. Rayner

ISBN 1 85617 216 3

Copyright © 1995 ELSEVIER SCIENCE PUBLISHERS LTD

All rights reserved
This book is sold subject to the condition that it shall not by way of trade or otherwise be resold, lent, hired out, stored in a retrieval system, reproduced or translated into a machine language, or otherwise circulated in any form of binding or cover other than that in which it is published, without the Publisher's prior consent and without a similar condition including this condition being imposed on the subsequent purchaser.

Other books in this series include:
*Hydraulic Handbook
Seals and Sealing Handbook
Handbook of Hose, Pipes, Couplings and Fittings
Handbook of Power Cylinders, Valves and Controls
Pneumatic Handbook
Pumping Manual
Pump User's Handbook
Submersible Pumps and their Applications
Centrifugal Pumps
Handbook of Valves, Piping and Pipelines
Handbook of Fluid Flowmetering
Handbook of Noise and Vibration Control
Handbook of Mechanical Power Drives
Industrial Fasteners Handbook
Handbook of Cosmetic Science and Technology
Geotextiles and Geomembranes Manual
Reinforced Plastics Handbook
Leak-free Pumps & Compressors*

Published by
Elsevier Advanced Technology
The Boulevard, Langford Lane, Kidlington, Oxford OX5 1GB, UK
Tel 010 44 (0) 1865-843000
Fax 010 44 (0) 1865-843010

Printed in Great Britain by BPC Wheatons Ltd, Exeter

Preface

The Pump Users Handbook places emphasis on the importance of correct interpretation of pumping requirements, both by the user and the supplier.

Completely reworked to incorporate the latest in pumping technology, this practical handbook should enable the reader to understand the principles of pumping, hydraulics and fluids and define the various criteria necessary for pump and ancillary selection.

Sadly, just before completion of this book the author, Ray Rayner, passed away. Elsevier Advanced Technology has endeavoured to complete this work to Ray's very high standards and would like to thank his family and former colleagues for any assistance they have given.

We hope that the Pump Users Handbook will live up to Ray's expectations and prove an invaluable aid when ordering pump equipment and in the recognition of fundamental operational problems.

The Publisher

Acknowledgement

The author acknowledged the time and effort that Robert Crawford, retired Engineering Manager of Peerless Pump Co. put into the review of this handbook. His suggestions and critique were very appropriate and current.

A unique series of dictionaries

- these books should be used by EVERY company seriously interested in doing business in the pumping sector in Europe - now the largest trading block in the world

- expertly checked for each and every entry for the 5 major European languages covered: English, French, German, Italian, Spanish

- For full details of EC legislation, standards and trading laws these titles are ESSENTIAL for ALL manufacturers and major specifiers of pumps and their components

EUROPUMP TERMINOLOGY SERIES

- Pump Applications ISBN 0 85461 071 5
- Pump Names ISBN 0 85461 089 8
- Component Names ISBN 0 85461 099 5

ORDERS & SUBSCRIPTION ENQUIRIES TO

Elsevier Advanced Technology
P O Box 150, Kidlington,
Oxford OX5 1AS, UK,
Tel: +44 (0)865 843848
Fax: +44 (0)865 843971

OR

In North America
Elsevier Advanced Technology
660 White Plains Road, Tarrytown,
NY 10591-5153, USA
Tel: +1 (914) 333 2458

REFUND GUARANTEE
Remember our money back guarantee. Should you wish to cancel your purchase, simply return the book within 28 days and we will refund your payment.

ELSEVIER
ADVANCED
TECHNOLOGY

ORDER FORM
☐ Please send me _____ copy/ies of the Pump Applications @ £100/US$160
[PIP61+B1]

☐ Please send me _____ copy/ies of Pump Names @ £69/US$110
[PIP62+B1]

☐ Please send me _____ copy/ies of Component Names @ £160/US$225
[PIP63+B1]

Payment
☐ Please find payment enclosed (please make cheques payable to Elsevier)
☐ Please charge my credit card:
    Access/Mastercard/Visa/Barclaycard/Eurocard/American Express
Card No. _____ Card Name _____
Expiry Date _____ Date _____
Signature _____
☐ Please invoice me. Company Purchase Order No. _____

Delivery Address
Name _____ Position _____ [JT:]
Organization _____
Address _____
Town _____ State _____
Post/Zip Code _____ Country _____
Tel _____ Fax _____
Nature of Business _____ [SIC:]

Sintered Silicon Carbide for Mechanical Seal Faces

EKasic® sintered silicon carbide components with excellent resistance to wear and corrosion in environmentally safe pumps.
A variety of grades available to optimize your pump performance.

For further information please contact us:
Elektroschmelzwerk Kempten GmbH
Kempten Plant
P.O. Box 1526, D-87405 Kempten, Germany
Phone +831 56 18 - 2 22 · Fax +831 56 18 - 3 57

REG. NO 3423-01

A company of
Wacker-Chemie

When Positive Displacement Pumps Just Will Not Do!

SUNDYNE® SUNFLO®

The Low Flow, High Head Specialists.

Sundstrand Fluid Handling and Sundstrand International S. A. have been serving the chemical, petrochemical, hydrocarbon and other process industries with single stage centrifugal, integrally geared, critical duty pumps since 1962. Technological advancements with speed, inducers and materials have established the Sundyne and Sunflo pumps as industry standards for reliability with the toughest process applications.

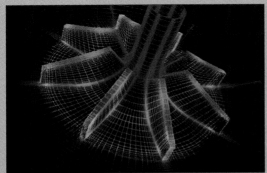

Today's marketplace demands the routine use of advanced design and analysis techniques to reinforce continuous improvement initiatives and to ensure the utmost product quality.

API Duties

The combination of low specific speed hydraulics with operational speeds to 25,000 RPM yields a design which transcends the common line between positive displacement and kinetic pumps. The result is superior low flow stability and multistage performance with the simplicity of a single stage centrifugal pump design.

Unparalleled **low NPSHR** performance through the continual refinement of a family of high suction specific speed inducers.

Medium Duties

Other Products Offered:
- HMD and KONTRO magnetic drive Seal/Less pumps
- SINE food grade pumps
- SUNDYNE/SUNFLO high speed centrifugal compressors
- SUNDYNE two stage, 2500 HP, integrally geared pumps
- SUNDYNE canned motor pumps

Performance:
- Flows to 225 m3/hr (1000 gpm)
- Heads to 1900 meters (6000 ft)
- Pressures to 150 barG (2160 psig)

ISO 9001 Quality Systems

Sundstrand Fluid Handling
Arvada, Colorado U.S.A. Telephone: 303-425-0800
Division of Milton Roy Company
Subsidiary of Sundstrand Corporation

Sundstrand International, S. A.
Brussels, Belgium Telephone: 322-721-5045
Division of Milton Roy Company
Subsidiary of Sundstrand Corporation

Contents

SECTION ONE – Basics
    Fluid Mechanics Principles 1
    Criteria for Pump Selection 23

SECTION TWO – Kinetic Pumps
    Special Effect Pumps 31
    Regenerative Turbine Pumps 35
    Centrifugal Pump Nomenclature
    Characteristics and Components 39
    Centrifugal Pump Types 111

SECTION THREE – Positive Displacement Pumps
    Rotary Pumps: Nomenclature, Characteristics, Components and
        Types 133
    Reciprocating Pumps: Nomenclature, Characteristics, Components
        and Types 153

SECTION FOUR – Pump Construction
    Materials and Corrosion 171
    Seals and Packing 183

SECTION FIVE
    Testing 191
    Installation and Start-Up 195
    Vibration 203
    Balancing 213

SECTION SIX – Drives
    Electric Motors 223
    Turbines, Engines, Gears, V-Belt Drives, and Couplings 227
    Variable Speed Drives and Speed Control 233

SECTION SEVEN – Applications
Water Pump Applications 239
Fire Pumps 257
Chemical Process Pumps 263
Food, Beverage and Pharmaceutical Pumps 283
Petroleum Production, Pipeline and Product Pumps 295
Pulp and Paper Pumps 311
Solids and Slurry Pumps 321
Waste Water/Sewage Pumps 335

SECTION EIGHT
Appendix 349

SECTION NINE
Buyers Guide 399
Trade Names Index 413
Advertisers' Names, Addresses and Contact Numbers 415
Editorial Index 417
Advertisers Index 427

SECTION 1

Basics

FLUID MECHANICS PRINCIPLES

CRITERIA FOR PUMP SELECTION

FLUID MECHANICS PRINCIPLES

The application of pumps is an engineering discipline that is based on principles of fluid mechanics. This chapter covers those principles that are independent of the type of pump and essential to a thorough understanding of the criteria that govern the performance of flow systems. They will be referenced throughout the text. The reader is encouraged to refer to them freely unless there is a full understanding of their meaning and use.

Definitions:

Fluid: A liquid that may contain or be mixed with solids, vapours, or gases.

Liquid: A pure fluid that contains no solids, vapours or gases.

Mass: (M), kgm (lbm) is a measure of the amount of matter in an object. It may be considered as that property of an object by which it exhibits inertia. An international standard prototype mass exists and all mass measurements can be done on an equal arm balance, traceable to that prototype.

Weight: (W), kg (lbf) is a measure of the gravitational pull on a body. At any given location on the earth, the weights of bodies are proportional to their masses. At sea level and 45° latitude a kgm = a kg_f* and a lb_m = a lb_f.

*The term kg_f should be avoided.

W = Mg (where g is the acceleration due to gravity, the units of g are meters/sec² (ft/s²)

Specific Gravity: (s) is defined as the ratio of the density of a substance to that of fresh water.

Specific Weight: (γ), N/m³ (lb_f/ft³) is defined as the weight per unit volume.

The specific weight of water is approximately 9795 N/m³, (62.3 lb_f/ft³) at 20°C, (68°F)

Specific Volume: (v), m³/N, (ft³/lb_f) is defined as the volume per unit weight. It is the reciprocal of the Specific Weight.

Density: (ρ) kg_m/m³, (lb_m/ft³) is defined as the mass per unit volume.

Atmospheric Pressure: p_{atm} is the force exerted on a unit area due to the weight of the atmosphere. Pa or N/m³, (lb/in²). When measured with a barometer it is called *Barometric Pressure*.

Standard Atmospheric Pressure is the pressure at sea level and zero degrees centigrade that is balanced by a column of mercury 76 cm high. It is equal to a pressure of 1,013,200 dynes/cm², 101.32 kpa, 1.0132 bar, 14.696 psia or 33.93 ft water.

Absolute Pressure: p_{abs} is the sum of atmospheric pressure plus gauge pressure.

Gauge Pressure: p_g is the pressure read by any instrument which has the atmospheric pressure acting in opposition to the pressure being measured, for example, a bourdon gauge.

Vapour Pressure: pv is that absolute pressure of a fluid at which it is in equilibrium, at a certain temperature. Any reduction in pressure or increase in temperature of the fluid will cause evaporation until a new state of equilibrium is reached due to the space above the liquid becoming saturated with the vapour. See Figure 1.1A and Table 1A in the Appendix. In the presence of a liquid and a confined space above that liquid, the molecules of the liquid will evaporate until an equilibrium point is reached. At this point, the number of molecules leaving the liquid and the number returning will be equal. This maximum vapour pressure of the liquid is the *saturated vapour pressure*. It is unaffected by the presence of other gases. For example if the confined space had air at atmospheric pressure of 76 cm mercury and the liquid was water at 20°C, then the pressure in the confined space would be 76 cm mercury plus 1.75 cm mercury which is the saturated vapour pressure of water at 20°C. The higher the vapour pressure the more volatile the fluid. This volatility must be taken into consideration on the suction side of pumps where acceleration however localized can cause cavitation.

Vacuum or Vacuum Pressure refers to pressures below atmospheric pressure and can be given in absolute pressure (a positive number), or gauge pressure (a negative number), or vacuum pressure an absolute number that must be subtracted from atmospheric pressure to get absolute pressure.

Head: h m (ft) is a term that is used interchangeably with pressure, since pressure can be

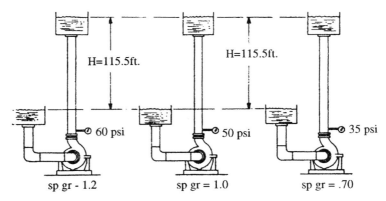

FIGURE 1.1 – Press/head relationships of identical pumps handling liquids of differing specific gravities

Always at your service!
KSB pumps - reliable and safe

KSB Aktiengesellschaft
67225 Frankenthal-Germany
Tel.: +6233/86-0

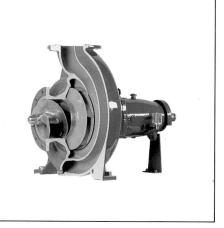

....*wherever pumping is needed!*
Pumpenfabrik ERNST VOGEL Aktiengesellschaft
A-2000 Stockerau, Ernst Vogel-Straße 2, Tel. 02266/604, Telefax 02266/65311, Austria, Europe

CHOPPER PUMPS   **PUMPING SOLUTIONS**

Pumping Solutions by
Vaughan

Vaughan Chopper Pumps handle the toughest industrial and municipal pumping applications, such as plant wastewater, scum, sludge transfer and digester recirculation, by chopping and pumping waste solids without clogging. Available in Vertical Wet Well, Submersible, Vertical Dry Pit, Vertical Recirculator, and Horizontal configurations, with flow rates of 10 M3/HR to 800 M3/HR.

Vaughan Chopper Pumps are guaranteed for *non-clogging* performance. Contact Vaughan Co. for your country representative for assistance in solving your toughest pumping applications.

THE VAUGHAN GUARANTEE

Vaughan

Vaughan Co., Inc.
364 Monte Elma Rd., Montesano, WA, USA 98563
1-206-249-4042, FAX: 1-206-249-6155, or
1-360-249-4042, FAX: 1-360-249-6155

"First in pumping solutions that Last"

FLUID MECHANICS PRINCIPLES

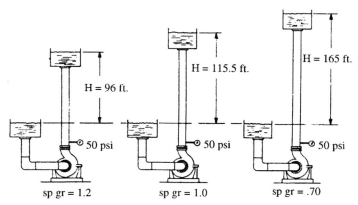

FIGURE 1.2 – Press/head relationships of pumps delivering the same pressure handling liquids of different specific gravity.

equated to the height of the balancing column of any liquid by dividing its height by the specific weight in the same units. It is a measure of the energy content of the liquid at the bottom of the column referred to datum. Head has an advantage in pumping systems over pressure in that the differential pumping head can be measured directly with no knowledge of the liquid. See Figures 1.1 and 1.2.[1]

Gauge Head: h_g is the head reading when opposed by atmospheric pressure. To convert gauge pressure to gauge head:

1 m water-h_g = 9800 p (p in *Pa*) [1 ft water-h_g = 0.4331p, (p in psi)]

Velocity Head: h_v is the kinetic energy at the cross-section of the flow path being considered.

$$h_v = \frac{U^2}{2g}$$

Where Velocity: U is obtained by dividing the flow by the cross-sectional area.

Elevation Head: z is the vertical distance of the point of measurement from datum.

Total Suction Head: hs is the sum of the gauge, velocity and elevation heads at the suction datum elevation plane.

$$h_s = h_{gs} + h_{vs} + h_{zs}$$

Total Discharge Head: hd the sum of the gauge, velocity and elevation heads at the discharge.

$$h_d = h_{gd} + h_{vd} + h_{zd}$$

Total Head: H is the measure of the work increase per unit weight of liquid imparted to the liquid by the pump. It is equal to the difference between the Total Discharge Head and the Total Suction Head

Cavitation: is defined from a fluid mechanics perspective as the formation and collapse of vapour filled cavities in a liquid due to dynamic action. From a metallurgical standpoint it is defined as the cavitation erosion attack of metal surfaces, caused by the collapse of cavitation bubbles on the surface of the metal and characterized by pitting.

Incipient Cavitation is the commencement of bubble formation and collapse.

NPSHA: m (ft) is the net positive suction head available at the first stage impeller datum available from the system. It is the total suction head at the pump impeller datum minus the vapour pressure of the liquid being pumped at that location.

NPSHR: m (ft) is the net positive suction head required by the pump to avoid more than a 3% loss in total head to the first stage of the pump at a certain capacity

Energy: J, N-m (ft-lb$_f$) is the potential a body has for doing work.

Internal Energy: u, is the energy stored within a substance due to the molecular activity and spacing. Generally we are concerned with the changes of this energy as opposed to the absolute value itself.

Potential Energy is the energy stored in that body by virtue of its spatial position.

Kinetic Energy is the energy of the body by virtue of its motion.

Flow-Work is the energy taken to push (force times distance) a unit weight across a section of the flow path.

Datum is the horizontal plane of a pump which serves as the reference point for Elevation Head measurements.

Equations:

There are three basic flow equations that are the basis for calculating most of the energy changes taking place in flow systems:

Continuity Equation - this equation states that the mass rate of flow is a constant

$$A1U1g1 = A2U2g2 = \text{Constant} \qquad \text{Eq. (1)}$$

Example: A pump with 25 cm (10 in.) suction flange and 20 cm, (8 in) discharge flange has a capacity of 200 l/sec (3170 gal/min) of water under operating conditions. What is the velocity in the suction and discharge flanges?

Since specific weight does not change and Q = AU then:

$$A = \pi d^2/4 = 0.786 d^2,$$

SI $A_1 = 0.786 \times 25^2 = 491 \text{ cm}^2$ USCU $A_1 = 0.786 \times 10^2 = 78.6 \text{ in}^2$

$A_2 = 0.786 \times 20^2 = 314 \text{ cm}^2$ $A_2 = 0.786 \times 8^2 = 50.3 \text{ in}^2$

$$Q = A_1U_1 = A_2U_2$$

SI 200 l/sec $0.001 \text{ m}^3/l = 491 \text{cm}^2$ $1 \text{ m}^2/10,000 \text{ cm}^2 \times U_1$.

$U_1 = 4.07 \text{ m/sec}$

$U_2 = 0.2 \times 10,000/0.314 \text{ m} = 6.36 \text{ m/sec}$

USCU 3170 gal/min $0.1337 \text{ ft}^3/\text{gal} = 50.3 \text{ in}^2/144 \text{ in}^2/\text{ft}^2$

$U_1 = 12.9 \text{ ft/sec}$

$U_2 = 424 \text{ ft}^3/\text{min}/0.349 \text{ ft}^2 \times 1 \text{ min}/60 \text{ sec} = 20.24 \text{ ft/sec}$

FLUID MECHANICS PRINCIPLES

Bernoulli's Theorem: This law states that in the absence of friction, when an incompressible liquid moves from one place to another the total energy remains constant but the make-up of that energy may change. Another way of expressing his theorem is to say that when a frictionless incompressible fluid flows the total head does not change.

$$P_1/\gamma + U1^2/2g + z_1 = p_2/g + U2^2/2\gamma + z_2 = \text{constant}$$

where p = pressure, γ = specific weight, U = velocity, z = elevation, g = gravitational acceleration constant.

The ***Steady Flow Energy Equation*** states that total energy remains constant even though its make-up may change. Friction can be shown as work. It is written as follows:

$$u_1 + p_1/\gamma + U_1^2/2g + z_1 + q + W = u_2 + p_2/\gamma + U_2^2/2g + z_2 \qquad \text{Eq. (2)}$$

Note that when there is no work done or heat exchanged (the latter is referred to as an *adiabatic* process) and the change in internal energy is not significant the equation matches that of Bernoulli.

The combination of terms $u + p/\gamma$ are often combined into a term called enthalpy which is available from tabulations of various liquids, vapours and gases such as Keenan and Keyes[2]. If there is a significant temperature rise in the fluid from 1 to 2 as may be seen in a high pressure ratio multi-stage pump then enthalpy should be used. For the vast majority of pump applications one can consider $u_1 = u_2$, and just calculate the flow work at suction and discharge.

Heat, q is positive into the fluid and in most pump applications will be negligible. For high temperature applications its significance should be tested. Work, W is positive for work into the fluid (pump) and negative out of the fluid (turbine). See Fig. 1.3.

Note that the Bernoulli Equation was not given an equation number. Because it is an equation for a specific case of the steady flow energy equation, it is recommended that the reader use the latter and go through the thought process of eliminating the terms that are not applicable.

Example 2: The same pump referenced in Example 1 and shown in Figure 1.3 is raising the pressure of the liquid from atmospheric, 0 kp_a (0 psig) to 303.2 kp_a (44 psig). The process fluid is at 20°C, (68°F). Datum is at the horizontal centre line of the pump. The discharge gauge is 30cm (1 ft) above datum. The suction gauge is 61 cm (2 ft) below datum. Calculate the work output from the pump.

$$u_2 + p_1/\gamma + U_1^2/2g + z1 + q + W = u_2 + p_2/\gamma + U_2^2/2g + z_2$$

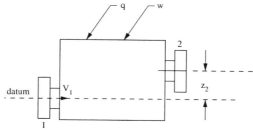

FIGURE 1.3 – Steady flow energy control diagram.

rearrange and let $u_1 = u_2$ and $q = 0$

$$W = (p_1 - p_2)/\gamma + (U_2^2 - U_1^2)/2g + (z_2 - z_1)$$

$$\gamma = 9807 \text{ N/m}^3 \ (62.3 \text{ lb/ft}^3)$$

SI
$$\begin{aligned} W &= (303.2 \times 10^3 - 0)/9807 + (6.4^2 - 4.8^2)/19.63 + .3 - (-.61) \\ &= 30.94 + 0.912 + 0.91 \\ &= 32.76 \text{ Nm/N} \end{aligned}$$

USCU
$$\begin{aligned} W &= (44 - 0)144/62.3 + (20.24^2 - 12.92)64.4 + 1 - (-2) \\ &= 101.7 + 3.62 + 3 \\ &= 108.3 \text{ ft-lb/lb} \end{aligned}$$

The above unit results, in pump output, when multiplied by the flow rate in units chosen are normally termed waterpower or liquid power and are expressed in kw (hp) units. Standard formulas exist for convenience.

SI   Wkw = QHs/367.5 where Q is measured in m^3/hr, H in m   (3)

or = QHs/6131 where Q is measured in kg/min, H in m

USCU   Whp = QHs/3960 where Q is measured in gal/min, H in ft   (4)

= QHs/33000 where Q is measured in lb/min, H in ft

Example 3: Calculate the Wkw (Whp) for the pump in the previous example.

SI
$$\begin{aligned} \text{Wkw} &= 9.8 \text{ QHs} \\ &= (200 \text{ l/s} \times 1 \text{ m}^3/1000 \text{ l} \times 3600 \text{ sec/hr} \times 32.76 \times 1)/367.5 \\ &= 64.2 \text{ kw} \end{aligned}$$

USCU
$$\begin{aligned} \text{Whp} &= \text{QHs}/3960 \\ &= 3170 \times 108.3/3960 \\ &= 86.7 \text{ hp} \end{aligned}$$

Example 4: Calculate the input power to the pump bkw (bhp) assuming the pump efficiency is 80%.

SI   bkw = Wkw/η = 64.2/0.80 = 80.25 kω

USCU   bhp = Whp/η = 86.7/0.80 = 108.4 hp

Often the power input to the pump driver is desired. This can be calculated if the driver efficiency is known.

Example 5: Calculate the driver input power in the above example. The driver efficiency is 92%.

$$\text{ikw} = 80.25/.93 = 86.3 \text{ kw}$$

$$\text{ihp} = 108.4/.93 = 116.6 \text{ hp}$$

The efficiency of the total pump unit is equal to the combined efficiencies of the pump (ηp) and the driver (ηd). In the case of motors this is quite often referred to as the *wire to water efficiency*.

$$\eta \text{ unit} = \eta_p \eta_d$$

FLUID MECHANICS PRINCIPLES

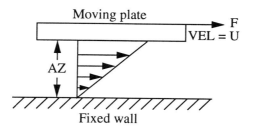

FIGURE 1.4 – Viscosity shear diagram

Viscosity - is the modulus represented by the ratio of

$$\frac{\text{shearing stress}}{\text{rate of shearing strain}}$$

A classic way to visualize viscosity is as follows:

Consider, as in Figure 1.4, a flat plate being pulled in a fluid filled channel parallel to and a distance Δz above the flat bottom of the channel at a constant velocity U. A thin layer will adhere to the bottom of the channel and will hence have zero velocity, whereas the thin layer that adheres to the bottom of the plate being pulled will have a velocity U. Assume that the fluid flows in laminar layers and parallel to the upper plate and bottom off the channel. If F is the force exerted on the plate and A is the area of the moving plate

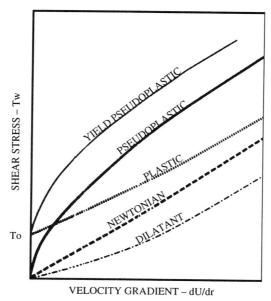

FIGURE 1.5 – Fluid classification of slurries.

then: $F/A = \tau$ = the shear force or stress involved. The rate of shearing strain = U/z.

$$\text{Viscosity} = F/A \,/\, U/z = Fz/UA = \mu$$

this is known as the *absolute* or dynamic viscosity and is commonly expressed in units of centipoise. One poise = 1 pascal second, or 1 gm/cm-sec.

Kinematic viscosity (υ) is defined as the ratio of absolute viscosity to density (ρ).

$$\upsilon = \mu/\rho$$

It is commonly given in units of centistokes. One stoke = one cm^2/sec although the use of Saybolt seconds for low and medium viscosity liquids and Saybolt Seconds Furol for high viscosity liquids is common in the US.

Newtonian Fluids are those where the viscosity is a constant for all shear rates and is independent of time, the ratio of shear stress to shear rate is a constant for all shear rates and zero shear rate only exists for zero shear stress. Water and mineral oil are *Newtonian Fluids* as are many coarse slurries. See Fig. 1.6.1.

A fluid that meets the above criterion only above some minimum shear stress, or yield point, greater than zero, is said to be *plastic*. Exampes are ketchup, gravel slurries, putty, moulding clay and sludge. See Fig. 1.6.3.

When the viscosity increases as shear rate increases and is independent of time, a fluid is said to be *dilatent*. Examples are clay, paints, printing inks and starch slurries. Some dilatent fluids will solidify at very high rates of shear. See Fig. 1.6.2.

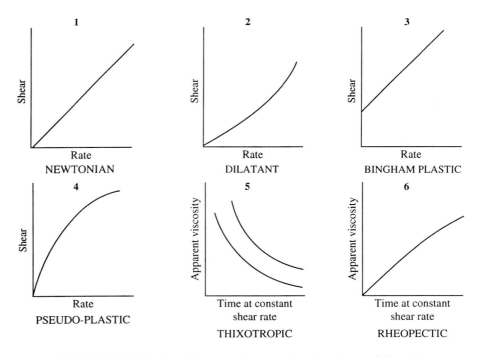

FIGURE 1.6, 1–6 – Stress/shear rate diagrams for various types of viscosity.

FLUID MECHANICS PRINCIPLES

When the viscosity decreases as shear rate increases, it is independent of time and zero shear rate occurs only at zero shear stress, a fluid is said to be *pseudo-plastic*. Examples are polymer solutions, water base emulsions, resinous materials and grease. See Fig. 1.6.4

When the effective viscosity decreases with time at a constant shear rate the fluid is *thixotropic*. Thixotropic fluids exhibit a hysteresis effect in that their viscosity at that point of time is dependent on the past history of the fluid. Examples are asphalt, grease, tar, gels and drilling mud. See Fig. 1.6.5.

When viscosity is constant for all shear rates at any given instant of time, but increases with time, a fluid is said to be *rheopectic*. To say this another way: if the apparent viscosity increases with time at a constant shear rate the fluid is said to be *rheopectic*. Gypsum suspensions are one example. See Fig. 1.6.6.

The reader is encouraged to also refer to Fig. 1.5[3] and 1.6[4] with these definitions.
Very strange things can happen with non-Newtonian fluids. Roco and Shook[5] point out that rheopectic and thixotropic fluids with time dependent shear resistance taken from pipelines do not show this time dependence whereas Couette viscometers do. They consider it likely that the pump action destroys the structure that can be re-established during shut down.

Pipe Flow

An understanding of the principles of incompressible flow in pipes or conduits is an essential prerequisite to the proper application of pumps in a pumping system.

In many applications one must consider not only the installed flow conditions in newly installed pipe, but must consider the effect of scaling and ageing of the pipe or process changes that may occur.

There are two types of flow laminar and turbulent. In the former visualize the flow as moving in layers or laminas. In the latter, there is transverse flow across these laminas which is imposed on the main flow. The larger the percentage of this transverse flow the flatter the flow profile becomes from that of the pure laminar flow. See Figure 1.7.

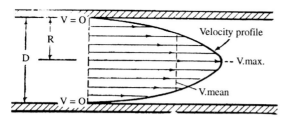

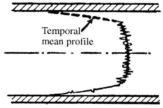

FIGURE 1.7

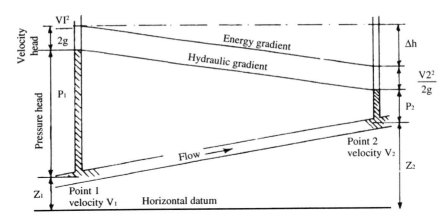

FIGURE 1.8 – Gradient diagram.

Laminar flow is found where the channel dimensions are relatively small, the velocity low or the viscosity high. The main resistance to the flow in these cases is the viscous shear effects. The velocity of the layer at the pipe wall is zero and it increases as the lamina approach the center of the pipe. See Figure 1.7.

Increasing the average velocity in the pipe, the size of the pipe, or decreasing the viscosity of the fluid can cause a steady decrease in the turbulence and a more uniform velocity profile. This reduces the viscous shear in the bulk of the fluid but the transverse flow in the form of vortices increases. Near the pipe wall the boundary layer becomes thinner due to the increased turbulence and the pipe roughness becomes a major factor by causing increased turbulence and energy loss.

A dimensionless number, the Reynolds number

$$Re = \rho UD/\mu \quad \text{where } D = \text{inside diameter} \qquad (3)$$

$$\text{or} \quad Re = UD/\nu \qquad (4)$$

is very helpful in correlating the main variable in pipe flow. Experiments[6] show that motion is laminar when $Re < 2000$ and turbulent when $Re > 3000$. The range of Re from 2000-3000 allows different conditions dependent on pipe entrance, roughness, disturbances etc.

Referring back to the steady flow energy equation 2 and letting pipe friction head loss (h) equal the work, we get:

$$h = (p_2 - p_1)/\gamma + z_2 - z_1 \text{ since } U_1 = U_2 \text{ and } q = 0$$

In the general case this can be depicted as shown in Fig. 1.8.

The Darcy-Weisbach formula is conventionally used for calculating head loss.

$$h = \lambda \, L/D \, U^2/2g \qquad (5)$$

where L = length of pipe

λ = friction factor or universal coefficient for head loss.

A bountiful supply of experimental data exists that allows one to determine the friction

FLUID MECHANICS PRINCIPLES

factor in circular pipes or tubes depending on the relative roughness factor and the Reynolds number. There also are computer software programs available of various degrees of sophistication in the technical market-place that contain the technical data of the type shown in this chapter and can solve these pipe flow problems.

Relative roughness = k/D, where k is the equivalent of the sand grain roughness of the inside diameter of the pipe. A chart of λ vs Re and k/D is shown in Figures (1.2A-1 and 1.2A-2)[7]. This is the same chart that is shown in ISO Test Specifications. The lowest curve in this chart may also be used for drawn copper tubing. Values of k/D for some common pipes and tubes are given in Figure 1.3A-2[8]. Note that below Re of 2000:

$$\lambda = 64/Re \quad (6)$$

Tables are available based on the specific pipe dimensions available in various parts of the world such as that shown for different schedule pipes common to North America in Tables 1.2A-1-2A-3[9]. These simplify the task even further. For those who need to do pipe friction loss calculations frequently, a set of such tables based on local pipe dimensions is recommended.

Non-Circular Pipes: Where pipes are non-circular the *hydraulic radius* is utilized.

Hydraulic Radius = r_h = cross-sectional area/wetted perimeter

$$r_h = D/4$$

Putting this into the Darcy-Weisbach formula (Eq. 5) gives us the form for non-circular pipes:

$$h = \lambda (L/4r_h) U^2/2g \quad (7)$$
$$Re = 4 U r_h \rho/\mu \quad (8)$$

For turbulent flow, Equations. 7, 8 can be utilized. For laminar flow, Lamb[10] gives theoretical results.

Aged Pipes: The amount of scaling or encrustation of pipes with age is dependent on many factors. Nevertheless, in the absence of any other data on ageing effects in a given application, empirical data is welcomed. One of the most widely used empirical formulas used for water under turbulent flow conditions is the Hazen-Williams formula.

SI $\quad U = 0.914 \times 10^{-2} C r_h^{0.63} S^{0.54}$ m/sec $\quad$ (9)

USCU $\quad U = 0.8492 C r_h^{0.63} S^{0.54}$ ft/sec $\quad$ (9)

where U = Average pipe velocity, m/sec
C = friction factor
r_h = hydraulic radius, mm, (ft)
S = head loss, m loss/m pipe length, (ft/ft)

Fig. 1.4A is a nomogram for the solution of the above formula. As age increases, lower C values are used. Tables 3A-1[11] and 3A-2 give some guidelines on the C values to be used for pipes of various ages.

Another empirically based equation is the Manning or Chezy-Manning formula: It can be used for closed pipes running full or partially full.

$$U = cr_h^{2/3}S^{1/2}/k$$

where S = Slope of the flow of h/ft of conduit   (11)

U = avg. velocity in m/sec (ft/sec)

r_h = hydraulic radius

n = the roughness coefficient

SI   c = 1

USCU   c = 1.49

Table 7A gives values of the roughness coefficient.

Viscous Liquids

Frictional losses of viscous fluids in Schedule 40 Pipe are given in Table 4A-1-4A-4 in the Appendix. Fig. 1.5A gives Reynolds numbers for various liquids and kinematic viscosities.

Pipe Fittings and Instruments

There is a ample supply of friction loss test data on pipe fittings and instruments available and empirical presentations. One of the more widely used sources is the Crane Company's Technical Paper 410[12], which in 1991 had had 25 reprintings since its presentation in 1955. Pressure drop through fittings can be correlated to the product of a coefficient of resistance, that has been found to be constant for geometrically similar fittings, multiplied by the velocity head of a matching size pipe.

$$h = KU^2/2g \qquad (10)$$

Table 5A-1-5A-4 shows the resistance coefficient, K for various valves and fittings.

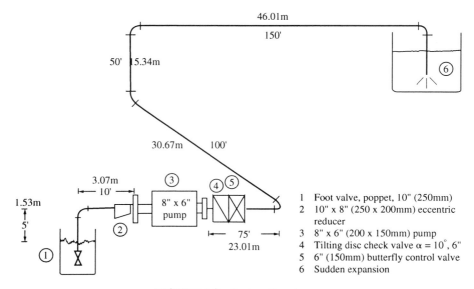

FIGURE 1.9 – System flow loop.

FLUID MECHANICS PRINCIPLES

Tables 6A-1-6A-3 give similar information. In the upper left hand quarter of the first page of 6A-1, the coefficient for three different inlets are shown. Particular attention should be paid to the big difference in relative pressure drop between these three, with the worst having 20 times the pressure drop of the best. This can be significant in energy savings and crucial in systems where the pump selection is sensitive to the suction conditions. In these figures, the symbol V = U.

Resistance coefficients for increasers and diffusers can be read off of Fig. 1.6A while those for reducers can be read off of Fig. 1.7A.

The loss through orifice, nozzle and venturi metres[13] is shown in Figs.1.8A through 1.10A in terms of a percentage of the meter differential.

Head losses through control valves, butterfly valves and shut-off valves are subject to considerable variation with design and the manufacturer's data on that valve should be used, if available, but data is given in Tables 5A and 6A for some valves and fittings from another source.

Example: 6:
SI

Calculate the head loss for the piping system shown in Fig. 1.9 at 0.057, 0.075, and 0.114 m3/s

Suction:
1. Entrance Loss
2. Static Lift
3. Foot Valve Poppet Type
4. 4.6 m-254 mm (id) pipe
5. 1-90° Long Radius Elbow
6. 1-Eccentric 254 mm/203 mm Reducer

Discharge:
7. Tilt Check Valve
8. Butterfly Control Valve
9. 114 m 154 mm (id) pipe
10. 15 m static head
11. 4- 90° Short Radius Elbows
12. Exit Loss

1. Entrance Loss
From Fig. 1.6A-1, K = 0.5 for square edged inlet
$U^2/2g$ at 0.057 m3/s = ? U = Q/A = (0.057 m^3/s) /0.785 (0.253m)^2 = 1.13m/s
$U_{0.075}$ = 1.13 (0.075/0.057) = 1.48m/s
$U_{0.114}$ = 1.13 (0.114/0.057) =2.26 m/s
$U^2/2g$ = 1.13^2/19.63 = 0.065 m
@ 0.075 m^3/s = ? = 1.48^2/19.63 = 0.111 m
@ 0.114 m^3/s = ? = 2.26^2/19.63 = 0.260 m
$h_{.057}$ = 0.5 (0.065) = 0.03 m
$h_{.075}$ = 0.5 (0.111 = 0.06 m
$h_{1.14}$ = 0.5 (0.260) =0.13 m

2. Foot Valve
   From Table 5A-3, K = 420 f_T. From Table 5A-1 f_T = 0.014
   K = 420 x 0.014 = 5.87 m
   $h_{0.057}$ = 5.87 x 0.065 = 0.38 m
   $h_{0.075}$ = 5.87 x 0.111 = 0.65 m
   $h_{0.114}$ = 5.87 x 0.260 = 1.52 m
3. Static Lift - 1.5m as shown
4. 4.6 m of 250 mm pipe

On Fig.1.3A-2 at the top find 25cm and drop vertically to Commercial steel pipe. Then move across horizontally to left and read 0.00018 Relative roughness. Now go to Fig. 1.3A-1. To determine the friction factor one must first ascertain the factor VD, (V= U).Then enter the top of Fig. 1.3A-1 with this value and drop vertically to the intersection with 00018. Then read to the left for the value of the friction factor.

$UD_{0.057}$ = 1.13 x 25 = 28.25   f = 0.0167
$UD_{0.075}$ = 1.48 x 25 = 37.0   f = 0.016
$UD_{0.111}$ = 2.26 x 25 = 56.5   f = 0.0153

h = f l/d U^2/2g
   $h_{0.057}$ = 0.0167 x 4.6/0.25 x 0.065 m = 0.02 m
   $h_{0.075}$ = 0.016 x 4.6/0.25 x 0.111 m = 0.03 m
   $h_{0.111}$ = 0.0153 x 4.6/0.25 x 0.260 m = 0.07 m

5. 1-90° 250 mm Long radius elbow
   Equivalent length in metres from Table 8A-5 = 5 m
   This is 5/4.6 of the values in 4 above.
   $h_{0.057}$ = 0.02 m
   $h_{0.075}$ = 0.03 m
   $h_{0.111}$ = 0.08 m
6. 1-Eccentric 250/200 mm reducer

Interpolating in Table 8A-5 for taper connectors is not advisable since these are increasers which can exhibit considerable turbulence if the angle of divergence exceeds approximately 15°. Using Table 8A-6 for a contraction ratio (d/D) of 0.5 an equivalent straight length of 8 is shown under the column for d (the smaller end).
This is conservative since our reducer has a ratio of 200:250 or 0.8. So arbitrarily reduce the figure to 5. Note that this is not going to result in a significant impact on the total head loss determination. This results in the same magnitudes as in item 5 above.
   $h_{0.057}$ = 0.02 m
   $h_{0.075}$ = 0.03 m
   $h_{0.111}$ = 0.08 m
7. 150 mm Tilt-check valve

Refer to Table 8A-4, flanged swing-check valve. The equivalent straight length is 40 m. For simplicity of calculation, proportion the results of 4 accordingly by multiplying them by the ratio of 40:4.6.
   $h_{0.057}$ = 0.02 x 40/4.6 = 0.17
   $h_{0.075}$ = 0.03 x 40/4.6 = 0.25
   $h_{0.111}$ = 0.07 x 40/4.6 = 0.61

FLUID MECHANICS PRINCIPLES

8. 150 mm Butterfly Control Valve
From Table 5A-3, K = 45f_T, f_T = ? Relative roughness from Fig. 1.3A-2 = 0.0003,
$UD_{0.057}$ = 1.13 x 15 = 16.95
   f_T = 0.018, K = 0.81
$UD0.075$ = 1.48 x 15 = 22.20
   f_T = 0.018, K = 0.81
$UD0.111$ = 2.26 x 15 = 33.9
   f_T = 0.017, K = 0.80

Alternatively, Table 6A-2 gives a K for 6 in (150 mm) butterfly valve as 0.9 and the difference is within the range that one should expect as one can see from Table 6A-3. Now lets take a short cut to finding the velocity heads by using the very convenient Table 2A. Dividing the metric pipe size in mm by 25 gives the USCU pipe size in inches. In this case 150/25 = 6 in. The corresponding gpm can be calculated by the conversion: 1 m^3/s = 15, 852 gpm (Table 9A-5).

0.057 m^3/s x 15, 582 = 900 gpm. From Table 2A-15
velocity head = 1.55 ft. or 1.55/3.05 m = 0.508 m.
   $h_{0.057}$ = 0.508 x 0.81 = 0.41 m
   $h_{0.075}$ = 0.41 x 0.075/0.057 = 0.54 m
   $h_{0.114}$ = 0.41 x 0.114/0.057 x 0.80/0.81 = 0.81 m

9. 123 m 150 mm steel pipe
As discussed in 8. above go to Table 2A-15, h_f in feet/100 ft of pipe (or metres / 100metres of pipe) = 5.05 m for 0.057 m^3/se (900 gpm).
   $h_{0.057}$ = 1.23 x 5.05 = 6.21 m
The gpm corresponding to 0.075 m^3/s =
   900 x 0.075/0.057 = 1200
   $h._{0.075}$ = 1.23 x 8.76(from table) = 10.78m
The gpm corresponding to 0.114 m^3/s is twice that of 0.057 m^3/s or 1800 gpm
   $h_{.111}$ = 1.23 x 19.4 = 23.86 m

10. 15.25 m Static head, as shown in Fig. 1.9.

11. 4-90° short radius elbows
From Table 6A-1 K = 0.29. From Table 2A-15
   $U^2/2g$ at 0.057 m^3/s = 1.55 x 0.305 = 0.473m
   $U^2/2g$ at 0.075 m^3/s = 2.76 x 0.305 = 0.842 m
   $U^2/2g$ at 0.114 m^3/s = 6.21 x 0.305 = 1.894 m
   $h_{0.057}$ = 0.29 x 0.473 = 0.14 m
   $h_{0.075}$ = 0.29 x 0.842 = 0.24 m
   $h_{0.114}$ = 0.29 x 1.894 = 0.55 m

12. Exit loss
From Table 5A-4 K = 1.0
   $h_{0.057}$ = 1.0 x 0.473 (See 11 above) = 0.47 m
   $h_{0.075}$ = 1.0 x 0.842 = 0.84 m
   $h_{0.114}$ = 1.0 x 1.894 = 1.89 m
The results are tabulated in Table 1 below:

TABLE 1

No.	Static head	$h_{.057}$ m³/sec	$h_{.075}$ m³/sec	$h_{.114}$ m³/sec
1		0.03m	0.06m	0.13m
2		0.38m	0.65m	1.52
3	1.5m			
4		0.02m	0.03m	0.07m
5		0.02m	0.03m	0.08m
6		0.02m	0.03m	0.08m
7		0.17m	0.25m	0.61m
8		0.41m	0.54m	0.81m
9		6.21m	10.78m	23.86m
10	15.25m			
11		0.14m	0.24m	0.55m
12		0.47m	0.84m	1.89m
Total	16.75m	7.87m	13.45m	29.60m

USCU Calculate the head loss for the piping system shown in Fig. 1.9 at 900, 1200 and 1800 gpm.

Suction:
1. Entrance Loss
2. Static Lift
3. Foot Valve Poppet Type
4. 15 ft. 10 in. Sch.40 pipe
5. 1-90°-10 in. Long Radius Elbow
6. 1-Eccentric 10/8 Reducer

Discharge:
7. Tilt Check Valve
8. Butterfly Control Valve
9. 375 ft. 6 in. Sch. 40 pipe
10. 50 ft. Static Head
11. 4-90° Short Radius Elbows
12. 1-Exit Loss

1. From Fig. 1.6A-1, square-edge inlet, K= 0.5, $U^2/2g$ = 0.208 @ 900 gpm, h_{900} = 0.5 x 0.208 = 0.104; $U^2/2g$ = 0.370 ft. @ 1200 gpm thus h_{1200}= 0.5 x 0.370 ft. = 0.185 ft.: $U^2/2g$ = 0.834 ft. @ 1800 gpm, thus h_{1800}= 0.5 x 0.834 = 0.417 ft.
2. Foot Valve Poppet Type, K = $420 f_T$ from Table 5A-3, f_T = 0.014 from top of Table 5A-1. K = 420 x 0.014 = 5.9 ft.; h_{900} = 5.9$U^2/2g$ = 5.9 x 0.208* = 1.22 ft.
    h_{1200} = 5.9 x 0.37 = 2.22 ft.
    h_{1800} = 5.9 x 0.834 = 4.92 ft.
    *from Table 6A-17

FLUID MECHANICS PRINCIPLES

3. Static Lift is 5. as shown in Fig. 1.9.
4. 15 ft. 10in sch. 40 pipe, 900 gpm: $U^2/2g = 0.208$ and h = 0.256 ft./100 ft or 0.256 x 0.15 = 0.038 ft.
   15 ft. 10in sch. 40 pipe, 1200 gpm: $U^2/2g = 0.37$ ft. and h = 0.703 ft./100 ft. or 0.703 x 0.15 = 0.105 ft.
   15 ft. 10in sch. 40 pipe, 1800 gpm $U^2/2g = 0.834$ ft. and h = 1.52 ft./100 ft. or 1.52 x 0.15 = 0.228 ft.
5. 1 = 90°-10 in. Long Radius Elbow, K = 0.014 from Table 6A-1 (flanged).
   $h_{900} = 0.014 \times 0.208 = 0.002$
   $h_{1200} = 0.014 \times 0.37 = 0.0052$
   $h_{1800} = 0.014 \times 0.834 = 0.116$
6. 1 Eccentric 10/8 reducer, from Table 5A, $\emptyset < 45°$ therefore use formula 1.
   $\beta = 8/10 = 0.8$
   $\beta^4 = 0.41$, Therefore $K_2 = 0.8 \sin(12/2)°(1-\beta^2) = 0.8 \times 0.105 \times 0.36 = 0.029$, Using Table 6A-16, $U^2/2g$ @ 900 gpm = 1.55, $h_{900} = 0.029 \times 0.518 = 0.015$ ft.
   $h_{1200} = 0.029 \times 0.920 = 0.027$ ft.
   $h_{1800} = 0.029 \times 2.07 = 0.06$ ft.
7. Tilt Check Valve, from Table 5A-2, $K = 80 f_T$ (assume $\alpha = 10°$). From Top of Table 5A-1 f_T for a 6 in. line = 0.015, thus K = 80 x 0.015 = 1.2, from Table 6A- 15,
   $h_{900} = 1.2 \times 1.55 = 1.86$ ft.
   $h_{1200} = 1.2 \times 2.76 = 3.312$ ft.
   $h_{1800} = 1.2 \times 6.21 = 7.452$ ft.
8. Butterfly Control Valve, from Table 6A-2, K = 0.9(note that Table 5A-3 gives K = 45f or K = 0.675, This discrepancy is a good illustration why K values should be obtained from the component manufacturer.
   $h_{900} = 0.675 \times 1.55 = 1.05$ ft.
   $h_{1200} = 0.675 \times 2.76 = 1.86$ ft.
   $h_{1800} = 0.675 \times 6.21 = 4.19$ ft.
9. 375 ft. Sch. 40 Pipe, 900 gpm, h = 5.05 ft./100 ft.* = 5.05 x 3.75 = 18.79 ft.
   1200 gpm, h = 8.76 ft./100 ft. = 8.76 x 3.75 = 32.85 ft.
   1800 gpm, h = 19.4 ft./100 ft. = 19.4 x 3.75 = 72.75 ft.
   * from Table 6A-15
10. 50 ft. static head, as shown in Fig.1.9.
11. 4-90° short radius elbows (flanged), from Table 6A-1 Regular 90° flanged elbow of 6 in. diameter has a K = 0.29. 0.29 x 4 = 1.16 ft. $h_{900} = 1.16 \times 1.55 = 1.8$ ft.
    $h_{1200} = 1.16 \times 2.76 = 3.20$ ft.
    $h_{1800} = 1.16 \times 6.21 = 7.20$ ft.
12. 1 exit loss, from bottom of Table 6A-4, K = 1. $h_{900} = 1.55$ ft.
    $h_{120} = 2.76$ ft.
    $h_{1800} = 6.21$ ft.

The results are tabulated in Table 2 below.

TABLE 2

No.	Static head	h_{900} gpm	h_{1200} gpm	h_{1800} gpm
1		0.10	0.19	0.42
2		1.2	2.2	4.9
3	5 ft.			
4		0.04	0.11	0.29
5		0.00	0.01	0.12
6		0.02	0.03	0.06
7		1.86	3.31	7.45
8		1.05	1.86	4.19
9		18.79	32.85	72.75
10	50 ft.			
11		1.80	3.20	7.20
12		1.55	2.76	6.21
Total	55 ft.	26.43 ft.	46.54 ft.	103.61 ft.

A plot of the data from the previous example is shown in Fig. 1.10. Note that at zero flow the curves start out at 55 ft. This is the static head of the system or z_2-z_1 for the condition of zero flow. The dashed curves on the left of the wide open butterfly valve curve represent curves of increased throttle positions of the butterfly valve, until at its fully closed position the dashed curve is straight up at zero flow. These curves are called system curves, because they portray the head loss imposed by the system at any particular flow condition.

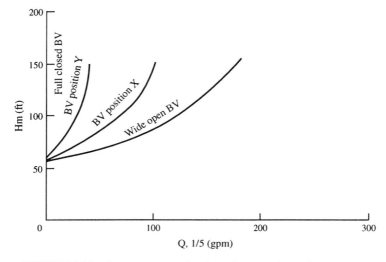

FIGURE 1.10 – System curves of various discharge butterfly valve positions.

FLUID MECHANICS PRINCIPLES

Open Channel Flow

Not all flow is in closed conduits. In many cases, the pumped liquid has a free surface. The Manning equation is used for this application: As previously covered it can also be used for closed pipes running full or partially full.

$$U = cr_h^{2/3} S^{1/2}/k \text{ where } S = \text{Slope of the flow} \quad (11)$$

or $\dfrac{\Delta z}{\Delta l}$ where l = length of conduit, m/m, ft/ft

U = average velocity in m/s (ft/sec)
k = the roughness factor

SI   c = 1
USCU   c = 1.49

Table 7A gives some roughness values.

Example: 7

A concrete flume has the dimensions of 4 m width, (13.12 ft.) It is 100 m, (328 ft.) long and has a slope of 0.004. The water depth is 4 m, (13.12 ft.). Calculate the flow rate.

SI $\quad U = C (r_h S)^{0.5} = 1 r_h^{1/6}/ k \times r_h^{3/6} \times .004^{0.5}$

$r_h = 4 \times 4/4 + 4 + 4 = 1.33$ m,
$k = 0.012$
$= 1 \times (1.33)^{2/3} \times 0.063 / (0.012)$
$= 1 \times 1.21 \times 5.27 = 6.38$ m/s
$q = AU = 16 \times 6.38 = 102.1$ m³/s

USCU $\quad U = 1.49 \, r_h^{2/3} \times 0.004^{0.5}/0.012$
$r_h = 13.12^2/13.12 \times 3 = 13.12/3 = 4.37$ ft
$= 1.49 \times 4.37^{2/3} \times 5.27$
$= 1.49 \times 2.69 \times 5.27$
$= 21.12$ ft/s
$q = 13.12^2 \times 21.12$
$= 3640$ ft³/s

Fig. 1.8 depicts the energy components and gradients. Note that frictional loss Δh is equal to the height loss or $z_2 - z_1$.

Secondary Flow in Elbows and Bends

When flow exists in a curved pipe or tube or even tanks centrifugal force sets up secondary flows that follow a spiral path to the origin or centre of curvature. See Fig. 1.11. The pressure at r_2 is greater than at r_1, increasing from the bend inlet to the *midpoint*.

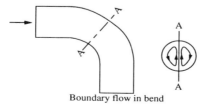

Boundary flow in bend

FIGURE 1.11 – Secondary flow in a bend.

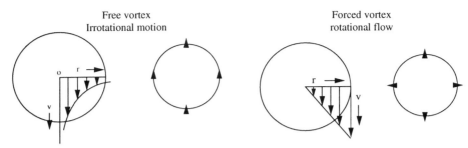

FIGURE 1.12 – Free vortex rotational flow.

FIGURE 1.13 – Forced vortex irrotational flow.

The combination of main and secondary flows sets up a twin spiral and the shape of the flow leaving the bend is similar to that shown. This can result in instrumentation errors, settling of slurries and higher pressure drop if not taken into consideration. Guide vanes installed in the curved elbow can reduce these secondary flows.

Free Vortex Flow

Circular flow in a horizontal plane with no work done on the fluid, is called a free vortex. In a free vortex, the velocity times the radius is a constant.

$$Ur = C \quad (12)$$

A free vortex is present in pump casings and most commonly seen in wash basins when they are draining. Its distribution is parabolic, Fig. 1.12

Forced Vortex Flow

A forced vortex has a relationship with flow and radius such that the velocity is equal to the product of radius times peripheral velocity or:

$$U = r\omega \quad \text{where } \omega = \text{peripheral velocity} \quad (13)$$

The highest velocity is thus at the maximum radius. The fluid rotates like a solid body, Fig. 1.13. Forced vortex flow is found in Vortex Pumps.

Note: In both types of vortex flow the pressure distribution is opposite that of the velocity distribution since the sum of velocity head and pressure head is a constant.

System Head Curve

In example 6, the head loss of a pipe system at three flow rates was calculated. These three points added to the static head plus the fourth point of the static head at zero flow, gave us a 4 point system head curve. A system head curve is a plot of the system flow rate by the system resistance and static head. Fig. 1.10 shows a series of system head curves representing the increased throttling of a discharge *butterfly* valve on a pump.

Energy Conservation

There are choices that can be made in the installation and pump selection that can conserve energy. The selection of the most efficient pumps and motors capable of doing the job is one area. Piping is another area. Pipe sizing is normally a compromise between first cost

FLUID MECHANICS PRINCIPLES

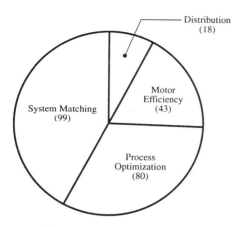

240 billion kWh/yr savings potential

- Motor Effieicncy Improvement
- Electrical Distribution Correction
- Motor-Drive/Mechanical System Matching (e.g., ASDs)
- Process Optimization

FIGURE 1.14 – Energy savings potential chart.

and operating costs but another area that is often overlooked is the optimal recovery of the kinetic energy in the pump discharge. Many times the discharge size of a pump is specified small, to allow the use of lower cost valves at the discharge and then expanded up by means of eccentric reducers.

The use of concentric reducers should be considered since they are much more effective at converting the velocity pressure to static pressure. The use of a conical 7° diffuser would result in a recovery of approximately 90%[14] whereas an eccentric reducer could have a estimated best case recovery of 55%. If the velocity head at the discharge flange is 3 m (10 ft.), then 2.7 m or 9 ft. would be recovered by the conical diffuser as opposed to the 1.65 m or 5.5 ft. recovery of the eccentric reducer. This is a difference of 1.05 m or 3.5 ft. Some of the unrecovered kinetic energy would be recovered later in the pipe, just as in a sudden expansion. Taking this recovery as being 50%, we are now showing 1.75 ft. unrecovered, or an equivalent power loss of almost 2%. Fig. 1.14 shows the result of a study by the U S Department of Energy, namely 240 billion kwh/yr potential savings in pump installations due to improvement in motor efficiency, electrical distribution correction, motor drive/ mechanical system matching e.g. adjustable speed drives, and process automation. While electrical distribution may or may not be controllable by the user, the other three more significant factors of high efficiency motors, system matching and process optimization are.

References

[1] Westaway, C. R. and A. W. Loomis, "Cameron Hydraulic Data " 16th Ed. pp 1-9, 1-10 Ingersoll-Rand (1979)

[2] Keenan, J. H. and F. G. Keyes, "Thermodynamic Properties of Steam" McGraw Hill (1936)

[3] "Valves Piping and Pipelines Handbook" 2nd Edition, Elsevier Advanced Technology, ISBN 85461-117-8

4. Hydraulic Institute Standards, 14th Edition (1983), pp 188,189
5. Shook, C. A. and M. C. Roco "Slurry Flow- Principles and Practice pp72 Butterworth-Heinemann (1991) ISBN 007506-9110-7
6. Reynolds, O. "An Experimental Investigation of the Circumstances Which Determine Whether the Motion of Water Will Be Direct Or Sinuous and of the Law of Resistance In Parallel Channels", Philosophical Transactions, Royal Society, London (1883)
7. Moody, L. F., "Friction Factor For Flow In Circular Pipes" ASME Transactions, Vol. 66. #8, (Nov. 1949), p671
8. Hydraulic Institute, "Engineering Data Book", p41, 2nd Ed. (1990).
9. Hydraulic Institute, "Engineering Data Book", pp51-76, 2nd Ed. (1990).
10. Lamb.H, "Hydrodynamics", 6th ED. Cambridge University Press (1932)
11. Salisbury, J.,Kenneth, Kent's Mechanical Engineers' Handbook-Power Volume,12th Edition, Wiley & Sons or Chapman & Hall, (1950)
12. Crane Co., "Flow of Fluids Through Valves, Fittings and Pipe", Technical Paper 410, (!988)
13. ASME, "Fluid Meters" 6th Ed (1971),Interim Supplement 19.5, pp 201,221 and 232
14. Shepherd, D., "Principles of Turbomachinery" The Macmillan Co. (1956), p155

Criteria for Pump Selection

There are a significant number of factors that should be considered when selecting a pump for a specific application. Otherwise the original goals of rating, low cost trouble free operation etc, may not be met. The number of applications of pumps, each with its own list of considerations, aside from any unique requirements, is itself limitless. And the number of different kind of pumps available to choose from can make selecting the right pump seem like an impossible task.

Fortunately, most pumps are used in repetitive applications and cost and process compatibility factors can quickly reduce the choices in less common applications. Proper and complete definition of the application, the system and the reasons for this pump will bring these criteria to the surface.

Classification

Fig. 2.1 shows a pump classification chart which breaks the classification down into two main categories; positive displacement and kinetic (or dynamic or rotodynamic). Positive displacement pumps are batch delivery, periodic energy addition devices whose fluid displacement volume (or volumes) is set in motion and positively delivers that batch of fluid from a lower to higher pressure irrespective of the value of that higher pressure.

Kinetic pumps are continuous delivery, continuous energy addition devices that build up kinetic energy in the rotating element or impeller and convert most of that energy into static energy to a point where the fluid delivery to the higher pressure level commences. Unlike positive displacement pumps the delivery is affected by the value of the discharge pressure that must be overcome. On the other hand, the kinetic pump will deliver an increasing amount of liquid as the discharge pressure is lowered, whereas the positive displacement pump delivery is fixed.

There are two main types of positive displacement pumps, reciprocating and rotary. The reciprocating being made up of piston, plunger and diaphragm types and the rotary composed of vane, gear, screw, lobe and the other types shown.

Centrifugal, regenerative turbine and jet pumps are the three main classifications of kinetic pumps. The centrifugal is by far the most common and has been estimated to make

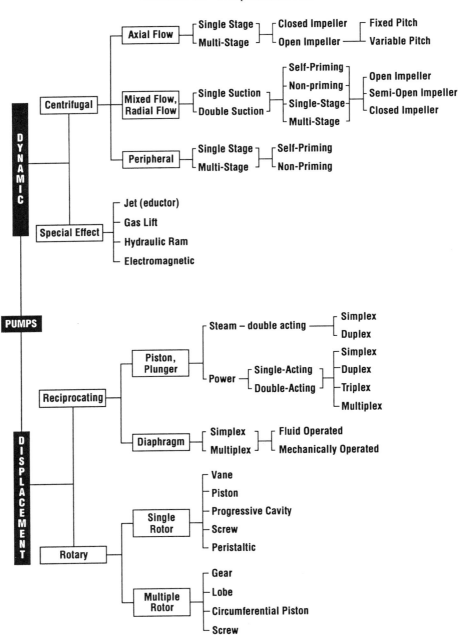

FIGURE 2.1 – Pump classification.

CRITERIA FOR PUMP SELECTION

up 90% of all pumps sold. Whether this figure is exact or not is not important. The commanding volume of the centrifugal is. There are two main types of centrifugal pumps, the diffuser type turbine pumps and the volute type.

Rough selection criteria

Economically, lowest cost criterion suggests selection consideration in the following order; centrifugal, rotary, reciprocating unless the head is over 6000 m (20, 000 ft), two phase flow of over 25% gas, viscosities over 600 centistokes (3000 ssu) or low shear requirements are dictated by the process, in which case there is a good possibility that the centrifugal may not meet the requirements. Fig. 2.2 shows the head-capacity envelopes of rotary, reciprocating and centrifugal pumps.

Pump selection can be affected by the following: head and capacity variability requirements of the system; elevations of the components; characteristics of the fluid such as viscosity, specific gravity, volatility or vapour pressure, corrosiveness, regulatory leakage restraints, sensitivity to shear, abrasiveness, size and settling characteristics of particulate carried, stability, and amount of entrained gases or vapours; the system layout; the system modes of operation such as continuous or intermittent operation, fluctuations in head and capacity; NPSHA-NPSHR; acceptability of downtime for maintenance and repair; ageing of equipment and system (effects of scaling of pipes and wear of pump components); future expansions; etc.

Positive displacement pumps are self-priming and can be used for metering. On the other hand they need relief valves (or the pressure against a high resistance such as a closed valve could exceed the design pressure retaining capability of the pump or system), pulsation dampening and accumulators plus additional NPSHA because of the acceleration that takes place on the suction side (The acceleration head loss can be 2–12 times the sum of all the other losses on the suction side during the suction stroke). Long discharge lines can also cause inertia problems that must be recognized and compensated for.

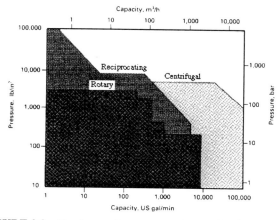

FIGURE 2.2 – Head-capacity range envelopes of various pumps.

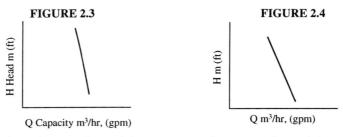

Reciprocating pump characteristic curve

Rotary pump characteristic curve

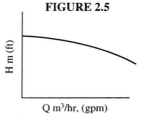

Centrifugal pump characteristic curve

Kinetic Pumps can be higher in efficiency than all but the most efficient reciprocating pumps.

In the ideal case (no hydraulic losses), on a head *vs* capacity plot the kinetic pump would have a horizontal constant head characteristic with varying capacity, while the positive displacement pump would have a vertical, constant capacity characteristic with varying heads. Actually, these pumps have losses and deviations from the ideal. Reciprocating pump hydraulic losses are the least and centrifugal losses the greatest of the three. These characteristics can be a deciding factor in the selection of the proper pump. See Figures 2.3-2.5. The rotary curve, Fig. 2.4, has a greater deviation from a vertical line than the reciprocating curve, Fig. 2.3, because rotary pumps generally have higher losses from leakage to the lower pressure than reciprocating pumps do.

Further specifics peculiar to each type of kinetic or positive displacement pump will be covered in the specific section on these pumps.

Pumps of both types can be placed in parallel with each other and in series. This is an important consideration since multiple pumps may be dictated for reliability and lowest cost operation reasons. For example: the operation alternatives of a cooling system circulating pump are such that loads as low as 30% for extended periods in the spring and fall and rising loads to occasional peaks of 115% of the calculated full load requirements of the system are expected, further, the risk of lack of capacity due to a pump shutdown is to be minimized. One could select four identical pumps, running in parallel. These pumps could be each good for 33% of the design full system load. This will allow 1-3 pump operation with a spare standing by and running time of the four pumps could be kept equalized. For the overload operation each pump could be run out to 38% of total flow

CRITERIA FOR PUMP SELECTION

(especially with centrifugals) or the fourth pump brought on the line for these few occasions.

System/pump interaction

The intersection of the system-head curve with the pump characteristic curve is an operating point. If this intersection is at the full flow condition of the system then it would be called the rated point. Actually, there could be one or several operating points specified in a pump order if the pump is expected to operate on other system head curves for significant periods of time. A present/future situation is one example. The opening of additional flow loops or stations is another. Many other possibilities exist. Refer back to Fig. 1.10. The parabolic curves for the different butterfly valve openings are examples where intersection with the pump characteristic curve, such as Figs 2.3-2.5 could occur. It may be appropriate to imagine the imposition of each of these latter curves onto Fig. 1.10 and note that the *capacity range* of the positive displacement pumps is nowhere near as great as the centrifugal, at least at a constant speed. This is as one would expect, and similarly the operation at different heads at a constant capacity would not be as well accommodated with a centrifugal as with a positive displacement pump. More on speed control later.

Atmospheric pressure effects

Atmospheric pressure effects must be considered carefully, Fig. 11A shows the effect of elevation on the local atmospheric pressure. The effect can be substantial. Even at sea level the atmospheric pressure changes significantly and during storms the pressure can drop two metres (six feet) or more. When this happens you have lost that much lift capability or NPSHA unless you have a closed system.

NPSHA-four standard cases

NPSHA-four standard cases present themselves over and over again. Fig. 2.6 gives a pictorial perspective of these four cases that apply to all types of pumps. Cases 1 and 2 cover supply from open tanks, below and above the pump. Cases 3 and 4 similarly cover supply from closed tanks, below and above the pump.

Case 1: Open Suction Supply Above Pump

$$NPSHA = Pa + Z + (-Vp) + (-hf)$$

where Pa = atmospheric pressure
Z = elevation from top of liquid in tank to c/l of pump
Vp = vapour pressure of the pumped liquid entering the suction at the maximum temperature expected.
hf = friction loss at the flow rate being considered in the suction pipe.

Case 2 : Open Suction Supply Below Pump

$$NPSHA = Pa + (-Z) + (-Vp) + (-hf)$$

Note: in this case where the supply is below the pump, Z is often called the "suction lift".

FIGURE 2.6 – Calculation of system Net Positive Suction Head Available for typical suction conditions.

NPSH Available is a function of the system in which the pump operates. It is the excess pressure of the liquid in feet absolute over its vapor pressure as it arrives at the pump suction. Fig. 2.6 shows four typical suction systems with the NPSH Available formulas applicable to each. It is important to correct for the specific gravity of the liquid and to convert all terms to units of "feet absolute" in using the formulas.

6A – Suction supply open to atmosphere – with Suction Lift

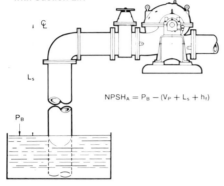

$NPSH_A = P_B - (V_P + L_s + h_f)$

6B – Suction supply open to atmosphere – with Suction Head

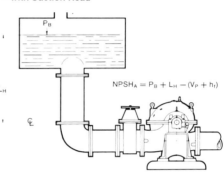

$NPSH_A = P_B + L_H - (V_P + h_f)$

6C – Closed suction supply – with Suction Lift

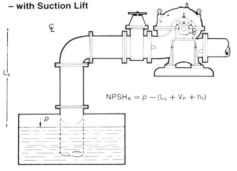

$NPSH_A = p - (L_s + V_P + h_f)$

6D – Closed suction supply – with Suction Head

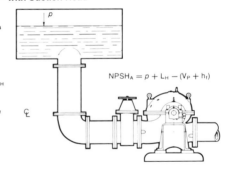

$NPSH_A = p + L_H - (V_P + h_f)$

P_B = Barometric pressure, in feet absolute.
V_P = Vapor pressure of the liquid at maximum pumping temperature, in feet absolute.
p = Pressure on surface of liquid in closed suction tank, in feet absolute.

L_s = Maximum static suction lift in feet.
L_H = Minimum static suction head in feet.
h_f = Friction loss in feet in suction pipe at required capacity

Case 3 : Closed Suction Supply Above The Pump

$$NPSHA = P + Z + (-Vp) + (-hf)$$

P = pressure in the tank

Case 4 : Closed Supply Below The Pump

$$NPSHA = P + (-Z) + (-Vp) + (-hf)$$

CRITERIA FOR PUMP SELECTION

Except for a simplified flow path depicted vs a more realistic and complex one that complicates the calculation of hf, these four cases cover most of the possibilities. Once the NPSHA is calculated the pump selected must have an NPSHR of a smaller magnitude. The difference between them is called the NPSH Margin.

SECTION 2

Kinetic Pumps

Kinetic Pumps as shown in Fig. 2.1 are made up to Centrifugal, Regenerative Turbine and Special Effects Types. Special Effects Pumps listed include Reversible Centrifugal and Rotating Casing (Pitot) Types. Electromagnetic, Eductor (Jet), Gas Lift, Hydraulic Ram and Pulsometer should be added to that list.

SPECIAL EFFECT PUMPS

REGENERATIVE TURBINE PUMPS

CENTRIFUGAL PUMP NOMENCLATURE, CHARACTERISTICS AND COMPONENTS

CENTRIFUGAL PUMP TYPES

Special Effect Pumps

Special effect pumps are for the most part used for specific applications and in small quantities. The Eductor or Jet Pump is probably the exception and sees the most usage of the group. Some of these pumps have been applied to their specific applications for over a century. The electromagnetic being the most recent addition.

Pulsometer

The Pulsometer is a steam-actuated expulser pump used in draining excavations, quarries, mine shafts, etc. They are simple, light weight, and do not require a foundation or lubrication and will handle water containing semisolids. The Pulsometer has a central suction air chamber, two pumping chambers, discharge and suction valve chambers and a foot valve. After the two pumping chambers are primed, steam is turned on and admitted to one of the two chambers forcing the contents of that chamber out into the discharge line. As the chamber liquid level reaches a low level the pressure drops. This forces the automatic steam valve to shift and admit steam to the second chamber. This all occurs while the vacuum in the first chamber is causing it to fill with a new charge. Capacities range from 100 to 2500 gpm. They have been used with high sand percentages, for clay slurries and sulfite paper stocks up to 8%. It finds little use today.

Hydraulic ram pump

The hydraulic ram pump has been used for over a century for raising water where the only power available is from a waterfall of *2 metres (6 feet)* or more. A portion of the water entering the suction can be raised to a height several times that of the waterfall. Or a smaller portion can be raised to an inversely proportioned higher height. For example: 1/6 the volume of water entering the suction can be raised to a height fives times that of the fall, or 1/12th of the water can be raised to a height ten times that of the waterfall. This pump also is seldom used today. See Fig. 18.3.

Rotating casing (Pitot) pump

The rotating casing (Pitot) pump is sort of a Hero's Turbine in reverse. The circular casing rotates while a stationary pitot pickup on the vertical centerline at the casing inside diameter, with its opening facing against the rotation, picks up this high peripheral speed liquid and converts its kinetic energy to head before discharging it. These pumps can reach heads of over 600 m, (2000 ft.) and capacities of 70 m³/h, (300 gpm). They are small low cost high speed units for this application. They are sensitive to grit abrasion due to the high velocities involved and to entrained gases and vapours which collect at the axial centre line of the pump. Maximum efficiency is approximately 40%. Sales volume is low.

Reversible centrifugal pump

The reversible centrifugal pump is merely a centrifugal pump with straight radial blades. Since they have no curvature the pump may be operated in either direction of rotation with no noticeable change in the conditions of service.

Air lift pump

The air lift pump has been used for lifting water or oil from wells. It consists of an air pipe inside of a discharge pipe. Air or other gas is discharged from the bottom of the air pipe. The result is a flow motivated by the reduction of specific gravity of the water or oil and the buoyancy of the air bubbles.

Water eductor or jet pump

The water eductor or jet pump has also been around for a long time and it is commonly used today in series with a centrifugal pump on domestic wells. It is good for small capacities (20-40 l/min or 5-10 gpm) and lifts up to approximately 40 me*tres* or 125 *feet*. Jet pumps use the discharge of a high velocity jet into a suction chamber to create a vacuum and educe a fluid into the chamber where it combines with the fluid from the jet. The motor-driven centrifugal pump at the top of the well recirculates a portion of its discharge flow back down into the well to the jet pump converging nozzle. This is the motivating fluid. The eductor, see Fig. 3.1, converts the static energy in this incoming fluid to kinetic energy in the converging nozzle and then it entrains the well water from the inlet pipe. The result is a combined flow at intermediate velocity which passes through a diverging nozzle where the significant part of the kinetic energy is converted back to static pressure. If the head loss in the discharge line between the eductor and the centrifugal pump is significant then it must be considered in the selection of the eductor. The Steady-Flow Energy Equation,

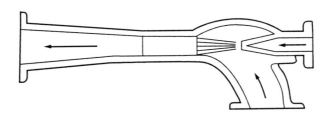

FIGURE 3.1 – Cross section of eductor.

SPECIAL EFFECT PUMPS

FIGURE 3.2 – Centrifugal pump combined flow diagram for residential well water supply.

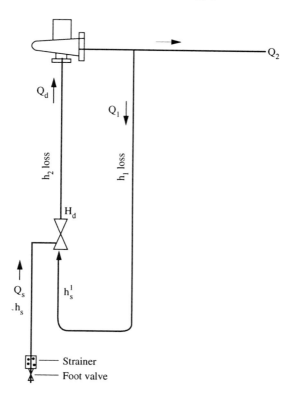

or in this case, the more specific Bernoulli Equation (with no work or heat exchange) can be used. Refer to Chapter 1. A flow diagram is shown in Fig. 3.2. For the flow through the converging nozzle we get the following relationship.

$$h_1 + \frac{U_1^2}{2g} = h_s + \frac{U_n^2}{2g}$$

where U_n = the velocity at the discharge of the converging nozzle and h_s = the head at the suction port of the eductor. Since U_1 is negligible compared to U_n, the formula becomes:

$$\frac{U_n^2}{2g} = h_1 - h_s$$

This term is commonly called the *operating head* The amount educted is obtained by the continuity of momentum equation:

$$m_1U_n + m_sU_s = (m_1 + m_s)U_t$$

where U_t is the velocity at the throat of the diverging nozzle. Across the diverging nozzle:

$$h_s + \frac{U_n^2}{2g} = h_d + \frac{U_d^2}{2g}$$

where h_d and U_d are at the discharge of the diverging nozzle and eductor, since U_d is very small relative to U_t, we can ignore it and we get:

$$\frac{U_t^2}{2g} = h_d - h_s$$

This term is commonly called the *discharge head*. The ratio of the operating/discharge heads, rh, is the ratio of the two velocity heads, or the two head differences, as follows:

$$rh = \frac{\frac{U_n^2}{2g}}{\frac{U_t^2}{2g}} = \frac{U_n^2}{U_t^2} \text{ or } \frac{h_1 - h_s}{h_d - h_s}$$

this term is called the head ratio. Another term used is the *mass ratio, rm = ms/m1*. From the continuity of momentum equation, since U_s is negligible; ms/m1 = Un/Ut–1 or $r_h^{0.5}$–1, thus

$$r_m = r_m^5 - 1$$

Since the specific gravities of all the flows are the same, then the *volumetric ratio* is also equal to rm.

The efficiencies of eductors range up to approximately 30%.

Electromagnetic pump

The electromagnetic pump found its impetus in the nuclear and aerospace fields. It is used for the circulation of liquid metals and other media of high conductance. They are expensive and large. A non-magnetic pipe has a magnet placed on it such that the lines of force are radial to the pipe. When energized this creates a force in the conducting fluid, causing it to flow. These pumps have no moving parts which is a distinct advantage in these services. Efficiencies are low.

References
[1] Marks, Lionel S., Mechanical Engineers Handbook, 5th Edition, McGraw-Hill,(1951) pp1830-1831
[2] Pumping Manual, 6th Edition, Trade and Technical Press (1979), pp 135-137
[3] Salisbury, J. Kenneth, Kent's Mechanical Engineers Handbook - Power Volume, 12th Edition, (1950), John Wiley & Sons, Inc and Chapman & Hall, Ltd., p5-82

REGENERATIVE TURBINE PUMPS

Regenerative Turbine Pumps are used for clean non-abrasive fluids with relatively high head and low flow requirements. The impeller is generally a solid disk with impulse type buckets cut around its periphery or radial blades on the upper part of the walls near the periphery. The latter is referred to as the side entry type. The flow is acted upon by a chain of impulse turbine type pulses as each vane imparts energy to the fluid being pumped. The energy imparted by these impulses from the impeller blade increases as the liquid makes its way along the casing passage. See Figs. 4.1 A, B, and C.

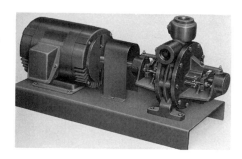
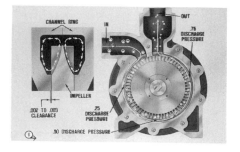

FIGURE 4.1 – Regenerative turbine pump.

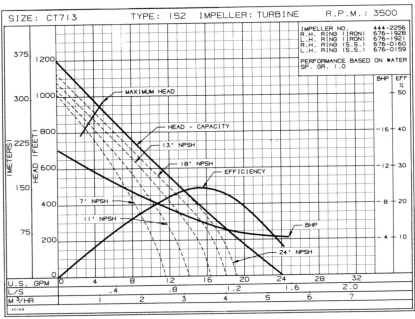

FIGURE 4.2 – Regenerative turbine pump performance map (3500 rpm).

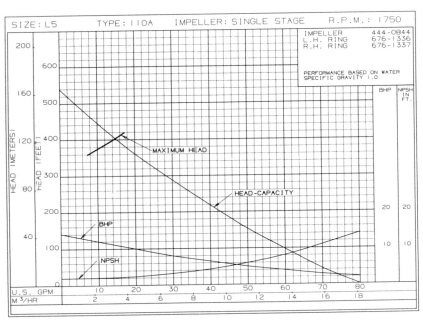

FIGURE 4.2A – Regenerative turbine pump performance map (1750 rpm).

REGENERATIVE TURBINE PUMPS

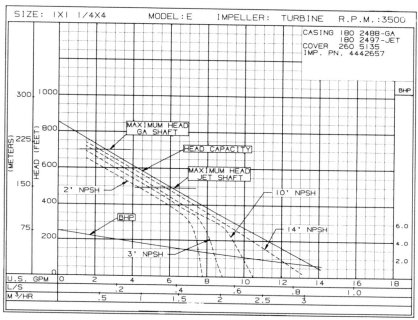

FIGURE 4.2B – Regenerative turbine pump performance map (3500 rpm).

The pressure at zero flow (shut off) can be several times that at the full load rating or the best efficiency point of the pump. In fact, the characteristic (head/capacity) curve is often steep enough that the driver power consumption increases inversely with the flow. The steepness of the characteristic curve and the rising BHP curve with decreasing flow can be seen in Figs. 4.2 A, B and C.

The casings are concentric (equal cross-sectional area), as opposed to the volute type.

Regenerative turbine pumps can be self-priming if the casing traps enough liquid to separate the suction and discharge sections with a seal. They are available for heads up to *350 metres (1100 feet)*/stage and capacities to 100 *litres/min* (250 gpm).

The regenerative turbine pump follows the affinity laws of centrifugal pumps with capacity being proportional to the speed and head proportional to the square of the speed. Refer to Chapter 5 for further information on affinity laws.

Centrifugal pumps are listed as having specific speeds from 10–385 (500–20,000 NA). Regenerative turbines fill the void up to *10 (500)*. They are capable of handling a relatively high percentage of gas or vapour without choking, e.g. 20%.

CENTRIFUGAL PUMP NOMENCLATURE, CHARACTERISTICS AND COMPONENTS

Centrigual Pumps, as stated previously, make up the vast majority of the pumps that are in use and those being sold each day. In this chapter, the types, fundamentals and characteristics of these pumps will be covered. Subsequent chapters will cover centrifugal pump types and their use in specific applications.

Nomenclature

The Hydraulic Institute nomenclature for centrifugal pumps, is covered in its 'Standards' publication and Elsevier's 'Europump Terminology'. They provide diagrammatical information (see page 42). These cross-sectional views and accompanying ballooned parts lists are extremely good reference items and will be used throughout this text. Their descriptions provide differentiation as commonly used by pump users, manufacturers and parts suppliers.

Rotation

The convention for rotation of centrifugal pumps is as follows: Imagine yourself sitting on the driver looking toward the pump. If the top of the shaft moves from left to right, then the pump rotation is clockwise. If it moves from right to left then the rotation is counter-clockwise.

Characteristic Curves

Characteristic curves of centrifugal pumps are their head, kW (bhp), NPSHR and iso-efficiency curves plotted against capacity for a given rpm. They are extremely important in the pump selection process.

Head/Capacity

Head/capacity curves come in several forms: a single curve that might represent the

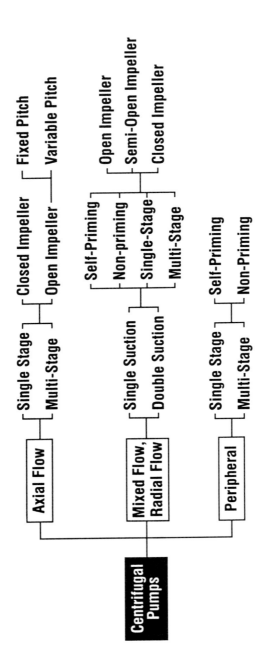

Classification of centrifugal pumps

CENTRIFUGAL PUMP NOMENCLATURE, CHARACTERISTICS & COMPONENTS 41

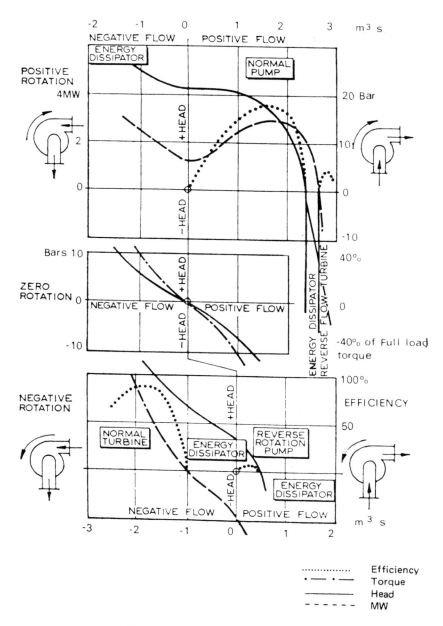

Characteristics of centrifugal pump.

A unique series of dictionaries

- these books should be used by EVERY company seriously interested in doing business in the pumping sector in Europe - now the largest trading block in the world

- expertly checked for each and every entry for the 5 major European languages covered: English, French, German, Italian, Spanish

- For full details of EC legislation, standards and trading laws these titles are ESSENTIAL for ALL manufacturers and major specifiers of pumps and their components

EUROPUMP TERMINOLOGY SERIES

- Pump Applications ISBN 0 85461 071 5
- Pump Names ISBN 0 85461 089 8
- Component Names ISBN 0 85461 099 5

ORDERS & SUBSCRIPTION ENQUIRIES TO

Elsevier Advanced Technology
P O Box 150, Kidlington,
Oxford OX5 1AS, UK,
Tel: +44 (0)865 843848
Fax: +44 (0)865 843971

OR

In North America
Elsevier Advanced Technology
660 White Plains Road, Tarrytown,
NY 10591-5153, USA
Tel: +1 (914) 333 2458

REFUND GUARANTEE
Remember our money back guarantee. Should you wish to cancel your purchase, simply return the book within 28 days and we will refund your payment.

ELSEVIER ADVANCED TECHNOLOGY

ORDER FORM
☐ Please send me _____ copy/ies of the Pump Applications @ £100/US$160
[PIP61 + B1]
☐ Please send me _____ copy/ies of Pump Names @ £69/US$110
[PIP62 + B1]
☐ Please send me _____ copy/ies of Component Names @ £160/US$225
[PIP63 + B1]

Payment
☐ Please find payment enclosed (please make cheques payable to Elsevier)
☐ Please charge my credit card:
    Access/Mastercard/Visa/Barclaycard/Eurocard/American Express
Card No. _____ Card Name _____
Expiry Date _____ Date _____
Signature _____
☐ Please invoice me. Company Purchase Order No. _____

Delivery Address
Name _____ Position _____ [JT:]
Organization _____
Address _____
Town _____ State _____
Post/Zip Code _____ Country _____
Tel _____ Fax _____
Nature of Business _____ [SIC:]

CENTRIFUGAL PUMP NOMENCLATURE, CHARACTERISTICS & COMPONENTS   43

THOUSANDS OF PUMPS NEVER SEIZE BECAUSE **GRAPHALLOY®** ENABLES PUMPS TO SURVIVE:

°RUNNING-DRY
°FLASHING PUMPAGE
°STAND-BY SLOW ROLL
°FREQUENT START/STOPS

Self-lubricating, non-galling GRAPHALLOY® provides the margin of safety to keep your pumps operating. Eliminate metal-on-metal rubs, galling, and seizures. Call or write for case histories.

 **GRAPHITE METALLIZING**
CORPORATION
1050 Nepperhan Avenue, P.O. Box 110, YONKERS, NY 10702 U.S.A.
Tel: 914-968-8400 · · FAX: 914-968-8468

ISO 9002 CERTIFIED

SEAL-LESS, BUT DIFFERENT!
constant flow as pressure rises

Hydra Cell pump features

Positive displacement, so flow is little influenced by pressure variations ★ Long life ★ Low maintenance ★ Shaft or Direct Drive ★ Low energy consumption ★ Very compact ★ Wide choice of materials ★ For water, chemicals, slurries, abrasives, hot and cold liquids etc.

HYDRA-CELL PUMP RANGE

Model	l/min	bar
D3	1-10	0-70
D10	2-32	0-70
H25	4-75	0-70
D40	20-150	0-80
D11	2-15	0-100

WANNER INTERNATIONAL LTD
Grange Court, Grange Road, Tongham, Surrey GU10 1DW, England
Tel: +44(0) 1252 7812234 Fax: +44 (0) 1252 781235

maximum diameter impeller performance for that pump casing as could be provided with a proposal by the pump manufacturer (see Fig. 5.2); Oak Tree Curves for various diameters of impeller to be fit into the casing, (see Fig. 5.3) or Oak Tree curves for various speeds in the case where speed adjustment or control is available, (see Fig. 5.4). These latter curves got their name from the resemblance of their iso-efficiency lines to the rings in an oak tree cross-section and are often found in vendor's sales manuals.

FIGURE 5.1 – Volute pumps.

Kw (bhp)

Kw (bhp) curves are shown in Figs. 5.24. There is one curve for each diameter or speed curve, representing the power required with that diameter or speed at any given capacity. Note the peaking ("non-overloading") power curve in Figs. 5.2 and 5.3 (C and D). These are generally found with the steeper HQ curves as opposed to the rising power curve of Fig. 5.7 with flatter HQ curves.

NPSHR

NPSHR curves show the amount of NPSHR required at any specific capacity. See Figs. 5.2 and 5.3.

Iso-efficiency

Iso-efficiency curve plots allow the interpolation of efficiency values at any head/capacity

CENTRIFUGAL PUMP NOMENCLATURE, CHARACTERISTICS & COMPONENTS 45

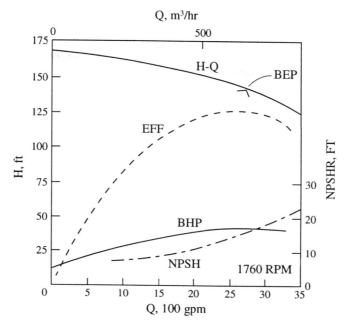

FIGURE 5.2 – Characteristic curves.

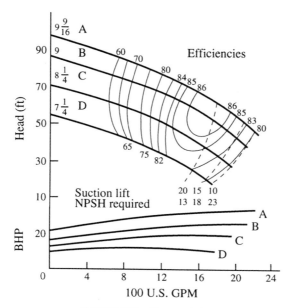

FIGURE 5.3 – Characteristic curves.

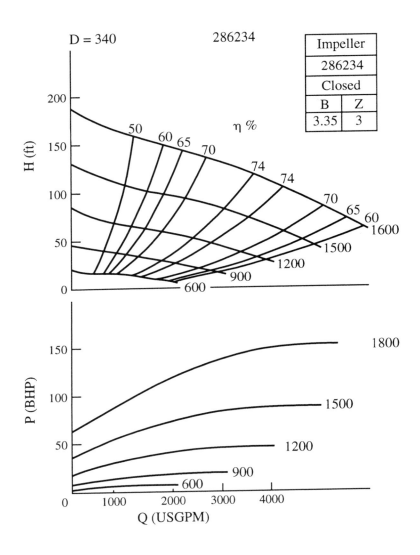

FIGURE 5.4

coordinate point. See Figs. 5.4 and 5.5. The BEP or best efficiency point is either specifically marked, as in the case of the single head capacity curve, (Fig. 5.2), or it can be found from interpolation within the iso-efficiency curves, as in Figs 5.2 or 5.3. This is the point at which the pump was designed and at which operation the losses are minimal.

CENTRIFUGAL PUMP NOMENCLATURE, CHARACTERISTICS & COMPONENTS   47

Multi-stage centrifugal pump – exploded view.

48   PUMP USERS HANDBOOK

The rating point for the application that the pump will be installed in should be as close to this point as possible.

Non-Overloading Diameter vs Motor Size data is shown in Fig. 5.5.

The above four figures were selected to show the different ways that characteristic data can be presented.

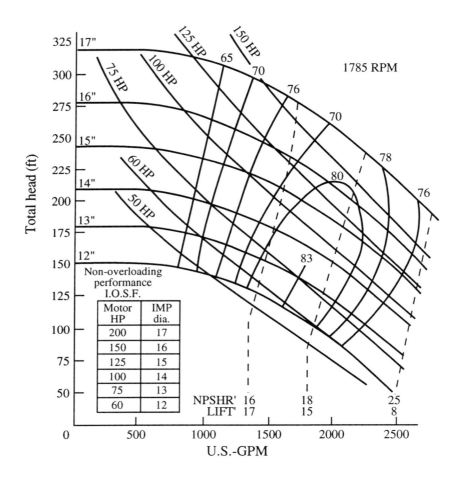

FIGURE 5.5 – Oak tree curve.

CENTRIFUGAL PUMP NOMENCLATURE, CHARACTERISTICS & COMPONENTS

Shape

Shape of H/Q characteristic curves is important to their compatibility with the application the pumps will be used in. A flatter curve, Fig. 5.6, will generally provide the most economical choice in terms of capacity per weight of pump and highest efficiency, whereas a steeper curve, Figs. 5.3, 5.5 or 5.7, will more likely provide non-overloading operation. That is to say the driver will not be overloaded due to a peaking power curve as opposed to a continuously rising power curve. A steep curve is also required for parallel operation of pumps, a subject which will be covered later in this chapter. Another important consideration that shows up in the H/Q characteristic curve is stability. Fig. 5.8, shows a curve with a negative slope as it approaches shut off. Such a condition, where there

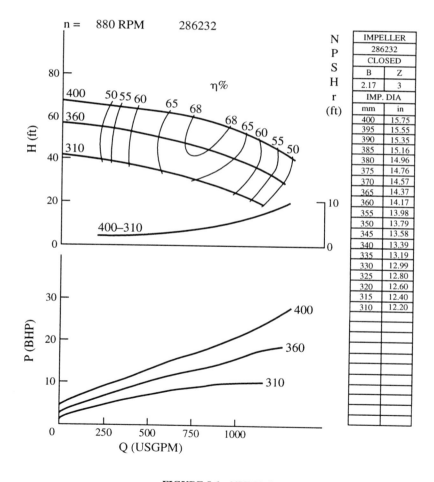

FIGURE 5.6 – NPT 33-4

CHEMICAL EQUIPMENT®

The Magazine Of New Technology

- Published monthly for today's CPI engineer, plant manager and purchasing professional.
  - For 30 years, the CPI's most respected source for product information.
- Concise coverage of innovative pumps, valves, filtration & separation equipment, free literature and more!

RETURN COMPLETED FORM BELOW TO SUBSCRIBE

Name _____ Title _____
Company _____
Address _____ City _____
Country _____ Postal Code _____
Signature _____ Date _____

Payment: MasterCard/Visa or check drawn on US bank.
☐ 1 year (12 issues): $70   ☐ 2 years (24 issues): $115

Visa # _____ Exp. Date _____
MasterCard # _____ Exp. Date _____

Mail to: Subscription Dept., CHEMICAL EQUIPMENT, P.O. Box 650, Morris Plains, NJ 07950-0650, USA. **Fax:** (201) 898-9281.

HYUNDAI PUMPS are running everywhere

Application for

- Power Plant (Thermal, Combined-Cycle and Nuclear)
- Desalination Plant
- Water Supply and Transfer
- Irrigation
- Drainage
- Dry-Dock De-watering
- Dredger
- District Heating Water Supply
- Cooling water for Petro-Chemical Plant

Hyundai pumps are the result of the accumulated technology and guarantee the highest efficiency.

Hyundai is running everywhere all over the world. Even in the hot desert, Hyundai's products are operating continuously and serve most favorably to all clients. Hyundai Pumps provide you with super efficient performance which translate into lower operating costs. Intelligent design, durability and simple maintenance have made Hyundai Pumps reliable worldwide.

HYUNDAI HEAVY INDUSTRIES CO., LTD.
Engine & Machinery Division
1 Jeonha-Dong, Dong-Gu, Ulsan, Korea.
Tel. (82) 522-30-7381~7
Fax. (82) 522-30-7694

If you want the best quality and economic selection, please do not hesitate to contact Hyundai Pump.

All Metric Sizes
Non-lubricating
High Temperature
Abrasives   Service
Downtime
Toxic Materials   Solidification

Solutions
WITH STANDARD DESIGNS

Corrosives   Shaft Deflection
Price   High Speed
Delivery   Limited Space
Installation   High Pressure

Solutions
WITH CUSTOM DESIGNS

Cartridge Split Seals PATENT PEND. SPLIT DESIGN -BY FAR THE SIMPLEST DESIGN AVAILABLE
Environmental Seals STANDARD AND PAT. PEND. DESIGN TO SATISFY TOUGHEST ENVIRONMENTAL LAWS
Cartridge Seals WELDED BELLOWS; SINGLE AND MULTIPLE SPRING; STATIONARY; LOW COST DESIGNS
Welded Metal Bellows SPECIALISTS IN MANUFACTURE AND APPLICATION
Conventional Spring Seals STANDARD AND CUSTOM DESIGNS
Complete Factory Support IMMEDIATE SERVICE AND RESPONSE
Seal Repair ALL TYPES AND BRANDS

ACTIVELY SEEKING AGENTS, DISTRIBUTORS, AND MANUFACTURING PARTNERS THROUGHOUT THE WORLD

INC.

CALL FOR RUSH RESPONSE
(802)878-8307 or 1-800-426-3594 in U.S.A.

1 JACKSON ST., P.O. Box 184 ESSEX JCT., VT USA 05453-0184
(802) 878-2479 FAX

CENTRIFUGAL PUMP NOMENCLATURE, CHARACTERISTICS & COMPONENTS 53

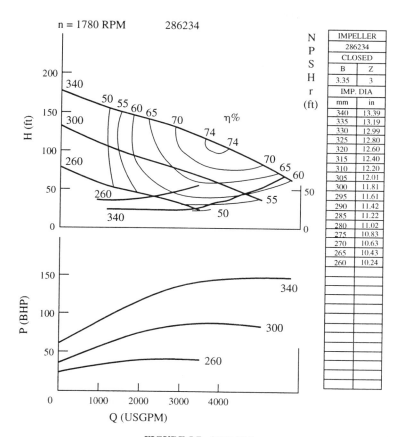

FIGURE 5.7 – NPT 42-8.

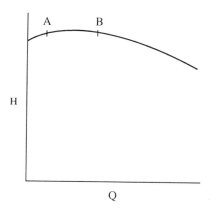

FIGURE 5.8 – Characteristic HQ curve.

are two different capacity points (A and B) that a pump could operate at for a given head, can be a concern if pumps will be run in parallel without individual throttle valves. On the other hand, if capacity control is to be obtained by throttling, the effect of a steep system head curve is to allow only the flow at the intersection of the throttled system curve with the otherwise unstable H/Q Curve. Stepanoff[2] gives further rules that determine instability under these conditions.

Affinity Laws

Affinity Laws are the relationships by which head, capacity and power required vary with speed. These laws are:

$$Q \propto N \quad \text{or} \quad Q_1/Q_2 = N_1/N_2$$
$$H \propto N^2 \quad \quad H_1/H_2 = N_1^2/N_2^2$$
$$P \propto N^3 \quad \quad P_1/P_2 = N_1^3/N_2^3$$

For corresponding points, the efficiency remains approximately constant over reasonable speed changes.

Diameter modifications follow the affinity laws within limits and with more tendency for inaccuracy on some types of pumps than others. A general set of formulas can be used:

$$Q \propto ND \quad \text{or} \quad Q_1/Q_2 = N_1D_1/N_2D_2$$
$$H \propto N^2D^2 \quad \quad H_1/H_2 = (N_1D_1)^2/(N_2D_2)^2$$
$$P \propto N^3D^3 \quad \quad P_1/P_2 = (N_1D_1)^3/(N_2D_2)^3$$

(NPSH varies as the square of the impeller diameter ratios similar to head, however, due to the occasional unreliability of these NPSH calculations, it is recommended that tests be used wherever possible.)

If speed or diameter is constant they can be cancelled out. It is hard to say just how much diameter correction can be applied and still stay close to these relationships because the sensitivity varies with the type of pump. Ross[3] recommends that impeller trims less than 80% of original diameter be avoided and on pumps of specific speeds of *2500* to *4000* that this be limited to 90%. Many catalogue pumps are sold with cut downs from the maximum diameter greater than these valid recommendations as can be seen from the oak tree curves in vendors' catalogues. The difference here is that the pump vendors have original test data and probably subsequent statistical data showing what these pumps will do at various diameter trims, and this is reflected in the catalogue curves. The changing of diameter after shipment may not enjoy the advantage of this experience unless time allows for the vendor to be contacted. The latter is very much recommended.

Example 1

Assume the performance of a *3500* rpm pump with a 9 in. (225 mm) diameter impeller, (D_1) is known (see Fig. 5.9), and that it is desired to find the performance of an 8½ in. (212

CENTRIFUGAL PUMP NOMENCLATURE, CHARACTERISTICS & COMPONENTS

mm) impeller (D_2) at the same speed. This can be calculated from the 9 in. diameter curve as follows:

1. Divide 8.5 by 9
2. Multiply by 240 (maximum capacity) and get 227 gpm
   i.e. $Q2 = Q1 \times D2/D1 = 240 \times 8.5/9 = 227$ gpm
3. Head varies as the square of the impeller diameters. Since the head (H) at 240 gpm on the 9 in. diameter curve is 287 feet, we get:
   $H2 = H1 \times (D2/D1)^2 = 240 \times (8.5/9)^2 = 256$ feet
4. The power of a pump varies as the cube of the ratio of the impeller diameters, or:
   $BHP2 = BHP_1 \times (D2/D1)^3 = 35.5 (8.5/9)^3 = 29.9$ hp.
5. NPSHR varies as the square of the ratio of the impeller diameters as did head, or:
   $NPSHR2 = NPSHR1 \times (D2/D1)^2 = 28.0 (8.5/9)^2 = 25.0$ ft.
6. Repeating this process enough times (minimum of three) will result in new curves, as shown in Fig. 5.9. Tabulated data from such calculations is shown below.

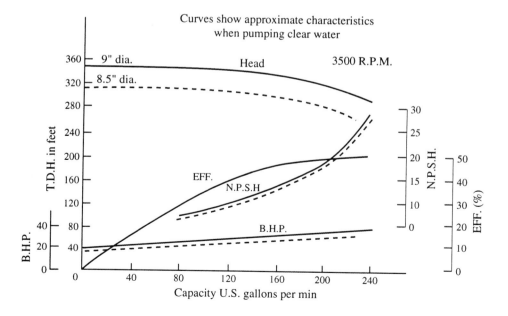

FIGURE 5.9 – Effect of diameter change.

TABLE 1
9 in. Diameter curve at 3500 rpm

GPM	Head	BHP	Eff.	NPSH
240	287	35.5	49.0	28.0
200	316	32.7	48.8	18.0
160	334	30.0	45.0	12.7
120	342	27.2	38.1	9.2
80	345	24.5	28.5	6.6
40	346	21.5	16.3	0
0	346	18.5	0	0

TABLE 2
8.5 in. Diameter curve at 3500 rpm

GPM	Head	BHP	Eff.	NPSH
227	256	29.9	49.0	25.0
189	282	27.5	48.8	16.1
151	298	25.3	45.0	11.3
113	305	22.9	38.1	8.2
76	308	20.6	28.5	5.9
38	309	18.1	16.3	0
0	309	15.6	0	0

Note that the efficiency remains constant as the capacity, head and horsepower are stepped down. For example, on the 9 in. curve the efficiency is 45% at the new point on the 8.5 in. curve. This can be checked by calculating the efficiency using the following equation:

(5.1) $\quad$ eff. $= 100 \ HQ/3960 \ bhp$ at 1.0 s.g.
$\quad\quad\quad\quad\quad = 151 \times 298 \times 100/(3960 \times 25.3) = 45\%$

Changing Rotative Speed from that shown on the standard curve can be estimated in the same manner. For example, to obtain a 9 in. (225 mm) curve at 2900 rpm ($N2$):

1. Since capacity varies as the ratio of the impeller speeds,
$Q2 = Q1 \ (N2/N1) = 240 \ (2900/3500) = 199$ gpm.

2. The Head of a pump varies as the square of the ratio of the impeller speeds or:
$H2 = H1(N2/N1)^2 = 287 \ (2900/3500)^2 = 197$ ft.

3. The horsepower of a pump varies as the cube of the ratio of the impeller speeds, or:
$BHP2 = BHP1 \ (N2/N1)^3 = 35.5 \ (2900/3500)^3 = 20.2$ hp.

CENTRIFUGAL PUMP NOMENCLATURE, CHARACTERISTICS & COMPONENTS

4. NPSH varies as the square of the ratio of the impeller speeds, or:
$$NPSH1 = NPSH2 (N2/N1)^2 = 28.0 (2900/3500)^2 = 19.2 \text{ ft}$$

This example and Fig. 5.9 courtesy of The Duriron Co., Inc.[4]

Impeller Modifications

Impeller modifications can be made which will adjust the performance of the pump to some degree.

Overfiling

Overfiling, as shown in Fig. 5.10 should be done after a cut down to bring the blade tips thickness at the new outside diameter back to their original dimension. This prevents losses by not allowing slip to increase and thus maintains the efficiency and the shape of the characteristic H/Q curve as closely as possible. Slip can be considered simplistically for purposes here as the result of blade geometry, that results in the impeller discharge flow

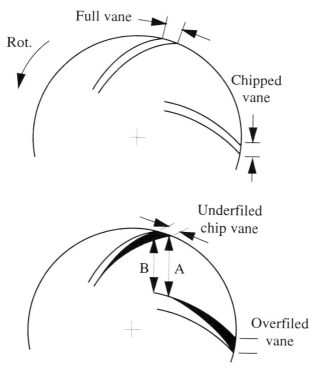

FIGURE 5.10 – Chip and underfile.

being at an angle that is less than the blade angle. This deviation can worsen as exit blade thickness is increased.

Underfiling, Undercutting, chipping or backing-off

Underfiling, Undercutting, chipping or backing-off the blade as shown in Fig. 5.10 can provide an improvement in efficiency as well as increased capacity and a slightly higher head from a flatter curve. As dimension A increases to B with the underfiling, the passage area increases and as a result the capacity increases. At the same time the outlet angle increases slightly with the resulting effect of slightly higher head and a flatter curve, The depth of the filing can be roughly established by the values in Table 3.

Any modifications to an impeller such as these should be followed by a re-balance of the impeller.

Wherever possible, the pump vendor should be consulted for his advice.

TABLE 3
8.5 in. Diameter curve at 3500 rpm

Diameter Range		Depth	
5–10"	13–25cm	1.5"	3.5cm
10–20"	25–50cm	3.5"	10cm
20–30"	50–75cm	5.0"	13cm
30–40"	75–100cm	6.0"	15cm

Actual vs Reference Curve Speed

Actual vs Reference Curve Speed is an area that needs closer attention from customers and sales representatives. Many times orders are received for a pump without driver, with the synchronous speed of the motor listed e.g. 3000 or 3600 rpm. This is especially true if the selection curves have been set up for synchronous speeds. This error is often caught in the order process at the pump vendor, but occasionally it is not. This has resulted in field problems where the motor has been overloaded. If the catalog curves are plotted at a more approximately correct speed such as 1475 (50 Hz) or 1770 (60 Hz) rpm for the range of motors covered, e.g. 25–50 kW (35–75p). Then the problem is ameliorated, but not necessarily eliminated. The actual driver full load speed could be 1488 (1786) rpm. Further error is added if the motor loading is less than its full load rating, e.g. 85%. Slip on an induction motor is proportional to load. If the motor selected has a full load rated speed of *1488* rpm then the slip is 12 rpm. At 80% load the slip is 0.8 x 12 or 10 rpm. The end result is an order for a pump has been entered, calling for *1475* rpm whereas the actual motor speed at the pump rating will be 1490 rpm. The effects are as follows:

CENTRIFUGAL PUMP NOMENCLATURE, CHARACTERISTICS & COMPONENTS

THE RIGHT GEAR FOR PUMPING

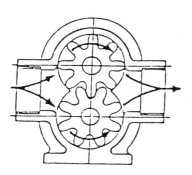

If you need standard or special purpose gear pumps or pumping systems, contact Albany.

- Many materials, inlcuding Hastelloy and PVDF
- Turbine, air, diesel and flameproof electric drives
- Wide range of industry standards including API 676
- Specialists in hot and viscous liquids — heated pumps
- Pump Gear Service — 400 gear pumps in stock

For details and quotations contact

Albany Engineering Co. Ltd.
Lydney, Gloucestershire
GL15 5EQ.
Tel: 01594 842275 Telex: 43363
Fax: 01594 842574

Quality Assurance to BS5750 Pt. 2. and ISO 9002

- DOSING PUMPS
- CENTRIFUGAL PUMPS

MAKERS OF STANHOPE, CROWN, & BARCLAY KELLETT GEAR AND CENTRIFUGAL PUMPS

Pumps with Built-In Reliability

For Acids Alkalis Corrosive and erosive fluids

Glandless Centrifugal Pumps - designed and manufactured to meet industry's highest standards

KESTNER

Engineering Co. Limited

Station Road, Greenhithe, *Telephone: 01322 383281*
Kent, DA9 9NG England *Fax: 01322 386684*

IWAKI MAGNETIC DRIVE PUMPS

Magnetically coupled process pumps

Main Features
- Sealess Construction
- Choice of Materials
- High reliability
- Easy maintenance
- Sizes from 1 LT/min — 1.5 m^3/min

Now offer models with dry running capability Ex Stock delivery

IWAKI PUMPS (UK) LTD
Unit 2 Monkmoor Ind. Estate
Monkmoor Rd.
Shrewsbury SY2 5SX
Tel: 01743 231363
Fax: 01743 366507

Actual Capacity, Qa = 1490/1475Q = 1.016 Q
Actual Head, Ha = (1.016)² = 1.033 H
Actual Power, Pa = (1.016)³ = 1.048 P

When one considers that the ISO Test Standards for Acceptance Class I[5] calls for an efficiency guarantee range of +/- 2.5% and we have already added 5% power due to a lack of understanding and improper specification of motor speed before the pump is even built the importance of ascertaining the actual speed the motor will run at can be seen. Hydraulic Institute Test Standard – 1.6[6] calls for a +5/– 0% head tolerance for one group of pumps. If the motor in this example had been sized close to rating, this error in impeller sizing due to incorrect speed information could have resulted in overloading and this has been the unfortunate experience of too many customers.

Specific Speed

Specific Speed[7] as used in pumps is based on a dimensionless ratio $NQ^{0.5}:(gH)^{0.75}$, any specific value of which describes a combination of operating conditions that permits similar flow conditions in geometrically similar hydrodynamic machines. The g is dropped and the result is dimensional but meaningful. In the case of SI, the units are consistent. In North American usage, the units are inconsistent.

(5.2) $\quad\quad\quad\quad\quad\quad\quad\quad\quad Ns = NQ^{0.5}/H^{0.75}$

where: N = rev/Min.
Q = m³/s, (gal/min)
H = m, (ft)

Converting Ns in SI units to USCU units as shown above: Ns (USCU) = 51.65 Ns (SI).

Another perspective of Specific Speed is that it is the speed at which an impeller would run if reduced in size to deliver 1 gpm against 1 ft. of head. Specific Speed is also called the Type #.

Fig. 5.11 shows the relationship between specific speed and impeller shapes and Fig. 5.12 is a plot of attainable efficiencies with well designed pumps *vs* specific speed. These curves alone show the tremendous value of the concept of Specific Speed. Specific Speed must be evaluated at the best efficiency point of the pump.

Radial Flow Pumps

Radial Flow Pumps are those which have no axial component to the discharge from their impellers. See Fig 5.11 for examples.

Mixed Flow Pumps

Mixed Flow Pumps are those which have an axial component to the discharge from their impellers. See Fig 5.11 for example.

Axial Flow Pumps

Axial Flow Pumps are those which have no centrifugal component to the discharge from

62 PUMP USERS HANDBOOK

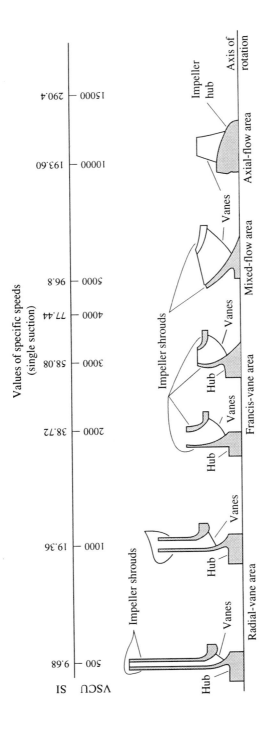

FIGURE 5.11 – Specific speeds.

CENTRIFUGAL PUMP NOMENCLATURE, CHARACTERISTICS & COMPONENTS 63

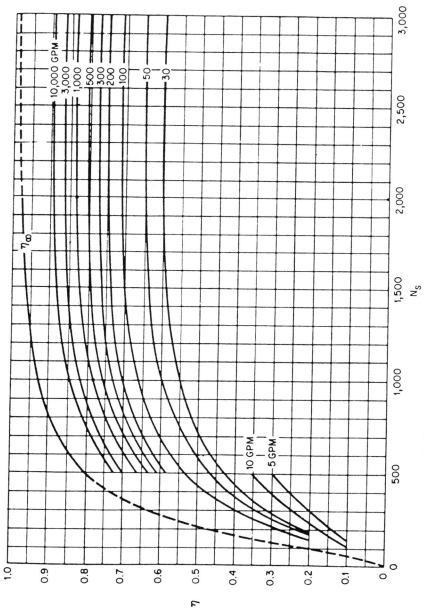

FIGURE 5.12 – Efficiency vs Specific Speed and Capacity.

their impellers and all flow is axial such as from a propeller or fan. See Figs. 5.11 and 5.14.

Suction Specific Speed

Suction Specific Speed, S (Suction Specific Speed Required) is another dimensionless ratio and is analogous to specific speed. It describes all the inlet conditions that produce similar flow conditions in geometrically similar inlet passages. Like specific speed it is defined at the bep of the pump.

$$(5.3) \qquad S = NQ^{0.5}/(NPSHR)^{0.75}$$

Note: Double Suction Pumps must have their total flow divided by 2 for use in this equation, so that you are looking at the inlet conditions of one of the two identical impeller inlets.

Suction Specific Speed Available

Suction Specific Speed Available, SA is the same formula except for the substitution of NPSHA for NPSHR. There should be an adequate margin between S and SA

$$(5.4) \qquad SA = NQ^{0.5}(NPSHA)^{0.75}$$

There are many cases where the practical limit on operating speed – for single and double suction pumps – can be exceeded with no ill effects and manufacturers experience should be taken into account. This is especially true on catalogue type pumps. Hydraulic design has progressed today to a point that calculated performances by a knowledgeable vendor can be accurate with much higher suction specific speeds than *8500*. This is the case with large engineered type pumps. S must equal or exceed SA, to significantly reduce the risk of cavitation. This is logical since NPSHA must exceed NPSHR to prevent cavitation and these two are now in the denominator of the SA and S terms. Suction Specific Speed has its limitations and will be covered in more detail at the end of this chapter.

Losses

Losses are shown for relative visualization in Fig. 5.13. The Euler Head curve is drawn based on the discharge angle of the impeller blades. Slip and the effect of uneven flow velocities are then taken into consideration. The result of this is the Ideal Head curve. The friction, shock and diffusion losses are then subtracted from the Ideal curve to give the actual H/Q curve. Likewise power losses are shown. The shock losses come from the turbulence set up as flow enters the impeller passages and discharges into the casing. Friction losses are a function of the amount of wetted surface in the vane and casing passages, the relative roughness and the velocity. Friction losses increase with capacity whereas leakage losses increase with head. The power loss of disc friction varies as the 5th

CENTRIFUGAL PUMP NOMENCLATURE, CHARACTERISTICS & COMPONENTS

(a)
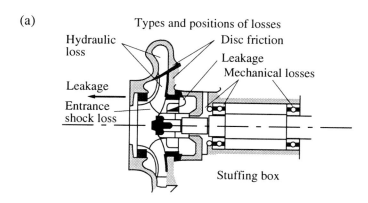

(b)
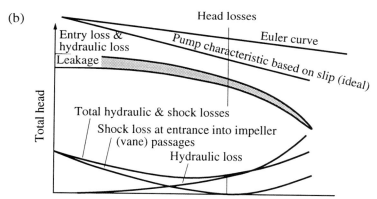

(c)
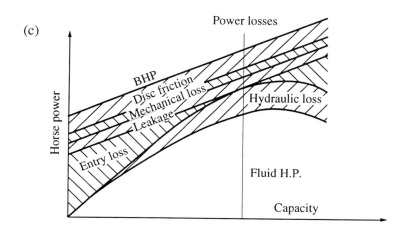

FIGURE 5.13 – Losses.

power of the diameter but is relatively constant with capacity as is mechanical loss. From a user standpoint the pertinence of this loss visualization is in the understanding of the effect of increased clearance and leakage with wear, the ramifications of excess head requirements in the specification over actual, and the effect on disc friction from the increased diameter or speed required, the increased friction losses from scale or corrosion and the result of increased shock losses from the dulling of blade inlet edges with erosion or impact damage.

Gases, Vapours and Priming

Centrifugal pumps are with few exceptions not designed to handle gases or vapours. 1% gas by volume can result in a discernible drop in head and capacity capability of the pump. 5–6% can cause a choking action where pumping actually ceases instantaneously and then tries to recover. If the suction is not flooded on start up by the system, most centrifugal pumps will not self-prime. A type of centrifugal specifically designed to keep the suction somewhat flooded and to allow startup under those conditions is called a self-primer. Even these units require an initial prime. These will be covered in more detail later. A slight amount of air (less than 1%) is sometimes purposely injected into the suction of a pump to quiet it down with no ill effects. Unlike the vapour bubbles which implode when they are pressurized, the gas bubbles will compress but not condense and provide an elastic cushion that muffles the shock of the vapour bubbles imploding.

Impellers

Impellers come in many configurations. In addition to the various centrifugal impeller shapes shown in Fig. 5.11, there are variations in the shroud (passageway cover) configurations, refer to Fig. 5.14.

The closed impeller has a front and back shroud or in the case of the double suction impeller two front shrouds (one on each side). It is the most commonly used design and generally the one that will result in the maximum efficiency at the design point (BEP). See Figs 5.21 and 5.23.

The open impeller is found in pulp and paper, chemical and pharmaceutical services as examples. It is generally used where stringy material might be found in the process fluid. It can be easily cleaned which is a factor in its widespread use in these applications. It has little back shroud area if any. Axial thrust is low. See Fig. 5.15.

The semi-open (or semi-closed) impeller See Fig. 5.19 is one which has a full or partial back-shroud. It is a stronger blade design from the web reinforcement. The axial thrust increases as the back shroud increases, until it is full. With a full shroud it can still handle stringy material and is readily cleaned.

The vortex impeller is covered in Chapter 6 under Vortex Pumps. Also see Fig. 5.22.

The single suction impeller can be any of the above (except Fig. 5.23. It is the most common impeller with its inlet on one side only. See Fig. 5.45.

The double suction impeller has an inlet on both sides having the symmetrical appearance of two single suction impellers back to back. See Fig. 12.2 and 5.23. Except

36th year of publication: all previous editions sold out

PUMPING MANUAL
9th Edition
By Christopher Dickenson

Widely recognised as the first source of reference on all aspects of pump technology and applications the Pumping Manual will enable you to...

Specify the right pump for the task
Design cost-effective pump systems
Understand the terminology
Ensure effective installation, operation and maintenance of all your pumping equipment
and much more!

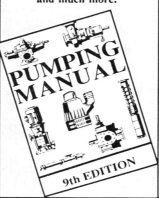

Section One: Introduction
SI Units
Pump Evolution
Pump Classification
Pump Trends

Section Two: Pump Performance and Characteristics
Fluid Characteristics
Pump Performance
Calculations, Type Number and Efficiency
Area Ratio
Pipework Calculations
Computer Aided Pump Selection

Section Three: Types of Pumps
Centrifugal Pumps
Axial and Mixed Flow Pumps
Submersible Pumps
Seal-less Pumps
Disk Pumps
Positive Displacement Pumps (General)
Rotary Pumps (General)
Rotary Lobe Pumps
Gear Pumps
Screw Pumps
Eccentric Screw Pumps
Peristaltic Pumps
Metering and Proportioning Pumps
Vane Pumps
Flexible Impeller Pumps
Liquid Ring Pumps
Reciprocating Pumps (General)
Diaphragm Pumps
Piston, Plunger Pumps
Self Priming Pumps
Vacuum Pumps

Section Four: Pump Materials and Construction
Metallic Pumps
Non Metallic Pumps
Coatings and Linings

Section Five: Pump Ancillaries
Engines
Electric Motors and Controls
Magnetic Drives
Seals and Packaging
Bearings
Gears and Couplings
Control and Measurement

Section Six: Pump Operation
Pump Installation
Pump Start-up
Cavitation and Recirculation
Pump Noise
Vibration and Critical Speed
Condition Monitoring and Maintenance
Pipework Installation

Section Seven: Pump Applications
Water Pumps
Building Services
Sewage and Sludge
Solids Handling
Irrigation and Drainage
Mine Drainage
Pulp and Paper
Oil and Gas
Refinery and Petrochemical Pumps
Chemical and Process
Dosing Pumps
Power Generation
Food and Beverages
Viscous Products
Fire Pumps
High Pressure Pumps

Section Eight: User Information
Standards and Data
Buyers Guide
Editorial Index
Advertisers Index

800 pages
1500 figures and tables
ISBN: 1 85617 215 5

For further details post or fax a copy of this advert complete with your business card or address details to:
Elsevier Advanced Technology, PO Box 150, Kidlington, Oxford OX5 1AS, UK
Tel: +44 (0) 1865 843848 Fax: +44 (0) 1865 843971

CIP4699

ELSEVIER ADVANCED TECHNOLOGY

PROFILE OF THE INTERNATIONAL PUMP INDUSTRY
MARKET PROSPECTS TO 1998

The 2nd edition of the highly successful Profile of the International Pump Industry is now available - fully revised and updated offering a whole new analysis of the pump industry

Do you know?

• Who the major players are, and how they are performing
• Who is active in acquisitions
• How technological developments will change the industry over the next five years
• What trends are developing in the main end-user markets

Contents include:

Executive Summary
World Pump Market by Country
International Pump Market
Pump Trade
Pump Production Data by Country
Market Analysis by End-User Industry
Technology Overview
Major Pump Suppliers
Profiles of 32 Leading Pump Manufacturers
Directory of Pump Companies
around 650 companies

☐ Please send me more information on **Profile of the International Pump Industry** 2nd Edition [PIP71.A1]

Price £545/US$872*
*Dollar price may change due to exchange rate fluctuations

Name (Mr/Ms) _____
Position _____ (JT:)
Organization _____
Department _____
Address _____
_____ Town _____
State _____ Post Code/Zip _____
Country _____
Tel _____ Fax _____
Nature of Business _____

☐ Please send me a copy of Advanced Materials & Engineering Catalogue [PBG10.A1]
☐ Please send me further information on Elsevier Market Research Reports

Please return to or call *Ryan Sheppard*
Elsevier Advanced Technology, PO Box 150, Kidlington, Oxford OX5 1AS, UK
Tel: +44 (0)1865 843828
Fax: +44 (0)1865 843971

In N. America (orders only)
Elsevier Advanced Technology, 660 White Plains Road, Tarrytown, New York NY 10591-5153, USA
Tel: +1 (914) 524 9200
Fax: +1 (914) 333 2444

To order, contact Ryan Sheppard on
Tel: +44(0)1865 843828
Fax: +44(0)1865 843971

CENTRIFUGAL PUMP NOMENCLATURE, CHARACTERISTICS & COMPONENTS

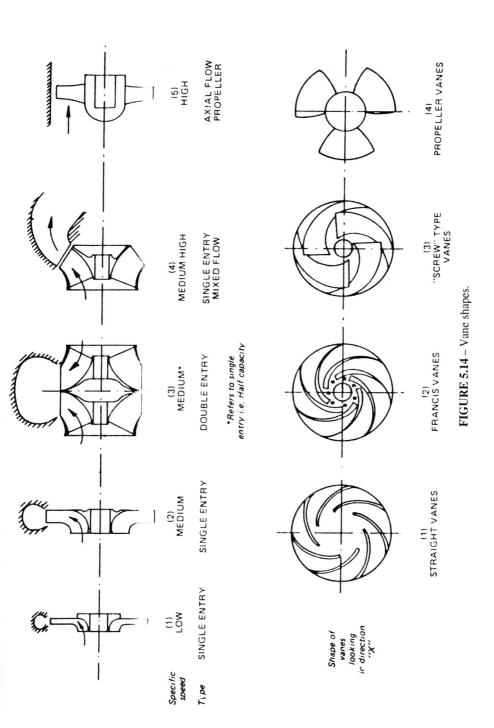

FIGURE 5.14 – Vane shapes.

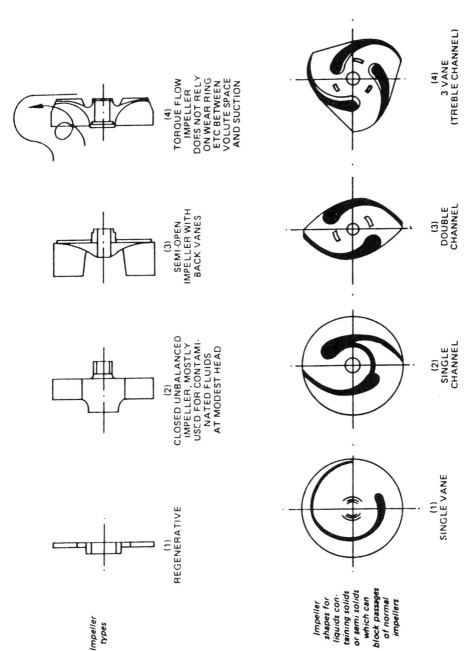

FIGURE 5.14 (continued) – Vane shapes.

FIGURE 5.15 – Open impeller.

for possible non-symmetrical geometry conditions in the impeller inlets, casing passages and wear ring clearances; and non-symmetrical flow conditions in the suction passages this unit has an inherent balanced thrust condition.

Casings

The volute casing

The volute casing has a cross-sectional collecting area that continually increases from the cutwater (or tongue) to the conical diffuser leading to the exit flange. See Figs 5.18 and 5.1. The tangential volute has a discharge centre line that is parallel to a tangent at the impeller o.d.. See Fig 5.1 . The centre line volute wraps around the impeller o.d. and then has a discharge axis that is radial to the impeller axis on its vertical centre line. It is also self-venting. See Fig 5.18. The centre line volute has come into common use in the process industries where piping and process changes are common and the pump is used to support piping loads. Its vertical centre line orientation splits the piping load evenly over the two pump casing feet and hold down bolts. The piping load on the tangential volute casing results in a downward force on the nearest leg and bolt to the discharge and a upward force on the farthest. The tangential volute can be expected to have an extra point of efficiency or two over an ideally designed centerline discharge due to the absence of the extra, constricted, 90° turn. The volute casing has a uniform casing pressure characteristic at the bep (best efficiency point) with close to zero radial load. On either side of the bep capacity, the radial load increases and can be substantial at low capacities.

The double volute casing

The double volute casing has a full or partial splitter vane splitting the flow in half. A full splitter covers the distance shown in Fig. 5.16 diagram 12 plus or minus 45°. Ross[8] points out that a full double volute will have 16% of the radial thrust of a single volute. The partial splitter may cover the second 180° of the collector plus an additional half of the axial length of the exit diffuser or as little as the lower left hand quadrant of the above figure. Tests[8] show that even this short splitter has a significantly lower (as low as 33%) radial thrust than

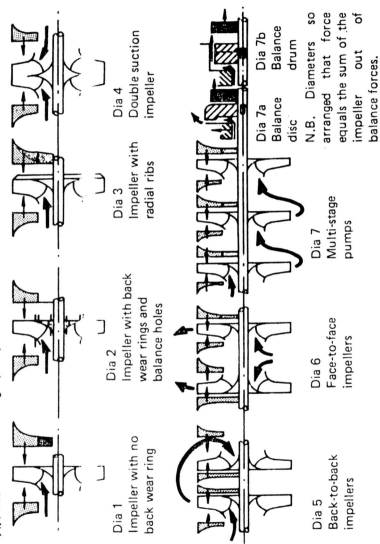

A. Axial thrust centrifugal pumps.

Dia 1 Impeller with no back wear ring
Dia 2 Impeller with back wear rings and balance holes
Dia 3 Impeller with radial ribs
Dia 4 Double suction impeller
Dia 5 Back-to-back impellers
Dia 6 Face-to-face impellers
Dia 7 Multi-stage pumps
Dia 7a Balance disc
Dia 7b Balance drum

N.B. Diameters so arranged that force equals the sum of the impeller out of balance forces.

CENTRIFUGAL PUMP NOMENCLATURE, CHARACTERISTICS & COMPONENTS

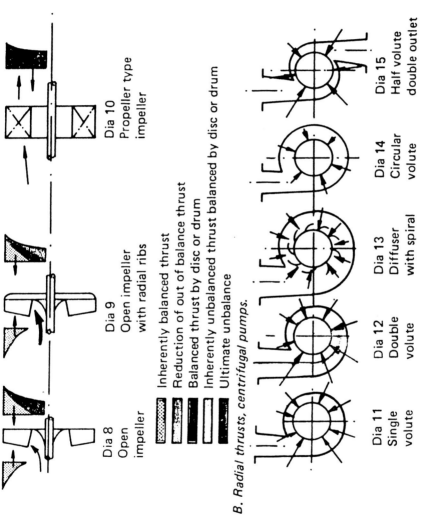

FIGURE 5.16 – Thrusts.

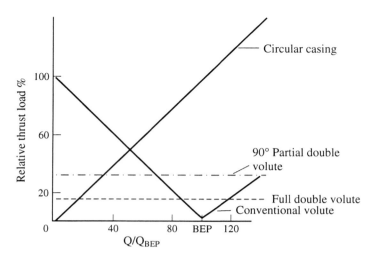

FIGURE 5.17 – Radial thrust vs volute type.

the single volute configuration. This feature is practical down to about 150 mm or 6 in. discharge connections on cast pumps. See Fig. 5.17 for relative thrust loads of the different configurations at one specific speed. Magnitude changes will vary with specific speed but the trends will be the same. Higher specific speeds show higher variations from shut-off to BEP. A slight loss penalty i.e. 1–2% is paid for the added friction generating surfaces at bep but this is generally compensated for by a similar efficiency improvement on either side of bep. Because of casting cleaning and surface finish problems, double volutes are not generally used below 150 mm (6 in.) discharge flange sizes.

The Circular or concentric casing

The Circular or concentric casing is advantageous for pumps in the 100–150 mm (4 in.–6 in.) range that can not accommodate the double volute for practical reasons. This is true in the area where radial thrust load considerations outweigh a 4% or so efficiency penalty. See Fig. 5.2. One good example where this design is used beneficially is in the wastewater pump area. The solids handling diameter requirements specified often force a 100 mm (4 in.) pump to be used on capacities that are hydraulically best handled by a 75 mm (3 in.) pump. This means pumps are being applied at full load ratings an undesirable distance to the left of the bep. The result is in very high radial loads and suction recirculation. From Fig. 5.17 one can easily see the advantage of the circular casing in this application. Actually, circular casings result in higher efficiencies than volutes when the specific speed is below approximately 12 (600) in addition to enjoying the reduction in radial loads.

The Diffuser Casing

The Diffuser Casing is used on vertical turbine pumps, where it is called a bowl, and is

CENTRIFUGAL PUMP NOMENCLATURE, CHARACTERISTICS & COMPONENTS   75

sometimes used in multi-stage pumps, (see Fig. 5.16) and low head propeller pumps. Radial thrust is low across the full range of capacity.

Wear Rings and Wear Plates
Wear (Wearing) Rings

Wear (Wearing) Rings are an economical solution to the problem of wear to casing, head, suction cover and impeller surfaces where they form their leakage joints between suction and discharge pressure. Without them the worn surfaces would have to be rebuilt by welding etc., or machined to take the wear rings at that time. By providing them as part of the original order, the replacement cost is minimized.

FIGURE 5.18 – Ahlstrom paper pumps.

Impeller Wear Rings

Impeller Wear Rings fit on the impeller at the eye and are of two types, radial and axial as shown in Figures 5.24 and 5.25 respectively. They may also be found in Europump Terminology and HI Standards..

Casing, Cover or Head Wear Rings

Casing, Cover or Head Wear Rings are the mates to the impeller rings mounted on the casing, suction cover or head depending on the type of pump. See the same figures as impeller rings above.

Wear Plates

Wear Plates are generally used for axial clearances such as shown in Fig. 5.25 for a Wastewater closed impeller and in Fig. 5.19 for a Pulp and Paper open impeller. In the latter case, as wear proceeds the wear plate may be moved axially to take up the excess clearance or in some cases the impeller is adjusted to the wear plate.

When performance drops as shown in Fig. 5.26 corrective action in the wear ring/plate area is required. Mating surfaces should have at least a 50 Brinnell hardness difference and low galling characteristics. Clearances are determined by the manufacturer of the pump

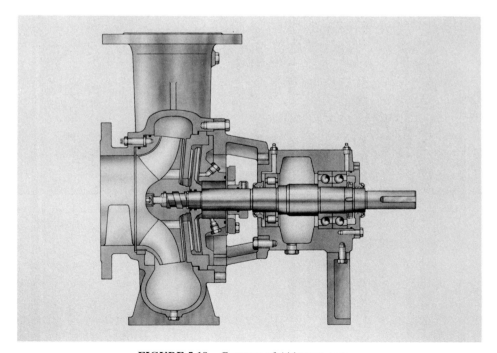

FIGURE 5.19 – Cutaway of Ahlstrom pump.

CENTRIFUGAL PUMP NOMENCLATURE, CHARACTERISTICS & COMPONENTS   77

BERESFORD PUMPS LTD

The Coventry based, British Pump Manufacturer, offers a competitive range of Pumpset for both the chemical + water industries.

- Materials used include, cast iron, bronze, stainless steel, polypropylene & PVDF etc. with duties ranging up to 18 cum/HR and heads to 40 m.

- Alcon Pumps (Part of B. Pumps) supplies self priming 2", 3", 4" and 6" in Bronze, Aluminium and Cast Iron. Petrol engine, Diesel engine and electric motor driven.

Beresford Pumps Ltd, Carlton Rd, Foleshill, Coventry CV8 7FL.
Tel: 01203 638484; Fax: 01203 637891.

With the addition of 6 extra chapters this renowned work provides all you ever need to know about centrifugal pumps in one book.

Centrifugal Pumps 4th Edition

By Harold H Anderson BSc (Hons), C Eng, FIMechE, Life Fellow ASME

This book will be of vital interest to all engineers and designers concerned with centrifugal pumps and turbines. Encyclopedic coverage of the subject enables you to...

- Find out how the various types of centrifugal pumps work
- Learn of the practical effects of design features and how to predict excessive flow
- Maximise pump performance and reduce cavitation through the correct choice of materials, ancillaries and design

... and much more, including the origination of the Area Ratio — now the universal design system!

CONTENTS

1. Fundamentals
Introduction
Units of the Systeme Internationale (SI)
General Review of Pumps
General Hydrodynamic Principles
Pump Performance and Dimensions
Theory of Centrifugal Pumps

2. Basic Data
Shape Number, Flow and Efficiency
Statistical Analysis of Pump Performance
Entry Conditions and Cavitation
The Determination of Pumps Speed and Dimensions for a Given Duty

3. Practical
Precision Manufacturers of Hydraulic Passage Surfaces for Fluid Machines
Pump Type Variations
Planning in a Range of Pumps
Losses in Centrifugal Pumps

4. Materials
Construction Materials
Mechanical Design of Centrifugal Pumps
Critical Speed, Vibration and Noise
Bearings
Pressure Vessel Aspects

Mechanical Seals and Other Shaft Sealing Devices

5. Characteristics
Approach and Discharge Conditions
Characteristics of Pumps and Pipe Systems

6. Installation of Pumps
Starting, Transient and Emergency Conditions
Testing and Test Codes
Monitoring and Maintenance of Pumps

7. Types
Single Stage Pumps
Multistage Spilt Casing Pumps
Multistage Cellular Pumps
Borehole Pumps
Cone Flow and Axial Flow Pumps
Severe Duty Pumps and Small High Speed Pumps

8. Application of Larger Power
Boiler Feed Pumps
Large Single Entry Pumps
Pumps for Storage of Energy and Other Large Power Duties

9. General
Submersible Pumps in General
Mine Pumps
Sewage and Similar Duties
Offshore Water Duties
Deep Sea Mining
Fluid Storage Caverns
Thermal Power Station Pumps including Nuclear
Offshore, Marine and Aerospace Pumps
Refinery, Chemical and Process Plant - Properties of liquids
Variable Flow
Electric Motors
Variable Speed
Oil, Water, Gas and Steam Turbines on Glandless Duties
Pumps for Various Specific Duties
Miscellaneous Aspects
Water Wheel
Past and Future

Appendices

Index

500+ pages, 150 photographs and diagrams   ISBN: 1 85617 231 7

For further details post or fax a copy of this advert complete with your business card or address details to:
Elsevier Advanced Technology,
PO Box 150, Kidlington, Oxford OX5 1AS, UK
Tel: +44 (0) 1865 843848 Fax: +44 (0) 1865 843971

CENTRIFUGAL PUMP NOMENCLATURE, CHARACTERISTICS & COMPONENTS

based on expected deflection, temperature rise and potential solids that may be contained in the pumped fluid, as well as scale and corrosion product build-up.

Balance Holes and Back-vanes

Balance Holes

Balance holes are provided to adjust the pressure in the stuffing box/seal chamber area. Pressure (or lack thereof) in this area can be a detrimental factor in: air in leakage along the shaft, excess process fluid leakage out and/or excess thrust. Ideally a slight positive pressure in the stuffing/seal chamber will be present at bep. These balance holes as the name implies tend to balance the pressures and reduce or eliminate the resulting differences mentioned before. Balance holes are shown in Figs. 5.19 and 5.20. There is a penalty to be paid in efficiency for their presence, in the order of 2 points on narrow impellers and less on the wider impellers. They are generally used in conjunction with pump out vanes. They also have a negative impact on NPSHR in the order of one or two feet.

Pumpout vanes

Pumpout vanes, sometimes called back vanes because they are found on the back shroud of the impeller are used to keep particulate in the fluid being pumped from freely flowing into the stuffing box or seal chamber. They serve as a pressure breakdown mechanism between the discharge pressure and the stuff box/seal chamber pressure. They are also shown in Fig. 5.20.

Dynamic seals

Dynamic seals are used as a pressure breakdown means as well as to alleviate or resolve pump sealing problems. They can eliminate product leakage or dilution and the need for a flush water supply to the seals. They can substantially reduce maintenance costs. Product leakage costs can be a concern if the product is expensive, has high cleanup costs or if there are environmental or safety concerns. Product dilution can be a consideration if seal flush water is required and must later be removed from the product. Dynamic seals can reduce or eliminate maintenance costs when abrasives, pulp or other such substances are kept out of the stuffing box during operation. Flush water is not readily available in many locations and making it available can be costly. A dynamic seal is shown in Figs 5.19, 5.20a and 5.20b. The shutdown condition of the dynamic seal has been its Achilles' Heel in the past because of the reliability of the static seals. Today, there are solutions such as elastic disc seals and mechanical seals that, where cost justified, can resolve this problem. Dynamic Seals, called expellers, set up an air/liquid interface from centrifugal force at some radius in their passageway that depends on the pressure difference between the stuffing box and discharge pressure that must be overcome. The higher this differential pressure the more beneficial the dynamic seal becomes. Suction pressure plays a large part in establishing this differential pressure. Dynamic seal expellers can be staged if one does not have the

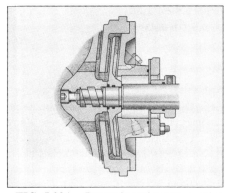
FIG. 5.20A – Dynamic seal pump, running.

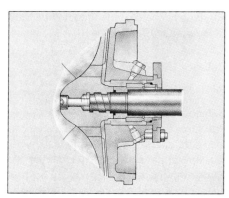
FIG. 5.20B – Dynamic seal pump, stopped.

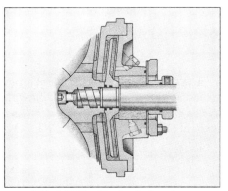
FIG. 5.20C – Mechanical seal pump.

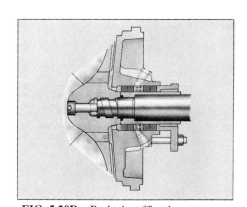
FIG. 5.20D – Packed stuffing box pump.

FIGURE 5.20 – Sealing alternatives.

capability to overcome the opposing back pressure. In such a case the liquid interface will be in the added expeller. Power/stage can run from 3/4 kw-7kw (1–10hp) for speeds up to *1800* rpm and diameters to 0.36 m (14 in.).

Stuffing Boxes and Seal Chambers

Stuffing Boxes and Seal Chambers are packing and mechanical seal chambers respectively. Figs 5.20d, 5.21, 5.22 and 5.23 show stuffing boxes and packing whereas Fig. 5.20c shows a mechanical seal chamber option for the same pump (the Hydaulic Institute Standards publication and Europump Terminology – see page 42 – show examples of stuffing boxes and seal chambers). Not too long ago manufacturers of pumps were using their creativity to come up with universal stuff box/chambers that would accommodate both packing and seals. The rapid progress of the seal industry in coming up with new seals

CENTRIFUGAL PUMP NOMENCLATURE, CHARACTERISTICS & COMPONENTS 81

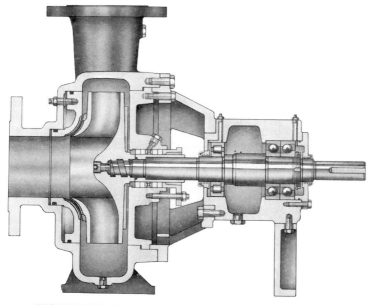

FIGURE 5.21 – Non-clogging wear resistant process pump.

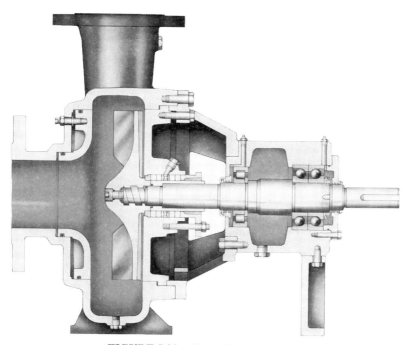

FIGURE 5.22 – Vortex impeller pump.

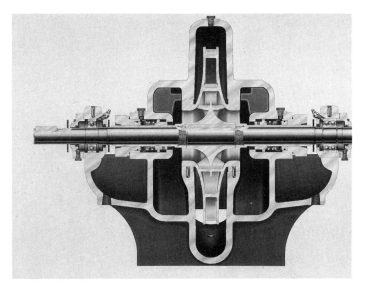

FIGURE 5.23 – Double suction horizontal split case pump.

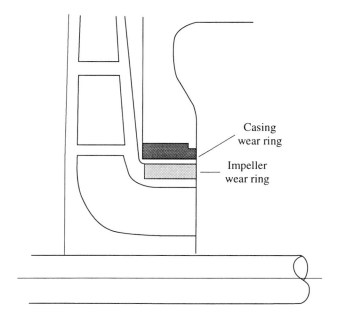

FIGURE 5.24 – Casing and impeller wear rings.

CENTRIFUGAL PUMP NOMENCLATURE, CHARACTERISTICS & COMPONENTS   83

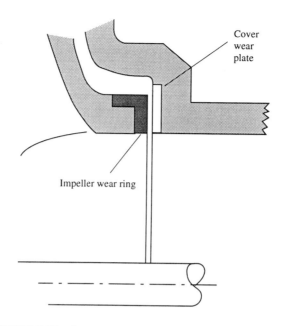

FIGURE 5.25 – Casing wear plate and impeller wear ring.

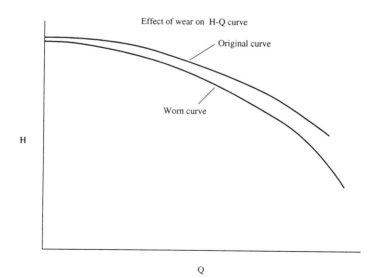

FIGURE 5.26 – New vs worn curves.

such as cartridge, bellows and tandem types meant that too often these universal chambers would not accommodate the new seals without special modifications. At the same time research into seal life was showing that modifications to the existing stuff box/chambers such as increased radial clearance at the top of the seal face, could have significant positive effects on seal life, and tighter environmental standards were resulting in improved pump designs from a standpoint of deflection of the shaft at the seal face. Fig. 5.27[9] shows four Seal chamber configurations, that were tested, and the test results are shown in Table 3A. The first shows the standard chamber that showed inadequate removal of heat, vulnerability of the seal to damage from abrasives and installation difficulties with double and tandem seals. The heat was mainly caused by the restricted throat on the left, which did not allow the heat generated by the seal to be carried off. The same was true for the second chamber where again the throat did not allow adequate mixing, however the seal face temperature rise dropped significantly due to the larger increased bore area above the seal. The third chamber had no throat and the result was that the seal chamber temperature rise dropped to only 1°F, but the seal face temperature rise went up 5°F more than the second due to the restricted conditions above the seal face. The tapered bore box was found to be as effective as a flush in cooling the seal faces with much less concern for abrasive damage and it also has good self venting capabilities. Large bore and taper bore boxes have now become ANSI B73.1 standard and an ISO Committee is looking at their standardization also.

TABLE 3A – Test results for Pump A seal chambers.

Seal Chamber	Seal Chamber Temp. Rise (°F)	Seal Face Temp. Rise (°F)	Seal Chamber Diff. Press. (PSI)
Standard	8	15	6
Enlarged I	10	9	6
Enlarged II	1	14	11
Tapered	1	5	18

Bearing Frames, Bearing Housings and Bearings

A *bearing frame* is a member of an end suction pump to which are assembled the liquid end and rotating element. A *bearing housing* is a pump component into which the bearings are mounted. In the case of the end suction pump the frame also serves as the bearing housing. Frames and/or bearing housings serve the purpose of providing bore alignment for the bearings, an oil or grease reservoir of adequate capacity, heat dissipation by convection or other means and protective shaft seals to keep out dirt and moisture and/or keep oil in. In addition, frames also serve as a mounting means for the wet end and provide a pedestal foot or feet to the base. Again, the Hydraulic Institute Standards publication shows frame/bearing housings for end suction pumps, some connected to the wet-end by an adapter bracket. Obviously, frames/housings must be sturdy enough to withstand the real life distortion forces they will be subjected to without distorting to a point that their

CENTRIFUGAL PUMP NOMENCLATURE, CHARACTERISTICS & COMPONENTS

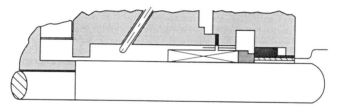

Standard stuffing box, Pump A.

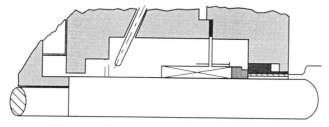

Enlarged cylindrical 1 seal chamber, Pump A.

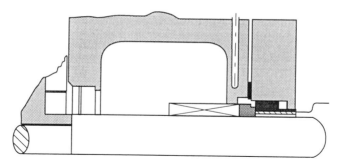

Enlarged Cylindrical II seal chamber, Pump A.

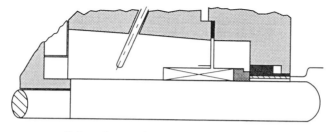

Enlarged tapered seal chamber, Pump A.

FIGURE 5.27

bearing alignment is threatened. *Bearings* for pumps are classified in many ways: *line vs thrust* describes the bearing's ability to handle radial or axial thrust loads and how the bearing positions the rotor. Figure 5.19 shows an inboard single row, cylindrical roller bearing that serves as a line bearing. Cylindrical roller bearings have no inherent axial thrust handling capability. The outboard bearing is made up of two angular contact ball bearings, back to back. These bearings can take both radial and thrust loads. The inboard ball bearing takes up thrust loads imposed in the direction of the driver and the outboard ball bearing takes up any thrust toward the inlet of the pump. A pump can have a single ball, line bearing on the inboard side and a double row ball angular contact bearing on the outboard side, which is a common arrangement. Frequently the thrust bearings handle a combination of both types of load. For example a thrust collar on the outboard side of the pump. *Anti-friction VS journal* refers to rolling VS sliding action, ball and roller types, VS the hydrodynamic cylindrical bearing that relies on forces in the lubricant in the clearance between the shaft and the cylindrical bearing to lift the shaft and keep it out of contact with the i.d. of the bearing. Ball bearings are the most common, with roller bearings showing up on larger shaft sizes where the axial load can be kept low. Rolling contact bearings are commonly used on catalogue type pumps up to approx. 1200 kW (1600 h.p.). *Self-aligning VS rigid* refers to the ability of a bearing to maintain its axis parallel to the shaft axis as the latter deflects. Fig. 5.28[10] shows cut-away views of 8 types of anti-friction bearings and balloons their component parts. The bearings most used on pumps are the single and double deep groove ball, the double row self-aligning and the single and double row angular contact.

Single Row Deep Groove Ball – Conrad Type: this bearing can carry significant radial loads and substantial thrust loads in either direction, even at high speeds.

Double Row Deep Groove Ball: similar to single row above but added row allows substantial increase in radial load.

Self-Aligning Ball Bearings have two rows of balls and a common spherical raceway which provides the self-aligning capability.

Angular Contact Ball Bearings can carry appreciable thrust loads in one direction either alone or with combined radial loading.

Corresponding Cylindrical Roller Bearings can offer higher radial load carrying capacity than their ball bearing counterparts. Rubber lined, Cutless bearings, see Fig. 5.29 are used in water service in applications involving sand and grit. These bearings were originally developed for maritime service. They are designed to provide a hydrodynamic wedge under the shaft and lift it clear of the bearing. They can operate dry for short start up periods. New plastic materials are being used similarly for specific applications that show promise of increased reliability and longer life. Bearing failures are one of the three highest failure areas on centrifugal pumps. These failures are caused by excessive operation at low flows, excessive axial thrust loads due to obstructions, improper upstream piping or wear, improper or inadequate lubrication, misalignment and piping strains, dirt and moisture.

CENTRIFUGAL PUMP NOMENCLATURE, CHARACTERISTICS & COMPONENTS

Bearing Terminology

The illustrations below identify the bearing parts of eight SKF basic bearing types. The terms used conform with the terminology section of the Anti-Friction Bearing Manufacturers Association, Inc. standards, and are accepted by anti-friction bearing manufacturers.

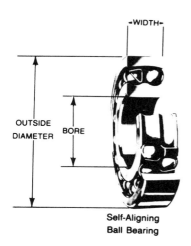
Self-Aligning Ball Bearing

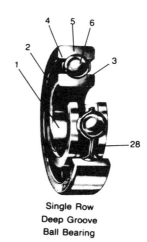
Single Row Deep Groove Ball Bearing

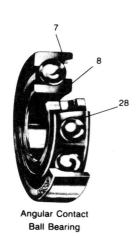
Angular Contact Ball Bearing

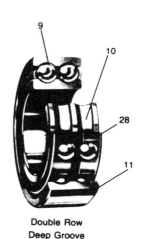
Double Row Deep Groove Ball Bearing

FIGURE 5.28 – SKF bearings.

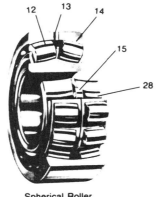

Spherical Roller
Bearing

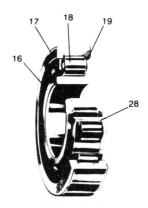

Cylindrical Roller
Bearing

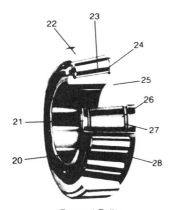

Tapered Roller
Bearing

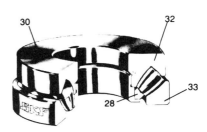

Spherical Roller
Thrust Bearing

1. Inner Ring
2. Inner Ring Corner
3. Inner Ring Land
4. Outer Ring Land
5. Outer Ring
6. Ball
7. Counter Bore
8. Thrust Face
9. Outer Ring Raceway
10. Inner Ring Raceway
11. Outer Ring Corner
12. Spherical Roller
13. Lubrication Feature
    (Holes & Groove) (W33)
14. Spherical Outer
    Ring Raceway
15. Floating Guide Ring
16. Inner Ring Face
17. Outer Ring Face
18. Cylindrical Roller
19. Outer Ring Rib
20. Cone Front Face
21. Cone Front Face Rib
22. Cup (Outer Ring)
23. Tapered Roller
24. Cone Back Face Rib
25. Cone Back Face
26. Undercut
27. Cone (Inner Ring)
28. Cage
30. Face
32. Shaft Washer
    (Inner Ring)
33. Housing Washer
    (Outer Ring)

FIGURE 5.28 (continued) – SKF bearings.

BOOKS FROM ELSEVIER

Heat Pumps for Energy Efficiency and Environmental Progress

Proceedings of the Fourth International Energy Agency Heat Pump Conference, Maastricht, The Netherlands, 26-29 April 1993

Edited by **J. Bosma**

©1993 616 pages Hardbound
Price: Dfl. 385.00 (US$ 220.00)
ISBN 0-444-81534-1

The 70 papers collected in this volume present an up to date review of the trends in heat pump technology. The heat pump is reviewed both as being part of a more comprehensive system, and as a refined device providing energy and greenhouse gas emission reductions. Its implementation in a system or process must be carefully considered at an early stage of design or development, and process integration is discussed in detail as a valuable tool for industry.

Pumps and Pumping

With Particular Reference to Variable-duty Pumps

By **I.I. Ionel**

Studies in Mechanical Engineering Volume 6

©1986 716 pages Hardbound
Price: Dfl. 490.00 (US$ 280.00)
ISBN 0-444-99528-5

"I suspect very strongly that a good number of A & E firms will find it an important addition to their reference libraries. ... the book is worth the asking price and should find many takers."
  Water & Wastewater

This book addresses itself to two major problems: the increasing demand for water and the increasing limitations on energy use. Within the general framework of problems regarding pumps and pumping installations, attention is focussed on variable-speed pumps. The book considers all aspects of pumps and pumping systems theory.

Dictionary of Hydraulic Machinery

In English (with definitions), German, Spanish, French, Italian and Russian

By **A.T. Troskolanski**

©1985 xvi + 736 pages Hardbound
Price: Dfl. 468.00 (US$ 267.50)
ISBN 0-444-99728-8
4,300 terms

"For those involved in the field of hydraulic machinery, particularly designers and researchers, this book should prove a valuable addition to their engineering bookshelf."
  World Pumps

Based on a Polish book of 1974 entitled Hydraulic Machinery - Basic Concepts, this revised and enlarged multilingual work is more a illustrated encyclopaedia than just a technical dictionary. Besides giving equivalent terms in the six different languages, precise and unequivocal definitions (in English) are provided, along with relevant formulae and classification tables. The approximately 4,300 terms are supplemented by 221 illustrations and 30 tables.

Elsevier's Dictionary of Water and Hydraulic Engineering

In English, French, Spanish, Dutch and German

By **J.D. van der Tuin**

©1987 xvi + 450 pages Hardbound
Price: Dfl. 431.00 (US$ 246.25)
ISBN 0-444-42768-6
5,117 terms

As its title suggests, this dictionary deals with water: water in relation to engineering projects designed to utilize it, to control it, or to defend u against it; water as a basic element of our environment, and water as the subject of a variet of physical phenomena.

ELSEVIER
An imprint of Elsevier Science

For further information on Elsevier's range of publications please contact:
Trade Relations Department, Elsevier Science B.V.
P.O. Box 211, 1000 AE Amsterdam, The Netherlands
Fax: (31) 20 6854 171

CENTRIFUGAL PUMP NOMENCLATURE, CHARACTERISTICS & COMPONENTS

FIGURE 5.29 – Water lubricated rubber bearings.

Bearing Protectors

Bearing Protectors (such as lip seals, Fig. 5.19 where drive end of shaft exits frame and bearing labyrinth seals Fig. 5.30[11]). The latter devices are generally rotating labyrinth seals as opposed to stationary labyrinths such as shown in Fig. 5.19 next to the lip seal. These devices can be very effective in keeping oil from leaking out and dirt or water from leaking in, especially water that is sprayed on a pump to hose it down. Because of their beneficial effects on bearing and seal-life, they are becoming quite common and are even being furnished as standard offerings on some types of pumps by the pump manufacturers.

Lubrication

Lubrication is normally one of three mediums, grease, oil and pumped product. The choice is a function of the environment the pumps are running in, concern for dilution or contamination of pumped product, ease of maintenance, reliability, first cost and operating costs. Grease is a common lubricant for anti-friction bearings supplied in the smaller catalogue pumps.

Oil lubrication

Oil lubrication is also widely used on an optional basis in small pumps as well as being

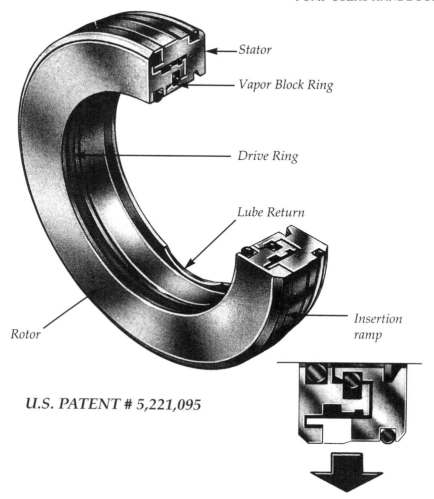

FIGURE 5.30 – Bearing isolator.

used as a standard on very large pumps. *Pumped product lubrication* is used where it is convenient such as in many vertical turbine installations and where use of another lubricant would raise the intolerable risk of contamination of the product being pumped.

Grease Lubrication

Grease Lubrication of anti-friction bearings has one recurring problem and that is over lubrication. Bearing cavities should be not be filled over approximately one third full. The results of overfilling is temperature build-up due to churning losses in the bearing that generally are nothing but alarming but can be damaging if not corrected. Oil lubrication

CENTRIFUGAL PUMP NOMENCLATURE, CHARACTERISTICS & COMPONENTS

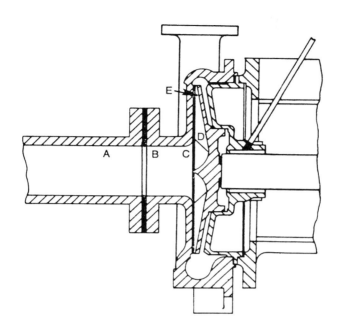

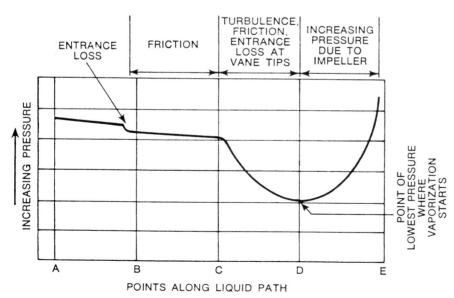

FIGURE 5.31 – Relative pressures at entrance to a centrifugal pump.

is found on larger pumps with hydrodynamic journal bearings where a pressurized system is found, and on some smaller pumps for specific industries such as the chemical and oil refining industries, or for occasional specific customers. When oil is called for catalog pumps are generally splash lubricated. Oil levels are maintained manually in most cases to the centre of the lowest ball position by watching the sight glasses. There are "constant level oiler" systems that do this automatically where customers call for them. An oil mist lubrication system can be the ultimate in providing the optimum film thickness of oil for anti-friction bearings when properly installed and operating, but these systems have a high first cost. It is extremely important to follow the pump vendors recommendations on the type of grease and oil as well as the viscosity of the oil and the maximum bearing housing surface temperatures. Too much grease or too high a oil level can cause bearing temperatures to soar.

Operation at Off-Design Conditions

Operation at Off-Design Conditions is the third of the three major causes of failure in centrifugal pumps. Inadequate NPSHA; increased vibration suction and discharge recirculation; excessive, constant or changing, radial and axial loads and high temperatures are some of the causes. This a complex subject, that is unfortunately not understood by many of those that select operate and even design pumps. It is, percentage wise, much less of a problem on low to moderate energy catalogue pumps (but there are many more of them) than engineered type pumps. On the higher energy engineered pump types, such as boiler feedwater pumps, it can be a critical one. Fig. 5.31 shows a centrifugal pump and

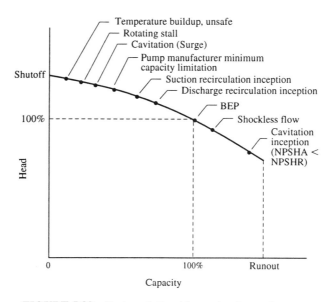

FIGURE 5.32 – Factor relationships on head capacity curve.

below it a curve of the pressure in the fluid being pumped as it progresses through that pump. Notice that there are three distinct losses: the sloped reduction due to friction, the stepped entrance loss and the turbulence and entrance losses at the vane tips until at a point D, just after the inlet tips of the blades a minimum pressure is reached. From this point on the pressure rises up to the discharge of the impeller (further rise in the volute collector is not shown). With this in mind consider the major goal in the application of a pump is to avoid vaporization of the pumped fluid. The NPSHR is equal to the total pressure shown at D in Fig. 5.31. The NPSHA must be adequate to prevent vaporization of the fluid at this lowest point of pressure under the worst conditions. Now look at a typical characteristic curve, Fig. 5.32, that shows the relative location on the curve of certain design and operating phenomena that pump users should understand and be aware of. Going from right to left, the first phenomenon is cavitation inception at slightly less than runout flow:

Cavitation Inception is the point at which bubbles begin to form due to localized conditions when the pressure is below the vaporization point of the fluid being pumped. It is very hard to determine this point of inception without a window and strobe light. Because centrifugal pumps are run at lower heads than design at certain times of the day, the year or the process cycle, they run at an operating point to the right of the rating point. Hydraulic designers of catalogue pumps design the impeller to provide the NPSHR at some compromise point to the right of bep. Designers of Engineered Pumps design the impeller to accommodate the majority of the points in a customer's specified duty cycle but like the catalogue pump designers will end up with a design point to the right of the BEP. This point is called the *shockless flow* point. It is generally between 1.1 and 1.3 times the BEP flow. At this point the flow angle entering the impeller at the eye of the impeller will equal the blade angle so that zero incidence exists. Cavitation damage rate increases much more rapidly on the right hand side of shockless flow capacity than on the left. Recent tests[12] show that erosion rate increases by a factor of 4 when the capacity is raised from 100% to 120% of the shockless flow. It is easy to see why at some point to the right of shockless flow, dependent on the NPSH margin (difference betwen the NPSHA and NPSHR), the margin would be used up and the vapour bubbles would form. Below shockless flow capacity the friction losses are dropping with the square power of capacity as opposed to rising on the right. Therefore going to the left you are using up less NPSHA although recirculation effects are opposing this trend.

NPSHR can be defined as that NPSH at which the pump total head resulting in the first stage has decreased by 3% due to low suction head and resulting cavitation. This is an arbitrary convention that is universally used as a standard unless specified otherwise. It has proven satisfactory in the bulk of cases on catalogue type pumps of low to medium specific speed. In the case of large boiler feed and large irrigation pumps and any of the high energy engineered type pumps the NPSHR criterion might better be lowered. The problem with the 3% criterion on such pumps is that the pump really starts cavitating at 0%. The effect of this cavitation on pump damage is non-existent on some pumps whereas on others it can be profound. Some engineered pumps have shown damage with NPSHA values between 0% and 1% head drop. Hydraulic designers are slowly gaining more insight into this area. Cavitation bubbles, for example, cause no damage if they are away from the surface metal when they are compressed and implode. The designers of engineered type pumps today work with designs that have bubble lengths between 12 mm (1/2 in.) and 100 mm (4 in.). They can control the length of the bubble to avoid cavitation damage with their blade design and then test with a strobe to measure the bubble size and confirm their calculations. The arbitrary 3% comes about because of the test problems involved in determining just where the pump is beginning to cavitate. See chapter on testing. Much more expense and capital costs, in terms of sophisticated instrumentation, are involved in tests when a lower pressure drop criterion is required. Cavitation inception is most likely going to start when some localized pressure drops to a point that vaporization takes place. This may be much sooner than would be justified by looking at the mean or average flow conditions. Suction piping is a major consideration, especially if it results in asymmetrical flow conditions entering the pump. Refer to the section on suction piping later on in this chapter. The *BEP point* is the design point of the complete pump.

Recirculation is a potentially damaging flow reversal at the inlet or discharge tips of the impeller vanes of a centrifugal pump as defined by Fraser[13]. It can occur in the presence of adequate NPSHA. He points out that "It is inherent in the dynamics of the pressure field that every impeller design must recirculate at some point-it cannot be avoided". He further states that discharge recirculation can be reduced in design but, only with an accompanying reduction in the rated efficiency of the pump; suction recirculation could likewise be reduced with an accompanying increase in NPSHR and that optimization of efficiency requires a reduction in the safety margin between the rated capacity and the discharge recirculation capacity. Fig. 5.33 shows an illustration of *suction and discharge recirculation.*

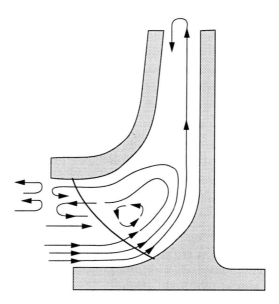

FIGURE 5.33 – Suction and discharge recirculation.

From his tests Fraser ended up with methods for calculating the incipiency of discharge and suction recirculation. Since that time these points of incipiency have been found to be conservative on many pumps, especially so on the catalogue types. In addition designers of engineered type pumps have found ways to significantly lower these incipiency points and still maintain good pump efficiency. Extensive surveys[14,15] have been conducted on pumps that have failed due to suction or discharge recirculation. The result has been recommendations not to exceed certain *suction specific speeds*. These recommendations while appropriate for the particular pumps and existing conditions in the respective surveys, have been inappropriately used on a more universal basis. As defined suction specific speed is an appropriate index for impellers of the geometrically similar inlets. Moreover, the units in these surveys were high energy units, used in refineries and utilities respectively. Other applications are not be as severe. Suction specific speed times eye velocity[16] or even-eye velocity squared would have been a more discerning parameter than suction specific speed alone.

The Hydraulic Institute set up a pilot survey on double suction, horizontal split case catalogue pumps of low to moderate energy because experience of its members on catalog pumps seemed to be in the order of magnitude of less than 1 suction recirculation failure in 1000, whereas on engineered pumps in the two surveys reported the failure rate was of the order of magnitude of 1 in 10. Results of this pilot survey on 192 different pumps with a population of over 65,000 was 13 total reported cavitation failures on 8 of the 192 pumps (less than the 1/1000 failure rate estimated). Pumps with suction specific speeds up to *300 (15, 500)* ran without cavitation failure (except for the 13 mentioned above). There are those who believe that Suction Specific Speed, S should have been defined at the shockless flow capacity of the impeller rather than the bep of the pump which includes the performance of the easing, wear ring leakage. Suffice it to say here that S alone is not a sufficient basis to determine the limits of pump operation. If it is to be used, then its own limitations must be recognized. This is especially true for engineered pumps being designed today[17]. There is a connection between discharge and suction recirculation when the vane overlap is low. Discharge and suction recirculation can combine and both start at the discharge incipiency point. This is likely to happen when eye diameter/impeller discharge diameter ratio exceeds 0.5. Ross[8] shows this climbing with specific speed to 0.65 at Ns = 58 (3200).

The *Pump manufacturers minimum capacity limitation* is the limiting flow that a pump manufacturer will show on his curves and documentation based on experience as to a safe minimum flow limit based on failures or in some cases vibration limit.

Cavitation (surge): Surging at the inlet is associated with low flows and reduced NPSH. A large spinning cavity in the inlet pipe grows and collapses. This is a high amplitude, low frequency phenomena in the 0–10 Hz range with accompanying cavitation.

Rotating Stall can occur at 1/3–2/3 of the shaft running frequency. One blade will stall then the next and so on.

Minimum allowable flow due to temperature rise occurs in the recirculation mode when the heat added to the fluid being pumped due to pump losses is greater than the amount of heat being carried away. When a pump operates in this mode for a certain (calculable) length of time, the temperature rise can reach the boiling point of the liquid in which case vaporization will occur and a potentially explosive and dangerous condition exists. Temperature rise, ΔT can be calculated from the total head and efficiency as follows:

(5.5) $$\Delta T = H(1/\eta - 1)/cp$$

where H = head in metres (feet)
 h = efficiency
 cp = specific heat at constant pressure, cal/gm–°C, (btu/lb–°F)

Example 1

Assume we have a pump whose characteristic curves are shown in Fig. 5.34. Plot the Temperature rise.

H and η are obtained from the characteristic curve for various capacities. Assume the fluid is water at 20°C, (68°F). Under these conditions cp = 1.0 .

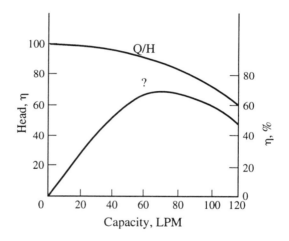

FIGURE 5.34 – HQ and η curves.

TABLE 4

Capacity, l/s	Head, m	η	ΔT – °C
120	60	48	0.15
80	84	67	0.097
60	90	66	0.11
40	94	50	0.22
20	98	26	0.64
10	99	14	1.39
5	99.5	6	3.59
4	99.7	4	5.5
2	99.8	2	11.3
1	99.9	1	22.8

CENTRIFUGAL PUMP NOMENCLATURE, CHARACTERISTICS & COMPONENTS   99

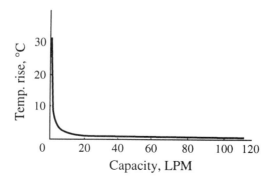

FIGURE 5.35 – Temperature rise vs. capacity.

One can see in Fig 5.35 that the temperature rises very rapidly once one gets below 10% flow. However this pump is a relatively low head pump. Look at another one with about twice the head and lower part load efficiencies and see what the effect is.

Example 2

Redo based on the pump whose characteristics are shown in Fig 5.36.

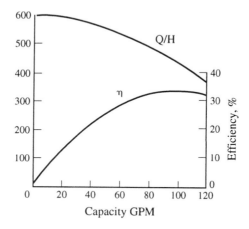

FIGURE 5.36 – HQ and η curves.

The results are plotted in Fig. 5.37. One can see that the higher head (lower specific speed) and lower efficiency unit is more sensitive to temperature rise at low flows, with the increase starting to accelerate as the flow dropped below 40 gpm. This is also an appropriate time to point out that this temperature rise can have an adverse affect on NPSHA. Due to leakage flows, the temperature of the fluid at the eye of the impeller may

TABLE 5

Capacity, gsm	Head, ft	η	ΔT – °F
120	360	32	0.98
80	500	33	1.30
60	540	28	1.78
40	575	21	2.78
20	600	11.5	5.93
10	605	6	12.2
5	610	3	25.4

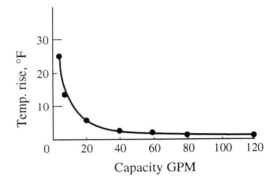

FIGURE 5.37 – Temperature rise vs. capacity.

be increased substantially because the leakage flow percentage to through flow may be high at these low flows. This is especially true on units with specific speed of 19 (1000) or less, and this could cause localized flashing.

Other low flow considerations

Power characteristic curves on high specific speed units rise with reduction in flow. Therefore care must be taken on such units that the motor is not overloaded. When the pumpage has entrained air, the ability of the pump to handle it is reduced at lower flows, and the result could be a choking condition (alternate pumping and no pumping).

Parallel – Series Operation

Parallel – Series Operation are common requirements but each has certain considerations that must not be overlooked. Parallel operation of two or more pumps requires consideration of adequate inlet piping sizing and take-off design (See Chapter on Sumps and Inlets, as well as adequate steepness of the characteristic curves for stability. Series operation of two or more pumps requires consideration of the pressure capability of the component parts of the pumps working at higher pressure as well as modifications such as break-down

CENTRIFUGAL PUMP NOMENCLATURE, CHARACTERISTICS & COMPONENTS

bushings or the capability to handle higher stuff box or seal chamber pressures. The steepness of the curves and the magnitude of the static head component are two of the major variables. Generally, pumps on systems that have high static head components would give better performance in parallel. System curves made up by a high percentage of friction losses will show higher flows through two pumps in series than two in parallel.

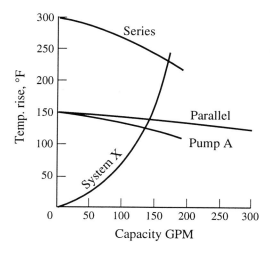

FIGURE 5.38 – Series vs. Parallel curves.

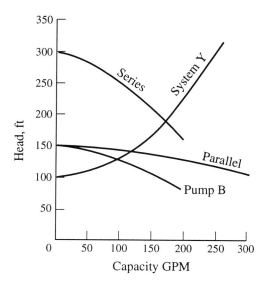

FIGURE 5.39 – Series vs. Parallel curves.

A plot of the power curves in such cases may show less power consumption in series also. However, series operation where parallel operation would meet the head requirements means giving up the benefits of having one or two pumps on depending on the load requirements. In series both pumps must be left on unless the head drops to less than half of the rated head. Figs 5.38 and 5.39 show the characteristic curves of two split-case, double suction pumps with flat and steep slopes respectively. Series and parallel curves for two identical A units and then two identical B units are superimposed on each, illustrating some of the points that have been made. Had the power curves been given, the question of which pump combination was expected to draw the most power could have been resolved. Add the heads of each pump at different capacities to obtain the parallel operation curve and the capacities at different heads to obtain the series operation curve. It is not necessary to stick with identical pumps for series and parallel operation. In the first case, the second pump in series should have approximately the same capacity as the first, but could have a different head. In the case of parallel operation it is desirable that the second pump has approximately the same head as the first, but the capacity could be different.

Intake Design: Sumps, Tanks and Suction Piping

The main source of hydraulic problems in pumps arises from improper design of the suction side of the system or intake. The function of an intake design is to supply an evenly distributed, non-rotating flow, free of entrained air and foreign- material with adequate submergence to the pump impeller.

Definitions

Intake: The structures into which liquid to be pumped is directed.

Sump: A wetted chamber with a free surface which receives liquid to be pumped and from which it is pumped.

Wet Pit, Wet Well: A sump having its bottom below the bottom elevation of its inlet, ground level or some other reference elevation.

Dry Pit: A non-wetted chamber housing pumping equipment located adjacent to a wet-pit.

Vortex: A rotation of a portion of the fluid.about its stationary or moving centre line.

In the case of *sumps and tanks* the leading statement in this intake design section means that their inlet should be below the minimum liquid level to avoid entraining air and as reasonably far away from the pump as possible to keep the flow as uniform as possible. Inlets with free-falling discharge to the sump or tank liquid level should be avoided as they cause entrapped air bubbles which can affect pump performance. The influent should not impinge against the pump, jet directly into the pump inlet or enter the sump or tank in such a way as to cause rotation of the liquid in the containment area. The volume of any tank or pit should be no less than that which would give two minute retention time or 3 or 4 starts per hour per pump. The latter depending on the motor winding's temperature limitations and ability to dissipate the energy absorbed in the form of heat (from the starting acceleration period) during the running and shut down periods between startups and still leave the minimum liquid level at the end of the run period. Once the volume of a sump or tank is ascertained, the next criterion is the dimensions. The Hydraulic Institute[17] has set up e.g. minimum and maximum criteria for sump dimensions and the latest edition of the HI Standards publication should be referred to for these. In the case of a single pump, the dimension S, a minimum dimension determines the width. Ideally, the dimension would be as deep as possible, but economics is the deciding factor. An average dimension is dependent on the type of pump. The pump manufacturers recommendations should be sought here. The edge of the suction bell should be close to the back wall. In those cases where this may not otherwise be possible a false wall should be installed. The shape and additions to sumps for multiple pumps are complicated. If at all possible the design should not have water flowing past one pump to get to the next unless certain criteria are met[18]. In addition to the Hydraulic Institute Standards, Duriron's Pump Engineering Manual[3], the Pumping Manual and the Pump User's Handbook are good references.

Note that sumps for solids bearing processes require some special considerations. The sump velocities, for example must be higher to keep the solids in suspension. Three ft/sec. is generally used as the minimum approach velocity. The sidewalls should be shaped to avoid solids settling in the corners.

Model Tests

If the sump design is not a relatively simple one that can clearly meet the guidelines shown here and in the attached references, it may be necessary to conduct a sump model test to assure adequate design. Agreement between the purchaser/user and the model test vendor as to what constitutes unacceptable flow conditions will reduce later misunderstandings. Fig. 5.40 shows a vortex classification system that may be used in such an agreement.

Tanks (including process vessels)

It is common to take the suction line off the bottom or side of a process or suction tank. General rules for sound intake design apply to suction tanks. In particular adequate submergence must be provided to prevent vortexing. One foot of submergence for each foot per second of velocity at the suction pipe inlet is recommended. A maximum velocity of 6 fps is suggested. Bellmouth or rounded inlets are recommended. If the recommended submergence cannot be obtained, the inlet pipe diameter should be increased or vortex breakers installed. Recommended breaker designs are shown in Fig. 5.41. Baffles should be placed between the inlet and outlet connections to prevent short circuiting.

Suction Piping

The function of good suction piping design is to provide uniform, non-rotating flow to the impeller avoiding air entrainment with a minimum of friction loss. In general, all piping should slope upward to the pump. Elbows should preferably be kept 10 pipe diameters

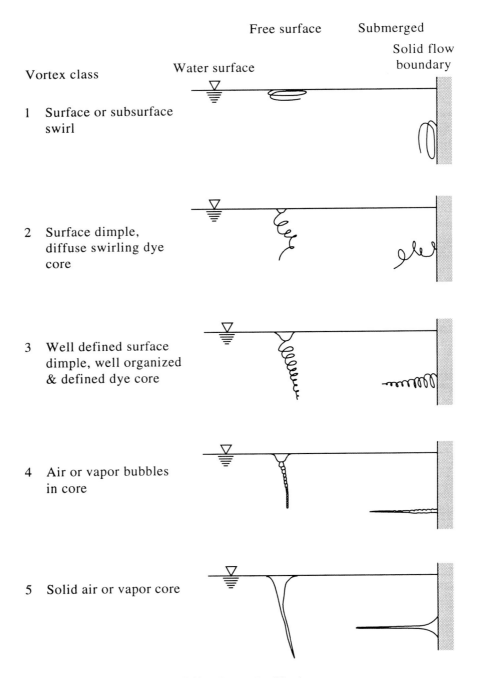

FIGURE 5.40 – Vortex classification system.

CENTRIFUGAL PUMP NOMENCLATURE, CHARACTERISTICS & COMPONENTS 105

away from the pump. On double suction pumps the elbows should only be installed in a plane perpendicular to the pump shaft and concentric reducers should be avoided in favour of eccentric reducers with the horizontal side on top. Long radius elbows and a larger pipe diameter than the pump suction with a reducer are recommended ways of increasing the NPSHA. Care must be taken with flange gaskets so as to avoid any protrusions into the suction pipe. This has resulted in problems of catastrophic proportion in numerous cases, especially in low NPSHR situations and flanges near or at the suction of the pump. From

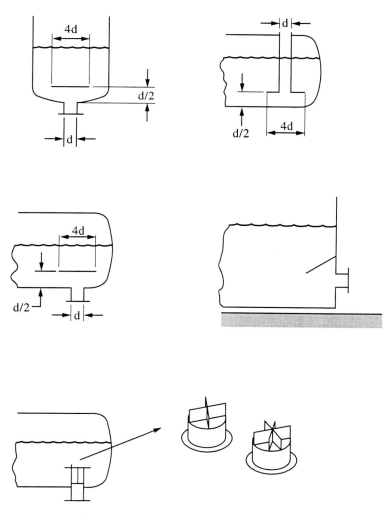

Vortex breakers in typical suction tanks

FIGURE 5.41 – Vortex breakers.

a mechanical standpoint the expansion joint closest to the pump should be anchored on the pump side to avoid passing pipe stresses and causing amplified vibration. The Technical Handbook[19] on expansion joints is highly recommended for reference.

References

1. Hydraulic Institute Standards, 14th Edition (1983), Hydraulic Institute.
2.
3. Ross, Robert R. and Lobanoff Val S.; Centrifugal Pumps- Design and Application, 2nd Ed. (1992) Gulf Publishing Co. ISBN 0–87201–200–X
4. The Duriron Co, Inc.,Pump Engineering Manual,5th Edition (1980)
5. Centrifugal, Mixed Flow and Axial Pumps – Code for Hydraulic Performance Tests for Acceptance Classes I and II, International Standards Organization
6. Hydraulic Institute Test Standards 1988 – Centrifugal Pumps 1.6, Hydraulic Institute.
7. Wislicenus, G. F., Fluid Mechanics of Turbomachinery, McGraw-Hill Book Co. (1947)
8. Lobanoff, Val. S and Robert, R. Ross: Centrifugal Pumps and Application, 2nd Edition, Gulf Publishing 1992, p55
9. Michael P. Davison, "The Effects of Seal Chamber Design on Seal Performance", Proceedings of the Sixth International Pump User's Symposium, Turbomachinery Laboratory, Texas A&M university, Jean C. Bailey, Editor, Copyright 1989.
10. Product Service Guide,190–710 (April 1992) , SKF
11. Maintenance Avoidance Program, Inpro Companies, Inc.
12. Schiavello, B., "Tutorial on Cavitation and Recirculation Troubleshooting Methodology" Proceedings of the Tenth International Pump Users Symposium, Houston, Texas (1993).
13. Fraser, W.H., " Flow Recirculation in Centrifugal Pumps" , Power and Fluids,(1982) Vol.8 /#2, Worthington
14. Hallem J. L., "Centrifugal Pumps: Which Suction Specific Speeds Are Acceptable?, Hydrocarbon Processing (April, 1982)
15. "Survey of Feed Pump Outages, EPRI, Electric Power Research Institute,(April 1978)
16. Rayner, R. E., "Understanding Suction Specific Speed", World Pumps (Feb. 1993)
17. Hydraulic Institute Standards, 14th Edition, 1983
18. Sanks, R.L.: C.E. Sweeney and G. A. Jones: "Self Cleaning Wet Wells For Constant Speed Submersible Pumps" Fifth Progress Report to EPA, (11/1/93)
19. Technical Handbook, Rubber Expansion Joint Division of the Fluid Sealing Association, Philadelphia, 5th Edition (1979).

CENTRIFUGAL PUMP NOMENCLATURE, CHARACTERISTICS & COMPONENTS

TABLE 6

Symptom	Probable Cause	Action
Pump does not deliver liquid	Impeller rotating in wrong direction.	Reverse direction of rotation.
	Pump not properly primed – air or vapour lock in suction line.	Stop pump and reprime.
	Inlet of suction pipe insufficiently submerged.	Ensure adequate supply of liquid.
	Air leaks in suction line or gland arrangement.	Make good any leaks or repack gland.
	Pump not up to rated speed.	Increase speed.
Pump does not deliver rated quantity	Air or vapour lock in suction line.	Stop pump and reprime.
	Inlet of suction pipe insufficiently submerged.	Ensure adequate supply of liquid.
	Pump not up to rated speed.	Increase speed.
	Air leaks in suction line or gland arrangement.	Make good any leaks or repack gland.
	Foot valve or suction strainer choked.	Clean foot valve or strainer.
	Restriction in delivery pipework or pipework incorrect.	Clear obstruction or rectify error in pipework.
	Head underestimated.	Check head losses in delivery pipes, bends and valves, reduce losses as required.
	Unobserved leak in delivery.	Examine pipework and repair leak.
	Blockage in impeller or casing.	Remove half casing and clear obstruction.
	Excessive wear at neck rings or wearing plates.	Dismantle pump and restore clearances to original dimensions.
	Impeller damaged.	Dismantle pump and renew impeller.
	Pump gaskets leaking.	Renew direction of rotation.
Pump does not generate its rated delivery pressure	Impeller rotating in wrong direction.	Reverse direction of rotation.
	Pump not up to rated speed.	Increase speed.
	Impeller neck rings worn excessively.	Dismantle pump and restore clearances to original dimensions.
	Impeller damaged or choked.	Dismantle pump and renew impeller or clear blockage.
	Pump gaskets leaking.	Renew defective gaskets.
Pump loses liquid after starting	Suction line not full primed – air or vapour lock in suction line.	Stop pump and reprime.
	Inlet of suction pipe insufficiently submerged.	Ensure adequate supply of liquid at suction pipe inlet.
	Air leaks in suction line or gland arrangement.	Make good leaks or renew gland packing.
	Liquid seal to gland arrangement logging ring (if fitted) choked.	Clean out liquid seal supply.
	Logging ring not properly located.	Unpack gland and relocate logging ring under supply orifice.

TABLE 6 – continued

Symptom	Probable Cause	Action
Pump overloads driving unit	Pump gaskets leaking.	Renew defective gaskets.
	Serious leak in delivery line, pump delivering more than its rated quantity.	Repair leak.
	Speed too high.	Reduce speed.
	Impeller neck rings worn excessively.	Dismantle pump and restore clearances to original dimensions.
	Gland packing too tight.	Stop pump, close delivery valve to relieve internal pressure on packing, slacken back the gland nuts and retighten to finger tightness.
	Impeller damaged.	Dismantle pump and renew impeller.
	Mechanical tightness at pump internal components.	Dismantle pump, check internal clearances and adjust as necessary.
	Pipework exerting strain on pump.	Stop pump and reprime.
Excessive vibration	Air or vapour lock in suction.	Stop pump and reprime.
	Inlet of suction pipe insufficiently submerged.	Ensure adequate supply of liquid at suction pipe inlet.
	Pump and driving unit incorrectly aligned.	Disconnect coupling and realign pump
	Worn or loose bearings.	Dismantle and renew bearings.
	Impeller choked or damaged.	Dismantle pump and straighten or renew shaft.
	Foundation not rigid.	Remove pump, strengthen the foundation and reinstall pump.
	Coupling damaged.	Renew coupling.
	Pipework exerting strain on pump.	Disconnect pipework and realign to pump.
Bearing overheating	Pump and driving unit out of alignment.	Disconnect coupling and realign pump and driving unit.
	Oil level too low or too high.	Replenish with correct grade of oil or drain down to correct level.
	Wrong grade of oil.	Drain out bearing, flush through bearings; refill with correct grade of oil.
	Dirt in bearings.	Dismantle, clean out and flush through bearings; refill with correct grade of oil.
	Moisture in oil.	Drain out bearing, flush through with correct grade of oil. Determine cause of contamination and rectify.
	Bearings too tight.	Ensure that bearings are correctly bedded to their journals with the correct amount of oil clearance. Renew bearings if necessary.

CENTRIFUGAL PUMP NOMENCLATURE, CHARACTERISTICS & COMPONENTS

TABLE 6 – continued

Symptom	Probable Cause	Action
	Too much grease in bearing.	Clean out old grease and repack with correct grade and amount of grease.
	Pipework exerting strain on pump.	Disconnect pipework and realign to pump.
Bearings wear	Pump and driving unit out of alignment.	Disconnect coupling and realign pump and driving unit. Renew bearings if necessary.
	Rotating element shaft bent.	Dismantle pump, straighten or renew shaft. Renew bearings if necessary.
	Dirt in bearings.	Ensure that only clean oil is used to lubricate bearings. Renew bearings if necessary. Refill with clean oil.
	Lack of lubrication.	Ensure that oil is maintained at its correct level or that oil system is functioning correctly. Renew bearings if necessary.
	Bearing badly installed.	Ensure that bearings are correctly bedded to their journals with the correct amount of oil clearance. Renew bearings if necessary.
	Pipework exerting strain on pump.	Ensure that pipework is correctly aligned to pump. Renew bearings if necessary.
	Excessive vibration	Refer to excessive vibration symptom.
Irregular delivery	Air or vapour lock in suction line.	Stop pump and reprime.
	Fault in driving unit.	Examine driving unit and make good any defects.
	Air leaks in suction line or gland arrangement.	Make good any leaks and repack gland.
	Inlet of suction pipe insufficiently immersed in liquid.	Ensure adequate supply of liquid at suction pipe inlet.
Excessive noise level	Air or vapour lock in suction line.	Stop pump and reprime.
	Inlet of suction pipe insufficiently submerged.	Ensure adequate supply of liquid at suction pipe inlet.
	Air leaks in suction line or gland arrangement.	Make good any leaks or repack gland.
	Pump and driving unit out of alignment.	Disconnect coupling and realign pump and driving unit.
	Worn or loose bearings.	Dismantle and renew bearings.
	Rotating element shaft bent.	Dismantle pump, straighten or renew shaft.
	Foundation not rigid.	Remove pump and driving unit, strengthen foundation.

NEW FOR 1995

If your system uses leak-free (or sealess) pumps or compressors this book will enable you to...

- Understand the various designs and properties of leak-free pumps
- Select the right pump or compressor for the job
- Find out how to measure flow without the leaks
- Comply with environmental legislation and ensure leak-free systems whatever the application

Contents include:

- A survey of leak-free centrifugal and positive displacement pumps
- Properties and design criteria for magnetic drives on pumps
- Zero-leakage pumps with permanent magnetic drive
- Leak-free centrifugal pumps in plastic
- Canned motor pumps: an important contribution to leakage-free operation
- Standardized chemical pump with canned motor in flameproof enclosure
- Canned motor and magnetic drive systems: a comparison
- Reciprocating metering pumps in leak-free design
- Leakage-free metering of fluids in fully automated processes
- Process diaphragm pumps
- Diaphragm compressors
- Liquid ring vacuum pumps and compressors with magnetic drive
- Leak-proof Roots vacuum pumps
- Buyers Guide
- Trade Names Index

For further details post or fax a copy of this advert complete with your address details to:

Elsevier Advanced Technology
PO Box 150, Kidlington,
Oxford OX5 1AS, UK
Fax: +44 (0) 1865 843971

OR

Call our order hotline

Tel: +44 (0) 1865 843842

250+ pages
Illustrations and tables

ISBN: 1 85617 230 9

Publication date:
October 1995

Price: £95
[PIP90.B1]

ELSEVIER
ADVANCED
TECHNOLOGY

Leak-free Pumps and Compressors Handbook

First Edition

Edited by G Vetter

CENTRIFUGAL PUMP TYPES

Centrifugal pumps

Centrifugal pumps come in many configurations. Figure 2.1 classifies them by mechanical configuration. The end suction configuration is shown in Figs. 6.3, 6.5 and 6.7-6.21. Its suction and discharge nozzles are perpendicular to each other, with the flow in the suction being along the centreline of the shaft for some distance. There is generally no shaft extension into the incoming flow.

The in-line pump configurations (see Europump Terminology or Hydraulic Institute Standards) have radial inlets as opposed to the end suction inlet. *Radial inlet* means the flow comes into the pump inlet perpendicularly and has to make a right angle turn coincident with the shaft before it enters the impeller. The between bearing unit configuration is the counterpart to the overhung configuration.

Figs 6.15-6.18 and 6.5-6.8 show vertical turbine type pumps. A submersible pump is one whose driver is submerged along with the pump. . A close coupled pump is one which has the impeller mounted on the motor shaft, as opposed to a separately coupled unit that has its own shaft separate from the motor's and connected with a rigid or flexible coupling. Some pumps have centreline support. This configuration is used where large temperature changes are expected and special provisions are required to maintain alignment. These large temperature swings do not necessarily have to be in the process fluid itself, but could be in the difference between shutdown and operating conditions with a high temperature process fluid. Refinery processes are a good example of this.

Some pumps have special features that give them a special designation, irrespective of application:

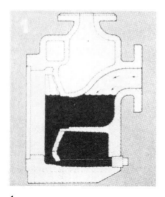

1.
For good priming a sufficient volume of liquid should be available in the casing. The liquid entrains air within the impeller.

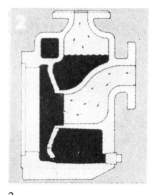

2.
The air/liquid mixture leaves the volute and enters the separating chamber. The air/liquid mixture separates in the chamber with the liquid 'settling' and the air venting out the discharge. The air-free liquid returns to the impeller through the bypass opening for re-entrainment. The air re-entrainment and removal cycle continues to reduce the pressure in the suction line.

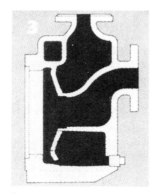

3.
The liquid rises in the suction pipe until the pipe is flooded. The pump then functions much like any end suction centrifugal pump. Once primed, the increased pressure in the volute reverses the flow through the bypass opening.

FIGURE 6.1 – Priming cycle for self-priming pump.

Maybe the best advice you were given and the secret to any good job, is to always start with a good primer.

The Stork **Fre-Flow**, is a tough self priming centrifugal pump created out of experience and innovation. Primers and pumps have one thing in common, both are crucial to a good job but both are seldom seen.

Aimed at your heavy duty operations, the **Fre-Flow** is there to handle your water, chemicals and contaminated liquids in sumps, pits and gulleys. There is no non-return valve to go wrong and an easily replaceable wearplate/wear ring is one of the many benefits. The **Fre-Flow** can prime up to 7m and is available in sizes up to 6" in a variety of materials.

STOP MESSING ABOUT WITH SUBMERSIBLES AND EJECTORS, AND GO WITH THE FLOW

Stork Pumps Ltd

Meadow Brook Industrial Centre, Maxwell Way, Crawley, West Sussex RH10 2SA.
Tel: (0293) 553495 • Fax: (0293) 524635

MACKLEY PUMPS

CLARKE CHAPMAN MARINE
PO BOX 9, SALTMEADOWS ROAD, GATESHEAD, TYNE & WEAR NE8 1SW
Telephone: 0191 477 2271 Fax: 0191 477 1009

Split Case and Multi Stage Ring Section Pumps: Specialising in a variety of liquid handling applications.

- Mine Dewatering
- Water Supply and Sewage Services
- Power Stations
- Chemical and Process Industries
- Irrigation and Fire Fighting
- Gas and Petroleum Industries

SCH 8/6
Diesel Driven Single
Stage Split Case Pump

INDUSTRIAL POWER GROUP

Rolls-Royce Industrial Power Group, NEI House, Regent Centre, Newcastle upon Tyne, NE3 3SB

New Line – Grindex's new dewatering pump program.

New Line means the best of thirty years' experience incorporated in a new more functional design plus a string of new features and improvements. For example:

SMART ™ (**S**urveillance of **M**otor **A**nd **R**o**T**ation) that monitors the motor temperature in three places and stops the motor in the event of a phase drop or other fault. Through the integrated ROTASENSE ™ system it also quarantees right motor rotation for optimal performance.

New Line also means: New dry run cooling valve, new patented shaft seal system, new power cable locking device and new larger junction box for easier access.

When others try to make the parts in their pumps cheaper to manufacture we have made a whole line of new and better pumps – Grindex New Line.

We try to serve you best

GRINDEX, Box 538, S-136 25 Haninge, Sweden. Telex: 17286 grindex-s. Fax: 745 53 28.

CENTRIFUGAL PUMP TYPES

Self-priming pumps

The unmodified versions of kinematic (roto-dynamic) pumps, with the exception of the regenerative turbine cannot pump if their suction sides are not "flooded". Even the regenerative turbine pump requires enough liquid to form a seal between the suction and discharge sections. Some types have an internal trap on the suction side that can be provided on an optional basis to perform this function. None of these pumps, however, can perform the function of a vacuum pump and expel enough air to let the water be drawn into the impeller. Adaptations are made internally to the pump or externally to purge the air out of the suction side.

A pump using external means such as a vacuum pump is not considered a self-primer. If internally provided eductors are used or the pump is shaped to perform the air removal the pump is considered to be self-priming. With the exception of the regenerative turbine this is accomplished by entrapment of air in the impeller or at its exit, the separation of this air in the stilling chamber (requiring a large free surface) and the return of the air free liquid to the exit of the impeller or the impeller itself where the cycle repeats itself until enough air is removed from the suction to allow normal pumping. In the process, the liquid is continually being recirculated in the discharge of the pump.

Suction and discharge recirculation have been used in the past, but today's pumps rely on discharge recirculation and no liquid is returned to the suction. Fig. 6.1 shows the priming cycle for pumps of this type. Fig. 6.2 shows a somewhat different self-priming

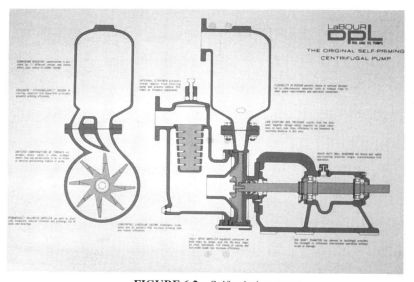

FIGURE 6.2 – Self-priming pump.

construction. In all of these cases some residual liquid must exist in the pump for the priming cycle to be effective, but designs generally hold enough liquid after the first charge that additional charges are not necessary under normal operating conditions.

Vortex pumps

Also called torque pumps are pumps with recessed impellers (see Figs 5.22 and 6.3). They are able to handle relatively large amounts of entrained gas and solids and stringy materials. A penalty is paid in power consumption because of the reduced hydraulic efficiency of the design which tops out at about 55%. Fig. 6.4 shows characteristic curves for a typical pump. These pumps are used in services where the pumpage has abrasive solids and/or entrained air. The severity of the abrasiveness of the solids that can be handled is a function of the ruggedness of the design, the materials and the use of sacrificial wear plates. They are found in use in waste water, mining, pulp and paper, chemical, food, industrial and agricultural applications.

Approximately 20% or less of the flow actually goes through the recessed impeller, but this flow provides the energy for the vortex motion in the main stream that generates the head/capacity characteristic.

Vertical radially split bowl, turbine, pumps

Also called bore hole and diffuser pumps, were originally developed for deep well applications. The depth of the water in the well was accommodated by lowering the pump into the well until the suction was adequately covered by water. The head was determined by the depth of this first stage, the losses pumping the water to the ground level and any

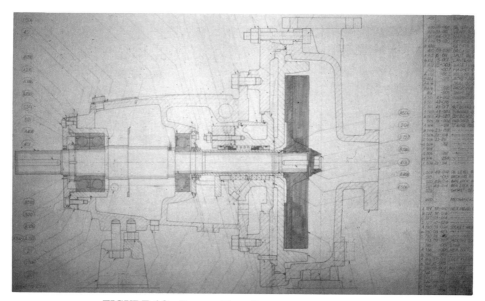

FIGURE 6.3 – Recessed impeller, vortex or torque pump.

CENTRIFUGAL PUMP TYPES

additional head that was imposed from that point. Enough pairs of impellers and bowls (stages) were then added to meet the head requirements and a segmented shaft connected the rotating assembly with the driver at ground level. A segmented column pipe carried its weight along with that of the bowls, inlet bell or suction pipe and strainer, line bearings etc. The limit in depth is determined by the maximum allowable elongation of the column relative to its original length, due to its weight and the down thrust. Below this limit a submersible configuration of deep-well submersible turbine pump is used. These pumps

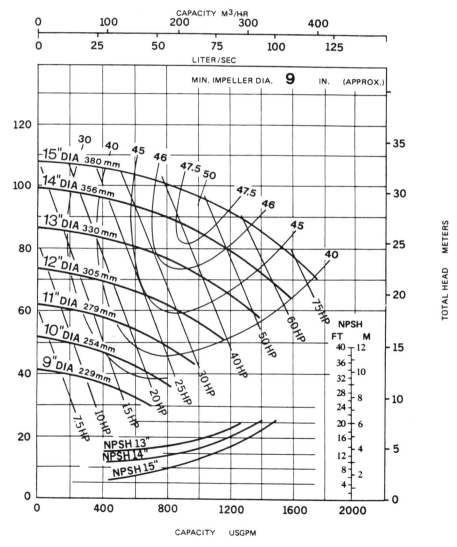

FIGURE 6.4 – Typical vortex pump, performance curve map.

have found extensive use in pumping from wells in agricultural applications. See Fig. 6.5. They are also used for pumping from open bodies of water such as ponds, lakes, rivers and oceans, as well as mine dewatering, sumps and caverns, power plants and oil field repressuring. They are especially well suited to low NPSH applications and since the first stage is submerged they are inherently self priming. This type of pump lends itself to a fine tuned solution of a specific application need. Not only can stages be added as required, but hydraulic designs can be selected over the range of specific speeds from *10 (500) to 230 (12 000)* and the maximum efficiency obtained. As in volute type pumps, the lower specific speed impellers (radial) are used for the higher heads; the mid-range specific speeds (mixed flow) for medium heads medium flows and the highest specific speeds, (axial-propeller) for high flow, low head applications. Figs 6.6[1] and 6.7[1] show impeller shapes for various specific speed and the characteristic head/capacity and power curves for each. Their range of speeds is generally below *3600* rpm. An outer casing is applied for discharge pressures above 150 kp_a (1000 psi).

Other advantages of this type of pump are the small footprint at floor or ground level and the ability to customize mechanical and metallurgical design options to the customer's needs. Disadvantages are submerged bearing system, remoteness that results in neglect

FIGURE 6.5 – Vertical turbine diffuser pump.

CENTRIFUGAL PUMP TYPES

from a monitoring stand point, special equipment that is required for installation and removal and high bay areas that may be required in building locations. Open and closed impellers are used with the latter preferred. Open impeller wear can result in the need replacement, whereas this is not generally the case with closed impellers. Also, open impellers require setting adjustment, that can be tricky, to optimize efficiency. The barrel or can pump as shown in Fig. 6.8, is popular in applications where additional NPSHA is required. This pump with its own sump can be sunk into a floor with minimal relative expense. Vertical turbine pumps are driven by induction or synchronous solid or hollow shaft motors or internal combustion engines through right-angle gear drives. It is common with variable speed drives to pass through a pump system natural frequency. This need not be a problem as long as provision is made to prevent continuous operation in this area.

Submersible pumps

Submersible pumps are pumps with connected motors that can be submerged into the sump, pit or well. In some cases the submerged, hermetically sealed motors are filled with oil and in other cases they are filled with air along with separate water cooling of the motor

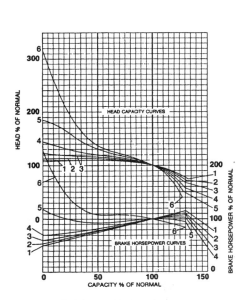

FIGURE 6.6 – Typical performance curve shapes for impellers of various specific speed design (see also Figure 6.8.)

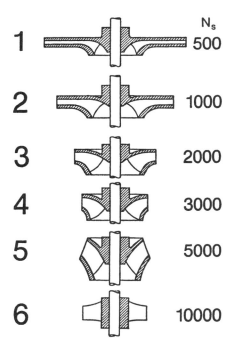

$$N_s = \frac{N\sqrt{Q}}{H^{3/4}}$$

FIGURE 6.7 – Impellers of various specific speed design.

FIGURE 6.8 – Barrel or can pump.

casing. See Fig. 6.9, a waste water pump. The smaller size pumps are furnished with air cooled motors requiring no coolant flow through the pump motor casing. The larger pumps are cooled with pumpage most frequently, utilizing the impeller back vanes as the pumping means. Seal chambers filled with oil separate the motor from the pump on some units. A submersible slurry pump with liner is shown in Fig. 6.10, a dewatering pump in Fig. 6.11 and a submersible propeller pump installation in Fig. 6.12, at Modesto, California, USA. These four pumps have a combined flow rate of 78 *MGD*. Diffuser and volute pumps of many types have been furnished in submersible configurations. Submersible pumps can be portable and are commonly used as contractors pumps for utility cleanup. Fig. 6.23 shows a cutaway of a portable submersible and Fig. 6.24 shows the vortex pump version of that same pump used for slurry transport duties, such as tank emptying. Further coverage of these pumps is included in Chapters 24 and 25.

Hermetic or sealless pumps

Hermetic or sealless pumps are used where there is a need to contain toxic, dangerous and/or valuable fluids[2]. They need and have no packing or mechanical seals and are sometimes called glandless pumps. The hermetics have low operating cost in some instances relative

CENTRIFUGAL PUMP TYPES   119

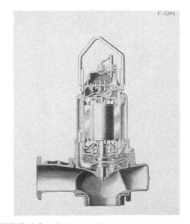

FIGURE 6.9 – Submersible waste water pump.

FIGURE 6.10 – Submersible slurry pump with liner.

FIGURE 6.11 – Submersible dewatering pump.

FIGURE 6.12 – In Modesto, California, treated wastewater plant effluent is lifted 16 ft into two massive storage ponds containing 8000 acre ft. Four Flygt submersible propeller pumps handle the flow at a combined rate of 78 MGD.

to mechanically sealed units[3]. Centrifugals are the most efficient of the hermetic types. They are available in two basic variations: *Magnetic Drive and Canned Motor*. The latter has been around since the twenties but has built up most of its usage in the last 40 years. The main difference between the two types is in the way the rotor is driven. The magmetic-drive rotor, see Figs 6.15-6.19, is driven by a set of permanent magnets located on one side of a containment shell or can, that drive another set of magnets on the inside of the shell. In the case of the canned-motor pump, the stator with its windings is isolated from the rotor by a sheet metal shell, or can, that is located in the air-gap of the motor. Government regulations on volatile organic emissions that have been issued around the world have resulted in a development effort in the pump industry of major proportions in the last 10

CENTRIFUGAL PUMP TYPES

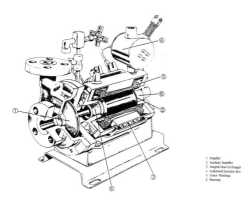

FIGURE 6.13 – Canned motor hermetic (sealless) pump.

FIGURE 6.14 – Canned motor hermetic (sealless) pump.

years, along with tremendous interest on the part of users. Canned motor unit units have been improved substantially. Their motor windings can now handle temperatures as high as 343°C (650°F)[4]. See Figs. 6.13 and 6.14. The hermetic magnetic-drive pump (Figs. 6.15-6.19) has seen its main growth during this time, assisted in great part by the advent of stronger permanent magnets and especially Samariun Cobalt magnets with their higher temperature capability, and a 8:1 strength ratio over their predecessors. Today a customer finds that he has a choice of many pump manufacturers to supply his needs in this area. It can also be said that the seal manufacturers at the same time have had a major development program of their own which has provided a magnitude improvement in seal reliability and reduced leakage rates to a point where zero leakage can be approached with conventional pumps. The seal less designs presently have a first cost of approximately two times that of a conventional pump and their failures can literally destroy the pump, but the new designs have a secondary containment that will avoid any catastrophic leakage for a period of several days. Condition monitoring equipment can also be furnished so that impending failures can be avoided. One has to balance the seriousness of a failure with the

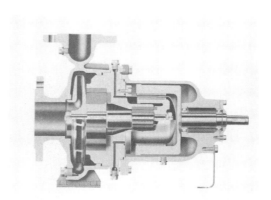

FIGURE 6.15 – Magnetic drive hermetic (sealless) pump.

 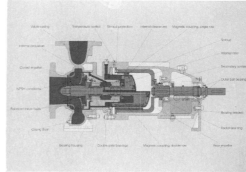

FIGURE 6.16A　　　　　　　　**FIGURE 6.16B**
Magnetic drive hermetic (sealless) pump.

FIGURE 6.17A　　　　　　　　**FIGURE 6.17B**
ANSI magnetic drive hermetic (sealless) pump.

CENTRIFUGAL PUMP TYPES

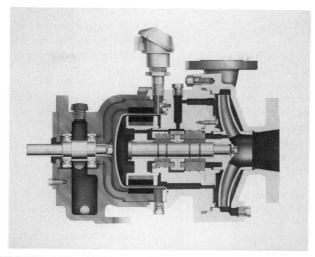

FIGURE 6.18 – ANSI magnetic drive hermetic (sealless) pump.

FIGURE 6.19 – Fibreglass ANSI magnetic drive pump.

economics to make a decision. In the case of many plants, where the total amount of emission has to be determined and reduced to the limit set by the new regulations, the solution is often a mixed one of retro-fitting some existing pumps with new seals and replacing others with new seal less pumps. Maintenance capabilities and costs can also be a deciding factor in the user's decision. The canned pump, see Figs. 6.13 and 6.14 has zero emission capability, tolerance to some dry running of the bearings, secondary containment and bearing wear monitoring capability built in. A main consideration with this pump is the expected life of the liner with a corrosive process fluid once the best material option has been selected.

Advantages and disadvantages of the two are as follows:

CANNED MOTOR PUMPS

Advantages
Compact
Quiet-no motor fan noise
Low installation costs:
  No heavy bed plate required
  Thin supports allow unit
    to move as an entity without
    creating internal distortion
  No special foundation requirements
No alignment required.
Low maintenance
Less parts
Experience with bearing monitors using
  power consumption monitors as
  backup has been excellent.
Secondary does not require
  a mechanical seal.
Lowest failure rate

Disadvantages
Special motor must be serviced by mfr.
Explosion-Proof applications require
  a testing agency (U.L.) label for
  pump and motor as a combined unit.
Motor heat must be dissipated to prevent
  flashing of pumpage
Whole pump replaced if windings fail
Capability for secondary containment
Thin can (0.015 in. to 0.018 in.)
Special instrumentation required to
  determine rotation.
Low tolerance to solids in pumpage due to
  close clearances between can and rotor
  or stator.
Maximum viscosity is 120 cp. Limit could
  be as low as 20 cp.
Higher temperature pumpage require
  isolation and external cooling of the
  stator cavity to prevent winding failure.
Cannot handle high temperatures as easily
  as magnetic drives.

MAGNETIC DRIVE PUMPS

Advantages
Use standard motors
No motor heat affects on pumpage
Can use ceramic shell
Can be repaired on-site
Magnets can tolerate heat better than
  canned motors windings.
Containment shell can be 5 times the
  thickness of canned motor shell
Can pump liquids up to 400°C (750°F).
Best tolerance for corrosive environments.
Pumped fluid must provide adequate
  bearing lubrication.
More tolerance to solids than canned
  motor pumps.

Disadvantages
Possibility of misalignment of motor and
  magnetic-drive shafts
Three sets of bearings: pump,
  magnetic-drive and motor shell.
Does not contain bearing monitors.
Cannot handle the pressures of canned rotor pumps.
Decoupling can be serious
Alignment can be tricky and costly.
Foundation and/or base required

Some types of pumpage handled by hermetic pumps are as follows: toxic, flammable, aggressive, poisonous, volatile, odorous or pungent, expensive, explosive, low vapour pressure, low surface tension and corrosive organic and inorganic chemicals, acids, bases, solvents, heat transfer fluids, (including refrigerants), liquefied and compressed gases, deionized and de-mineralized water, oils, brines and fluorocarbons[5]. Sealless pumps eliminate leakage across seal faces[6]. but, they must be properly selected and customized,

CENTRIFUGAL PUMP TYPES

if necessary, for the application. One user[8] of many canned and magnetic-drive units found that they even had to go back to conventional pumps on one application because they were too overzealous in applying hermetics. If they are not properly applied and maintained, they will fail, sometimes drastically. Training is something you will have to commit to immediately because these pumps are sensitive to human error and improper operation or maintenance.

To decide if a pump is right for your application, you need to understand their design, their weaknesses and be ready to provide the vendor with complete necessary information on design and off design conditions and the properties of the pumpage. Users point out that reliability is as important a factor as emission elimination, especially with difficult to seal liquids such as those that harden when exposed to air or water. Special designs are called for if the pumpage has poor lubricating qualities (common), poor heat absorption, is corrosive, has low vapour pressure or low or high viscosity.

If you have an application where the sealed pumps are failing repeatedly, you must determine if the problem is in the pump or system. Users have replaced sealed pumps with hermetics only to find that they failed too. Hermetic units generally use the pumpage as a lubricant and coolant to take away motor heat and eddy current losses: In the case of the canned motor pump and magnetic-drives with metallic cans; the eddy current losses and lastly, friction losses in the bearings of both types as well as impeller disc friction losses. Any system condition that will cause the the pumpage being circulated to boil can result in a failure. Bearing failures and running dry are two of the main causes of failures in hermetic pumps. Viscosities that are too high (even 30 cp could require special considerations for the small clearances and conduits to handle) can cause the coolant to boil. Viscosities that are too low can cause bearing failures due to inadequate hydrodynamic film in the sliding bearings of either type.

Suspended solids, fluids that polymerize and cavitation can also result in failures. Most of these failures are caused by misapplication or operator errors. Most users see a reduction in failures as their staff get used to the differences of operating these pumps vs the sealed types they have been used to.

Chemical pumps that comply with Din-24-256 or ISO-2858 for Europe and ANSI B-73 for the US are the most widely used equipment in the chemical industry. Most magnetic-drive units being offered will match the standardized dimensions of one of those two specifications. If a particular hermetic is being considered, the user should demand heat rise calculations be done on the cooling circuit. The temperature rise is a function of the losses being covered, the speed of the pump, the specific gravity and specific heat of the cooling fluid and the amount of fluid being pumped and circulated. If the temperature is too high then cooling means must be employed. Temperature rise is calculated as follows:

(1) SI Temperature Rise in cooling circuit (°C) = 0.86 x Drive Losses, KW/Recirculation coolant flow, mph^3 x S.G. x S.H.

(2) USCU Temperature Rise in cooling circuit (°F) = 5.09 x Drive Losses, HP/Recirculation coolant flow, gpm x S.G. x S. H..

Note: If the coolant is taken off the impeller discharge the full temperature rise of the coolant is the rise through the pump plus the rise in the coolant circuit. At minimum specified flow this would equivalent to:

SI $\Delta T = 0.86$ x KW Consumption at minimum flow/pump flow x S.G. x S.H.

or

USCU $\Delta T = 5.09$ x HP Consumption at minimum flow/pump flow x S.G x S.H.

The calculation should be made at least at operation, minimum flow and maximum flow conditions and the question as to what happens to the fluid on shutdown from the residual heat absorbed should be answered. Temperature control in the bearing circuit is crucial. You have to understand the temperature rise of the cooling fluid and how it responds to heat[7].

Containment shells can be metallic (e.g. stainless steel, hastelloy), ceramic (zirconia, silicon carbide) fibreglass or plastic (Peek, polyetheretherketone). The metallic shells result in eddy current losses with some metals having lower losses than others. For example, hastelloy has less loss than 316 stainless. Ceramics and plastics have no losses. One user[8], points out that peek was found to be very sensitive to human error and specifically the use of steam. Magnets can be Neobydium up to 120°C (250°F), Samarium up to 220°C (425°F), and above that Al-Ni-Co is used. Samarium also tends to resist the demagnetization of decoupling better than the others. Bearings are plain journal, sliding type of carbon in the older canned motor units and silicon carbide in the magnetic-drives. The direct sintered type of silicon carbide is slightly more expensive but offers more corrosion resistance. Magnets should be sealed from corrosive, aggressive fluids and should be conservatively rated especially if service with different pumpages can be expected. Caution should also be taken that oversized motors put on service on a magnetic-drive pump are not so large as to cause decoupling during start up. Because of the use of pumpage for cooling, the head capacity curve will fall below the curve for the same impeller and speed of a sealed pump, unless the seal losses are high such as with a double seal. Expect to see more internal auxiliary pumps in the near future to maintain a positive pressure in the cooling circuit and avoid flashing.

Monitoring

Monitoring of hermetic pumps is increasing. Measuring power instead of current has advantages[8]. Power monitoring is 10 times as sensitive at light loads as current sensing. Power sensors can be used with variable frequency and D.C., whereas current sensors can not. Thermocouple probes are not reliable for sensing dry running. Vibration measurements are not reliable for sensing dry running problems soon enough. Pressure and flow switches have been used satisfactorily. In spite of all the precautions and concerns these hermetic units do give acceptable zero emission service and after a learning curve lower maintenance costs than sealed pumps. Two good references are references 1 and 4. The future looks even brighter. Magnetic bearings will solve the bearing failure problem. Their development is moving along with the main barrier to their use now being cost. Magnets will continue to get stronger, and a new option of *Barrier Pump Design*, where a pressurized gas barrier is set up between the impeller and the magnetic-drive, shows promise. It allows ball bearings to be used instead of the sliding bearings.

CENTRIFUGAL PUMP TYPES

Non-metallic pumps

Fibreglass and plastic materials are now being used quite commonly to construct the wet-ends and bases of pumps for corrosive duties. See Fig. 6.19.

Sanitary Pumps

Standards have been set up in various countries on the pumps used in various food industries such as the 3A Standard in the U S for milk and milk products. This latter standard governs the shape of the food wetted and external surfaces, surface finishes, materials, and requirements of pump features such as openings as well as requirements on gaskets and static seals. These specifications take into account the CIP (clean in place) processes that are used to clean the pumps in between batches or at the end of specific periods of time.

Aseptic

Aseptic is another term that means germ-free, it is used in the canning, food, dairy, pharmaceutical and other industries where the emphasis is on bacteria-free operation to assure product freshness, flavour, colour and shelf-life. Centrifugal pumps are produced that fall under these descriptors.

Another aspect to pumps for the food, drink and pharmaceutical industries is the compatibility of the pump with the food or drink being processed. Many food slurries and drinks or beverages are non-Newtonian and may tolerate little shear for example. Centrifugals are widely used in the beverage area on soda pop, milk, beer and wine. They

FIGURE 6.20 – Sanitary centrifugal for use on product and CIP return in dairy plant.

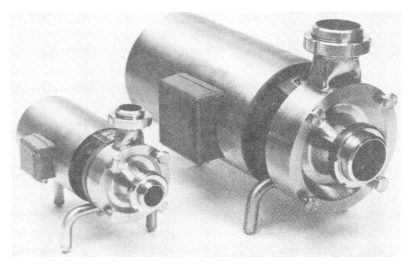

FIGURE 6.21 – Stainless steel centrifugal pumps for food processing.

are generally of a stainless steel metallurgy. They are also used in the ancillary processes of CIP, sterilization and transfer. See Figs 6.20 and 6.21 for some sanitary centrifugal pumps.

Grinder, chopper and cutter pumps

These specialty pumps in waste water and sludge service are used on unscreened pumpage containing rags, debris, or stringy material. The latter could clog a normal waste water pump. Generally, at the intake of sizable treatment plants there are bar screens to remove this type of material before it gets to the pumps. In some small plants this type of pump may be incorporated to chop, cut or grind up this material as well as pump it. *Cutter* pumps essentially have modified standard impellers with cutter knives attached. *Grinder* pumps use a cutter attachment on a vortex (recessed) impeller. *Comminutors* are just grinders that have no pumping capability. Fig. 6.22 shows a chopper pump.

Inducers

To quote from Lobanoff and Ross[9] An inducer is basically a high specific speed, axial flow, pumping device roughly in the range of Ns = 4, 000 to 9, 000 that is series mounted preceding a radial stage to provide overall system suction advantages. Inducers are essential enabling devices in the technological advancement of high speed centrifugal pump units. They reduce the NPSHR that would be required by the radial stage alone. Fig. 6-25 shows a popular high speed pump with step up gear and inducer.

Booster pumps

Booster pumps are not so much a distinct pump type as the use to which the pump is put. Instead of an inducer a separate pump preceding the main pump is called a booster pump. By raising the pressure at the suction of the main pump NPSHR problems can be satisfied.

CENTRIFUGAL PUMP TYPES

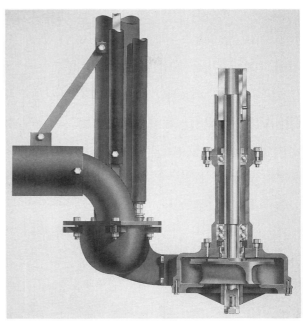

FIGURE 6.22 – Chopper pump.

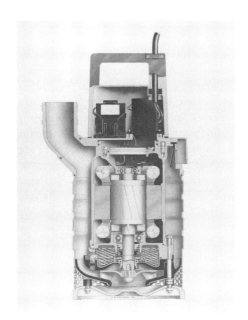

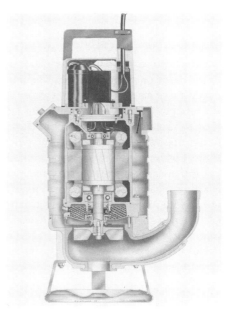

FIGURE 6.23 – Portable submersible pump.   **FIGURE 6.24** – Portable submersible vortex pump.

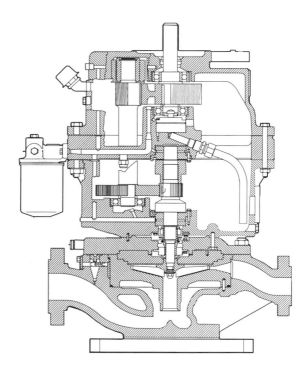

FIGURE 6.25 – High speed pump with step up gear and inducer.

High pressure multistage pumps

When the head required is greater than can be attained with a reasonable diameters and/or speed based on proven designs available the economical alternative that is available is the addition of stages on the same shaft. The application chapters on water (specifically boiler feed pumps for example) and Petroleum give numerous figures of multistage pumps with up to 15 stages. There are problems of axial thrust that must be overcome. One means is to use opposing banks of impellers with results similar to the double inlet split case designs. Casing design for pressure is another consideration. Axial split casings are used up to the economic restraints on their pressure capability design. Segmental, stage, pancake type casings then come into play. These individual stage casings are then bolted together with long tie rods and the casing seals provide the integrity. When this design reaches its design limit. The axial split or segmental casings are then put into barrel casings giving a dual casing construction. See Figs 18-6, 18-7 and 20-3 especially. Note the balance piston or balance disc in Figs 18-6 and the opposed impellers in Fig. 20-3. Balance drums are another means of balancing axial thrust.

References

[1] Greutink, H., "Vertical Diffuser Type Turbine, Mixed Flow and Axial Flow Turbine (Propeller) Pumps – A Brief Overview", Turbomachinery Laboratory, Mechanical Engineering Depart-

CENTRIFUGAL PUMP TYPES

ment, Texas A& M University, from Proceedings of the 5th International Pump Users Symposium, p 67, Copyright 1988.

[2] Hydraulic Institute, "Sealless Centrifugal Pump Standards" HI 5.1 – 5.6 (1992)

[3] Nasr, A., "When to Select a Sealless Pump", Chemical Engineering (May 26, 1986)

[4] Jaskiewicz, S and J. Cleary, "High Temperature Canned Motor Pumping", World Pumps (Feb. 1993)

[5] Cleary, J., "The Use of Sealless Pumps for Heat Transfer Service", Rocon Conference Proceedings, NJIT (Nov. 10-12, 1993).

[6] Zimmerman,G., "Going Sealless" Pumps and Systems Magazine (Mar. 1993).

[7] Cleary, J., "Sealless Pumps: The Effects of Heat", Pumps and Systems Magazine, Mar. 1993.

[8] "Reliable Mag Drive Protection: Should You Monitor Temperature, Flow, Current or Power?", Chemical Processing (June 1992)

[9] Lobanoff, V. and R. Ross, "Centrifugal Pumps Design and Application", 2nd Edition (1992), Gulf Publishing Co.

SECTION 3

Positive Displacement Pumps

As stated in Chapter 2, Positive Displacement Pumps are batch delivery, periodic energy addition devices, whose fluid displacement volume (or volumes) is set in motion and is positively delivered from a lower to higher pressure, irrespective of the value of that higher pressure. The reader is referred back to Chapter 2, Part 1 for other background information on Positive Displacement Pumps.

Classification: Fig. 2.1 is repeated here. This classification chart shows the positive displacement pumps broken down into two main categories; rotary and reciprocating. Rotary pumps are made up of vane, gear, screw, lobe, flexible vane, flexible tube (peristaltic), flexible liner, axial piston and circumferential piston types. Reciprocating pumps are made up of piston or plunger steam and power pumps and piston, plunger or diaphragm controlled volume pumps.

ROTARY PUMPS: NOMENCLATURE, CHARACTERISTICS, COMPONENTS AND TYPES

RECIPROCATING PUMPS: NOMENCLATURE, CHARACTERICSTICS, COMPONENTS AND TYPES

Multimedia Tools for Process Engineers

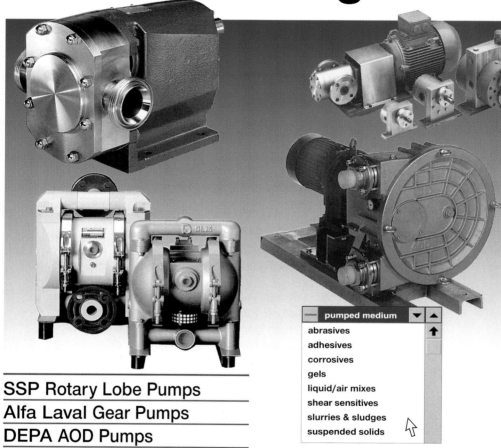

SSP Rotary Lobe Pumps
Alfa Laval Gear Pumps
DEPA AOD Pumps
Alfa Laval Peristaltic Pumps

Positive pumps from Alfa Laval move a multitude of often difficult media and handle the full span of process transfer tasks. They embody the most advanced designs and manufacturing quality you would expect from a world leader.

Choice of the optimum pump type, range, model and specification means exactly the right unit for your application with high efficiency, low maintenance and minimal pumping problems.

So when it's a question of positive pumping, call us for multimedia knowhow and the cost effective answer.

Alfa Laval Pumps Ltd
Birch Road Eastbourne East Sussex BN23 6PQ England
Telephone: 01323 725151   Telefax: 01323 730495   SU01

⋏ Alfa Lava

ROTARY PUMPS: NOMENCLATURE, CHARACTERISTICS, COMPONENTS AND TYPES

Rotary pumps

Rotary pumps make up the second largest group of pumps in terms of numbers. They also represent the second most economical selection, next to centrifugals. Most rotary pumps are self-priming and along with that have the ability to handle fluids consisting of liquids with entrained gas or vapour. Compared with the high pulsations and definitive batched flow of the reciprocating types, the rotary has a more continuous flow with lower pulsation levels. They are available in types that can handle fluids of extremely high viscosity. However, the most efficient speed drops as viscosity increases above a certain point. This is a function of clearance and the shear action. With high viscosity fluids the clearance is generally opened up by the manufacturer to reduce the power consumption and maintain the low shear effects on the product. Their capacity varies with speed but is affected by pressure to some extent due to its affect on slip in the low viscosity ranges but as viscosity increases this effect continues to diminish to a point. If, at some viscosity, the latter impedes the intake of the fluids into the displacement compartment then the capacity and efficiency will drop off. The operation of rotary pumps has been described[1] aptly as having a suck and squeeze action. They suck the fluid in and then squeeze it out. Rotary pumps are designed to operate with close clearances and wetted internal surfaces. Therefore they are sensitive to fluids containing abrasive solids. Because they are positive displacement pumps they should not for safety sake be run with a closed discharge. The preferred term for rotary units is pressure as opposed to head. Rotary pumps should be sized to provide the capacity required when handling the lowest viscosity expected. While their drivers are sized for the power requirements with the maximum viscosity expected. Rotaries depend on the lubricity of the pumpage. Fluids to be pumped with low lubricity should cause one to look at other alternatives first, for example – reciprocating pumps with replaceable liners. Rotary pumps should not be installed such that the the piping, motor or thermal

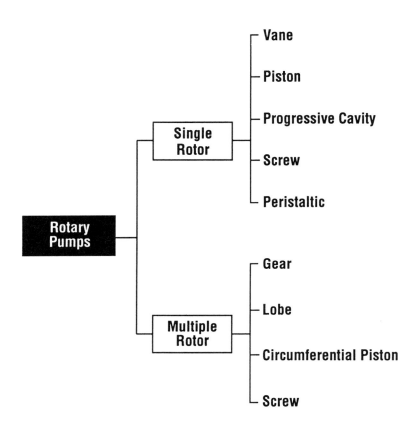

FIGURE 7.1 – Types of rotary pumps.

expansion will impose strains on their casings. Air in leakage through packing or mechanical seals should be monitored closely. Grease lubricated ball-bearings are common. Magnetic-drive pumps have sleeve bearings on the pump side of the drive assembly. Timing and intermediate gears will require oil lubrication. The presence of lubricant in gears should be checked at installation, before start up, as most are shipped without lubricant. Bearings are usually shipped with lubricant.

Nomenclature
Nomenclature for various rotary pump types is shown in Figs. 7.2 through 7.17 and further examples can be obtained from the Hydraulic Institute Standards publication and Europump Terminology.

When your process needs to be driven hard to handle viscous, slightly abrasive or non lubricating fluids then a Stork SRT internal gear pump is unlikely to be overtaken.

When it comes to good handling the SRT can be applied to all kinds of applications. The performance of the Stork SRT positive displacement pump works best when the installation is geared to the process.

There is no reason why your business should not get the best, ask Stork to help you get into top gear.

STORK®

Stork Pumps Ltd.

Meadow Brook Industrial Centre, Maxwell Way, Crawley,
West Sussex RH10 2SA.
Tel: (0293) 553495 • Fax: (0293) 524635

Fieldhead Business Centre
ListerHills, Bradford BD7 1LG
Tel: (0274) 742247 • Fax: (0274) 742228

NEMO®-Pumps.
The versatile Netzsch pumping concept for all branches of industry.

Quality pays for itself!

® **NEMO:** Registered trade mark of **NE**TZSCH **MO**HNOPUMPEN GMBH

Our Quality assurance system ist certified according to DIN ISO 9001 under the DQS-No. 22405-01.

Netzsch-Mohnopumpen GmbH
Liebigstraße 28, P.O. Box 1120
D-84464 Waldkraiburg, Germany
Phone 08638/63-0, Fax 67999
Telex 56421

NETZSCH

THE RIGHT GEAR FOR PUMPING

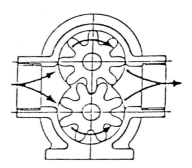

If you need standard or special purpose gear pumps or pumping systems, contact Albany.

- Many materials, inlcuding Hastelloy and PVDF
- Turbine, air, diesel and flameproof electric drives
- Wide range of industry standards including API 676
- Specialists in hot and viscous liquids — heated pumps
- Pump Gear Service — 400 gear pumps in stock

For details and quotations contact

Albany Engineering Co. Ltd.
Lydney, Gloucestershire
GL15 5EQ.
Tel: 01594 842275 Telex: 43363
Fax: 01594 842574

Quality Assurance to BS5750 Pt. 2. and ISO 9002

Certificate No. FM 20986

- DOSING PUMPS
- CENTRIFUGAL PUMPS

MAKERS OF STANHOPE, CROWN, & BARCLAY KELLETT GEAR AND CENTRIFUGAL PUMPS

ROTARY PUMPS: NOMENCLATURE, CHARACTERISTICS, ETC.

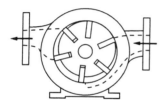

FIGURE 7.2 – Sliding vane pump (balanced).

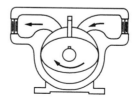

FIGURE 7.3 – External vane pump.

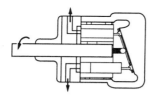

FIGURE 7.4 – Axial piston pump.

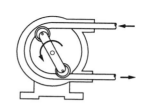

FIGURE 7.5 – Flexible tube pump.

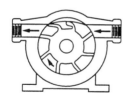

FIGURE 7.6 – Flexible vane pump.

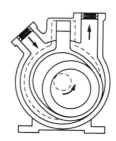

FIGURE 7.7 – Flexible liner pump.

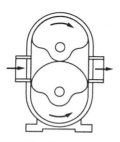

FIGURE 7.8 – Single lobe pump.

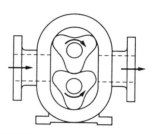

FIGURE 7.9 – Three-lobe pump.

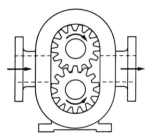

FIGURE 7.10 – External gear pump.

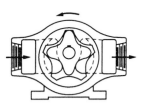

FIGURE 7.12 – Pumping principle of internal gear pump.

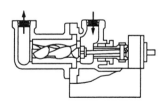

FIGURE 7.14 – Single screw pump (progressing cavity).

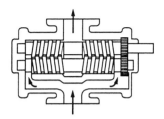

FIGURE 7.16 – Two screw pump.

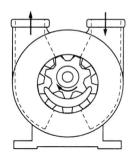

FIGURE 7.11 – Internal gear pump (with crescent).

FIGURE 7.13 – Circumferential piston pump.

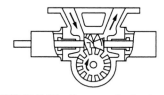

FIGURE 7.15 – Screw and wheel pump.

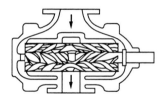

FIGURE 7.17 – Three screw pump.

ROTARY PUMPS: NOMENCLATURE, CHARACTERISTICS, ETC.

Definitions:

Displacement (D), m³/rev. (in³/rev) is the volume displaced per revolution of the rotor(s). A variable displacement pump is rated at its maximum displacement.

Slip, S m³/h, (gpm) is the quantity of fluid which leaks through internal clearances of a pump per unit of time. It is dependent on internal clearances, differential pressure, the characteristics of the fluid being handled and in some cases on the speed.

Metering effectiveness is the ratio of the pump's minimum volumetric efficiency to maximum volumetric efficiency, expressed as a percentage, over the specified operating range.

Capacity (Q) m³/h (gpm). The quantity of fluid delivered per unit of time, including any dissolved or entrained gases under stated operating conditions. In the absence of any vapour or gas, capacity is equal to the volume displaced per unit of time, less slip.

SI: $Q = 0.019\,DN - S$, m³/hr   Eq. 1
where D = Displacement, m³
N = RPM
S = Slip, m³/hr

USCU $Q = \dfrac{DN}{231} - S$, gpm
D = in³
N = RPM
S = gal/min

Power input (P) kW (hp) is the power delivered to the pump drive shaft.

Power output (P_u) kW (hp) is the power imparted to the fluid being pumped. It is less than the power input by the amount of losses in the pump.

SI $P_u = 0.0004352 Q p_{td}$, kw   Eq.2
USCU $p_u = 0.0005834 Q p_{td}$, hp
where p_{td} = Pressure differential of pump, kp$_a$ (psi)

Net Inlet Pressure p_n is the difference between the absolute pressure at the pump inlet and the vapour pressure of the liquid.

Net Inlet Pressure Required p_{nr} is that required by the pump manufacturer to avoid vaporization of the fluid and cavitation at the speed, pressure and fluid characteristics that will exist.

Overall efficiency, η_o is the ratio of the pump power output to the total power input, p_{mot}. It is also called wire to water and overall unit efficiency.

$$\eta_o = \dfrac{P_u}{p \times 100}$$

USCU $\eta_o = \dfrac{P_u}{P} \times 100$

Pump efficiency, hp is the ratio of the pump power output to the pump power input.

$$\eta_o = \frac{P_u}{P} \times 100$$

Pump volumetric efficiency, hv: is the ratio of the actual pump capacity to the volume displaced per unit of time.

SI $\qquad n_v = \frac{Q}{0.019DN} \times 100$

USCU $\qquad n_v = \frac{231Q}{DN} \times 100$

Rotation

The rotation of a rotary pump as viewed from the shaft end.

Characteristic curve

Typical HQ and Power Curves are shown in Fig. 7.18

Types

A Classification Chart is shown in Fig. 7.1.

Vane

There are two basic types of vane pumps, internal and external. See Figs. 7.2 and 7.3, The internal vane pump has the sliding vanes in the eccentric rotor or cam whereas the external type has the sliding vane in the stator. The vanes may actually be blades, buckets, rollers, slippers, etc. The pumps come with the hydraulic forces on the rotor balanced or unbalanced. The internal type comes in constant and variable displacement types. Figures

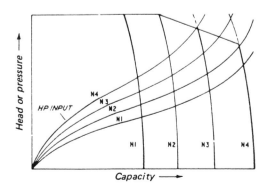

FIGURE 7.18 – Typical rotary pump HQ and power curve.

7.2 and 7.3 are constant displacement unbalanced types. These pumps are generally of light, compact construction; economical, good up to moderate pressures, and suitable for handling air and other gases as low pressure compressors and mid-range vacuum pumps. The internal vane pump vanes are thrown out by centrifugal force against the stator whereas the external vane rotor drives the sliding vane in the casing up with its cam action from its elliptical shape when approaching its major axis and they are forced back against the rotor by springs or other means when approaching the minor axis. Fluid is being aspirated into the pump on the suction side at the same time it is being discharged on the discharge side. Leakage occurs at the tips and sides of the vanes. Increasing the number of vanes can materially improve the volumetric efficiency. The maximum number of blades that can be used is determined by the viscosity. The higher the viscosity, the lower the number of blades. Some pumps have swinging vanes that are hinged on the rotor. Because of the space taken up by these vanes the pumps are generally larger than the sliding types and cannot accommodate as many vanes. The joints are prone to wear and the lower number of blades makes it hard for this type to compete on an efficiency basis with the sliding vane type. Vane pumps are self-priming. They give constant delivery and discharge pressure at constant speed and have minor pulsations. Blade wear is self compensating. They can pump in either direction and do not require check valves. However, they do require protection against closed discharge. Adjustable capacity vane pumps at constant speed are available. Vane pumps are used in large volumes on oil burners, and hydraulic drives. They are not used to any great degree in food processes because the wear products could contaminate the food. Fig. 7.19 shows a typical characteristic curve for a vane pump.

Axial piston

This pump is really a reciprocating pump that converts rotary shaft motion to an axial reciprocating motion of a piston. They may be fixed or variable displacement. They generally use either a port plate or check-valves in the ports. The pistons are oriented

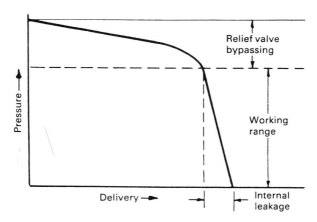

FIGURE 7.19 – Typical vane pump characteristic curves.

axially with the drive shaft[3]. See Fig. 7.4. Displacement is varied by changing the angle between the swash plate and the piston block face. They have high volumetric efficiencies, can handle pressures of 725 kpa (5000 psi) and can be run up to speeds of 6000 rpm.

Radial piston or radial plunger

These pumps are also reciprocating pumps that convert rotary shaft motion into to a radial reciprocating motion of the pistons or plungers. The displacement can be fixed or variable. Check-valve or valve spindles are used for porting. They have high volumetric efficiencies and can operate at pressures up to 1450 kpa (10, 000psi). Fig. 7.23 shows a radial piston pump.

Flexible tube or peristaltic pump

These pumps (Fig. 7.5) have a flexible tube of elastic material that is squeezed by a compression ring or shoe on an adjustable eccentric. Rotor rotation causes the hose to be compressed and the fluid in the hose pushed out of the pump. After the compression ring passes the hose returns to its circular cross-section, creating a lower pressure and aspiration of the fluid from the suction line into the space takes place. They are used for low pressures and low capacity applications. They are less sensitive to abrasives than other rotary pumps and hence are used to handle low flow streams of this type of material. It is also used for metering and low shear applications in such diverse fields as food processing, waste treatment[4] and paper production. Figs 7.24 A and B show the front and side views and Fig. 7.24C the rotor and hose views of a peristaltic pump. Fig. 7.25 shows a characteristic curve for a peristaltic pump. The critical component in determining the discharge pressure capability of a peristaltic pump is the hose. Once limited to 2-3 bar., these pumps are commonly being used for 15 bar discharge pressures today. One vendor limits the size of solid particles to third the tube internal diameter and points out that some dry running is acceptable, but the cooling of the tube is dependent on the pumpage. Some designs have a glycerin mixture in the body of the pump surrounding the tube for cooling purposes. Many of the designs available are reversible, and will work as efficiently in either direction of rotation. A shaft seal is not required for the pumpage, but pumps with a glycerine or other coolant surrounding the peristaltic tube could require a seal.

Flexible vane (impeller) pump

The flexible vane (impeller) pump has a cam shape between the suction and discharge ports on the inside diameter of its housing. An impeller with flexible blades rotates in this housing. When each blade passes the discharge port, it is depressed radially inward, which forces fluid to discharge. When it passes the cam and the vane straightens out radially again a slight vacuum is created which induces the fluid from the inlet into the displacement volume between it and the following blade. These pumps have a gentle action and are used extensively in the food industry. They are self-priming and are used on large capacity applications. The maximum pressure capability of the pump is in the order of 10 kp_a (70 psi) because of the non-rigid displacement volumes. This same property also eliminates the need for a relief device. Dry running for short periods (a minute or less) can be accommodated. Maximum operating temperatures are a function of the liner material, and

ROTARY PUMPS: NOMENCLATURE, CHARACTERISTICS, ETC. 141

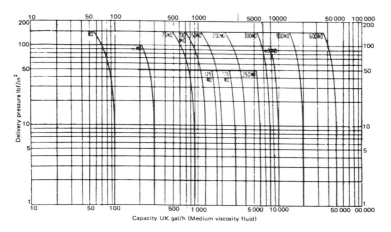

FIGURE 7.20 – Typical lobe pump HQ characteristic.

generally in the range of 80-100°C. Like lip seals the blade tips should be prelubricated before initial use. Some applications are food, beverage, milk and dairy products, pharmaceutical, chemical marine engine cooling, entrained air containing products etc. See Fig. 7.6.

Flexible liner pump

These pumps are constant volume eccentric cam squeegee pumps. The eccentric cam rotates within a flexible liner and squeegees the fluid trapped between the liner and the housing. No shaft seal is necessary with this pump because the liner isolates the pumpage from the moving parts. The pumping action is the same as the peristaltic pump and its usage would be similar. See Fig. 7.7.

Lobe pumps

Lobe pumps were one of the earliest types of rotary pumps to come on the market. They were used extensively for low pressure, high capacity pumps and blowers. They require timing gears since the lobe rotors do not drive themselves. They are used to pump fluids with solids in suspension and widely used in the food processing industry as sanitary pumps because they are easily cleaned and adapted to CIP. See Figs 7.8 and 7.9. Single lobe pumps are used for pumping large delicate products, such as strawberries, meat chunks, large curd cottage cheese, potato salad, etc. Because of their line contact and relatively large clearance area they have a high leakage flow and therefore are not suitable for high pressures. They are classified as gear pumps. The lobe pump resembles the external gear pump in action. See Fig. 7.20 for HQ characteristic. They require at least two product seals.

External gear pumps

(Fig. 7.10) utilize the meshing of spur, helical or herring-bone gears to form a series of expanding cavities at the pump inlet that are angularly displaced before being contracted

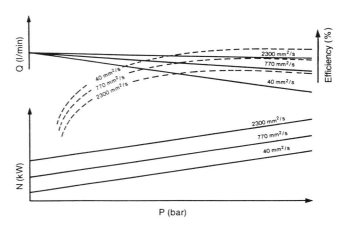

FIGURE 7.21 – Typical performance curve, screw pump.

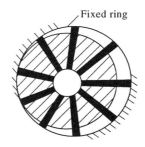

FIGURE 7.23 – Radial piston pump.

FIGURE 7.24A – Peristaltic pump, front view.

ROTARY PUMPS: NOMENCLATURE, CHARACTERISTICS, ETC.

FIGURE 7.24B – Peristaltic pump, side view.

FIGURE 7.24C – Peristaltic pump, view of rotor tube assembly.

at the discharge, where the gears mesh[5]. The fluid is carried between the gear teeth and displaced when they mesh. External gear pumps have all the gears cut externally. One gear is keyed to the shaft and the other is an idler. The higher the number of teeth the lower the leakage loss. Spur gear pumps operate up to approximately 600 rpm and herring bone gear pumps up to 1750 rpm. Helical gears are quiet, but produce an undesirable end thrust. They are especially suited to clean fluids with lubricating qualities. Characteristic curves are

TYPE SP/50

Inner diameter
pump hose Ø 50 [mm]

Minimum starting
torque 620 [Nm]

Capacity per revolution
2,9 [l]

Maximum discharge
pressure 1.500 [kPa]
(15 bar/220 psi)

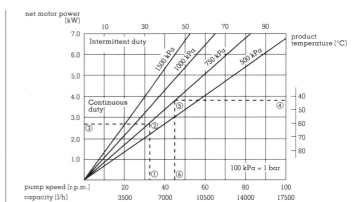

FIGURE 7.25 – Peristaltic pump characteristic curve.

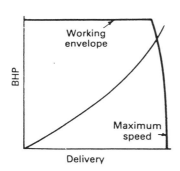

FIGURE 7.26 – Gear pump characteristic curve

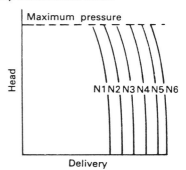

FIGURE 7.27 – Gear pump characteristic curve

shown in Figs. 7.26 and 27. Performance is generally given for a specific viscosity. These pumps are widely used for lube oil and fuel oil service and have found wide use in asphalt, bitumen and molasses applications and for metering, injecting and circulating in the food, chemical, pulp and paper and petroleum industries.

Internal gear pumps

These pumps (Figs. 7.11 and 12) have one rotor with externally cut teeth meshing with another rotor with internally cut teeth. They are available with and without the crescent. Generally the crescent design has two less teeth in the internal rotor than the external rotor. The internal gear designs have lower localized pressures and lower shearing effects on the fluid than the external type. Fig. 7.28 shows a heavy-duty internal gear pump in cross-section.

ROTARY PUMPS: NOMENCLATURE, CHARACTERISTICS, ETC.

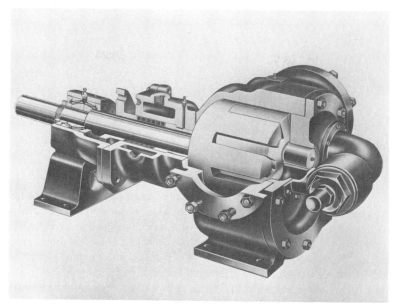

FIGURE 7.28 – Internal gear pump.

Circumferential piston pump

See Fig. 7.13. Fluid is carried from inlet to outlet in the space between the arc shaped pistons. In the external type the rotors must be timed separately, whereas the internal type requires no timing gears. The external type requires at least one piston, whereas the internal type requires at least two. Fig. 7.13 is an external type. A constant volume is delivered with a pulse less flow by these pumps. Fig. 7.35 shows a sanitary metal rotor rotary pump driven through a gear motor for medium to high viscosity products. Fig. 7.37 shows an aseptic twin single wing circumferential piston pump. The lower figures show the steam (or sterile solution) flow when sterilizing. Circumferential piston pumps have relatively low shear action and good metering effectiveness.

Screw pumps

See Figs. 7.14 through 7.17 and 7.36. Fluid is carried in spaces between the threads and is displaced axially as they mesh. The single screw pump is called a *Progressing Cavity Pump*. See Fig. 7.29A. These pumps have long corkscrew like rotor with single external

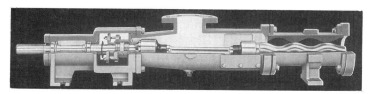

FIGURE 7.29A – Progressive cavity pump.

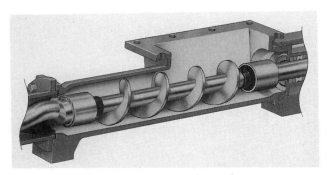

FIGURE 7.29B – Progressive cavity pump.

threads that operates inside a double threaded helix stator or liner. The rotor threads are eccentric to the shaft axis. Fluid is trapped in a series of pockets between the meshing threads move axially toward the discharge. Since the rotor rotates in an eccentric motion, it is connected to the drive shaft by a connecting rod with a universal joint at each end or a flexible shaft. The flow has little in the way of pulsations or shear and low velocity. They should not be run dry. Seals and relief devices are required. This type can handle a very wide range of viscosities, and those with an elastomeric stator can tolerate abrasives more than other types of rotaries. Their temperature limitation is dependent on that of the elastomeric liner used. Efficiencies are good because of the fit of the elastomeric stator against the rotor. These units are widely used in food, pharmaceutical and waste water treatment processes. They are self-priming. A cutaway view is shown in Fig. 7.29 along with balloons and descriptions of the components. The multiple screw pumps have one rotor driving the other or others. They are capable of operating at higher rpm's than other rotary pumps and are often built with opposed screws to reduce, or in the case of the two screw pump, eliminate thrust. They can be timed. See Fig. 7.21 for typical performance curves of a screw pump. The rotor operates eccentrically to the driver shaft. Limits in capacity come from the universal joints that are used. The pin type joint is common in the small and medium sizes with the cardan joint (automobile universal) being used in the larger sizes. Fig. 7.29B shows a hopper for material input into the suction of a progressing cavity pump.

Sine pump
This pump is a of unique construction where a sinusoidal impeller made up of two complete sine curves dividing it into four separate compartments. As the rotor turns these compartments progress through liners, providing a positive displacement of fluid from suction to discharge. A sliding scraper gate prevents return of the product past the discharge to the suction side of the pump. The four compartments never change in volume, eliminating compression in the pump and damage to the particulate in the fluid. The suction port's volumetric area is constant and both sides of the rotor are fed simultaneously so there is no pulsation, axial thrusting and negligible effects of viscosity within the pump. The pump has negligible slip and can purportedly pull a 9 metre (30 feet) lift. It has USDA and 3A approval for sanitary processing applications. Its low shear capabilities and gentle

ROTARY PUMPS: NOMENCLATURE, CHARACTERISTICS, ETC.

handling of sensitive products has been shown in many applications in the food and cosmetic industries. Fig. 7.33 shows the sinusoidal rotor and Fig. 7.34 shows the performance curve for this line of pumps.

Hermetic rotary pumps

See Figs. 7.30, 31, 32 and 36 which show Hermetic magnetic-drive internal gear, external gear vane and screw pumps that are presently being marketed. It is anticipated that the hermetic rotary offerings will increase substantially over the next few years.

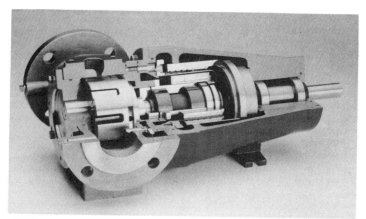

FIGURE 7.30 – Cutaway cooling system, mag drive internal gear pump.

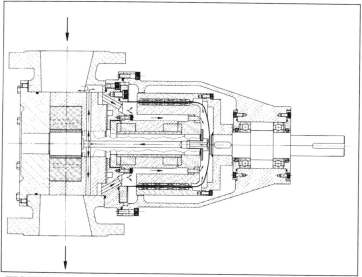

FIGURE 7.31 – Cutaway cooling system, mag drive internal gear pump.

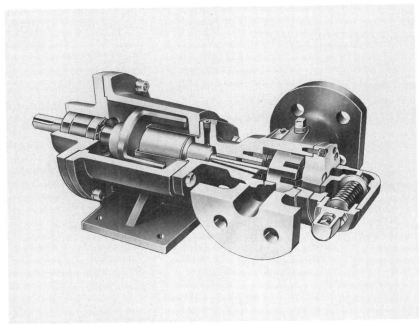

FIGURE 7.32A – Internal gear pump.

FIGURE 7.32B – External gear pump.

ROTARY PUMPS: NOMENCLATURE, CHARACTERISTICS, ETC.

FIGURE 7.32C – Vane pump.

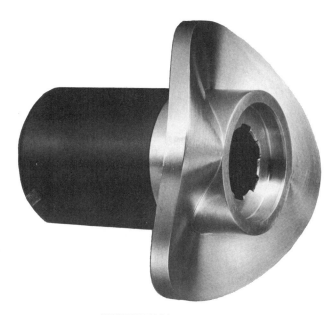

FIGURE 7.33 – Sine wave pump rotor.

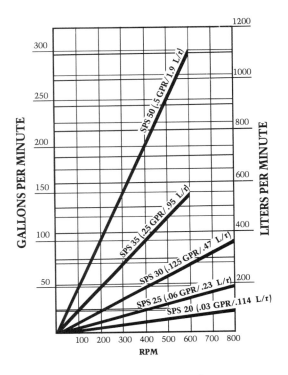

FIGURE 7.34 – Sine wave pump performance curve.

FIGURE 7.35 – Sanitary rotary gear motor drive pump for medium to high viscosities.

ROTARY PUMPS: NOMENCLATURE, CHARACTERISTICS, ETC. 151

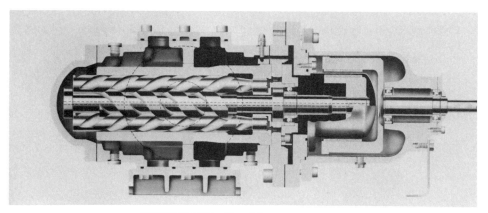

FIGURE 7.36 – Mag drive screw pump.

Aseptic PD Pumps

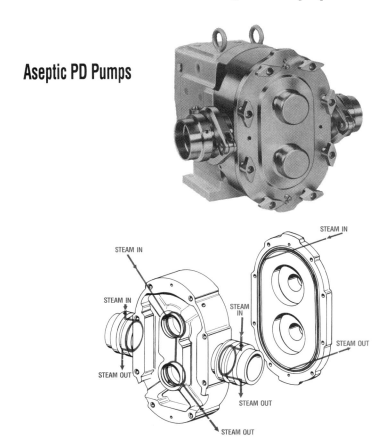

References

[1] Dickenson C., The Pumping Manual, 8th Edition, Elsevier Advanced Technology (1992), pp201.
[2] Hydraulic Institute Standards, 14th Ed.(1983).
[3] Kauffman, J., "Pumps and Their Application", Machine Design, pp180-184, (Mar.14, 1968).
[4] Dickenson, C., Pumping Manual, 8th Edition (1992) pp267
[5] Cunningham, E., "Fluid Handling Pumps – Reference File", Plant Engineering, 1982.

RECIPROCATING PUMPS: NOMENCLATURE, CHARACTERISTICS, COMPONENTS AND TYPES

Reciprocating Pumps

Reciprocating pumps are positive displacement pumps that fall into three types: power, controlled volume and steam. Refer to Fig. 8.1. These pumps have an uncontested niche in the very high pressure, low capacity areas as shown in Fig. 2.2. In applications such as steam generating plants, where there is steam readily available, the direct-acting steam pump still finds a place. The advantages of reciprocating pumps over their centrifugal

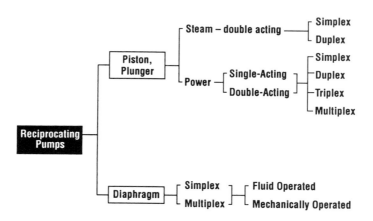

FIGURE 8.1 – Types of reciprocating pumps.

counterparts lie in their capacity, head and speed flexibility and competitive constant efficiency over the range of operation in the low specific speed range of centrifugals. When this advantage overcomes the smaller footprint, lower initial and maintenance costs and lack of pulsations and vibration of the centrifugals, the reciprocating unit is selected. They can handle variable pressures without speed adjustment or the use of energy

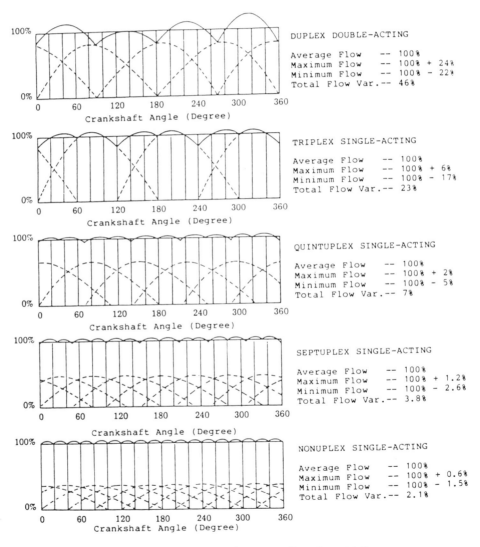

FIGURE 8.2 – Reciprocating pumps flow characteistics.

SPECK Pumps (UK) Limited

Manufacturers of

**PUMPS FOR HIGH PRESSURE
WATER APPLICATIONS**
Flows 5-250 litres/min, Pressures to 500 bar

Our triplex pumps can be supplied with brass, bronze or stainless steel heads. We also stock a wide range of accessories for the water jetting market ie. guns, lances, nozzles, unloader valves, etc.
For more information contact:

SPECK Pumps (UK) Ltd.
Units 11&12 Wycombe Industrial Mall
West End Street
High Wycombe
Bucks HP11 2QY, UK

Tel 01494 523203   **Fax:** 01494 441542

Certificate No. S0560

High-pressure water generation with optimum pressure-litre ratio

The new WOMA central valve pumphead series from 150 - 480 HP sets standards in reliability, robustness and flexibility. A series with a wide performance range of pressure and volumetric flow with operating pressures up to 2500 bar.
The pumpheads, which are free from load-alternation, have coaxially arranged valves. Thanks to their minimum dead volume and optimum valve kinematics they offer high volumetric efficiency. The metallic sealing of the valves and conversion set components as well as the central water guide ensure optimum hydraulic conditions, maximum operational reliability and easy maintenance.
The outstanding quality of the WOMA central pumphead is easy service and a wide range of conversion sets. It offers economity, efficiency, a wide flexibility in operation and minimum shutdowns.

Water as a tool for a cleaner environment

WOMA Apparatebau GmbH
Werthauser Straße 77-79
47226 Duisburg
Telefon (0 20 65) 3 04-0

SEALLESS PUMPS

WARREN RUPP SOLVES TOUGH PUMPING PROBLEMS.

Air-Operated Diaphragm Pumps for safe and easy pumping of:

Abrasive slurries.

Heavy viscosity. **Caustics and acids.** **Toxic and aggressive fluids.**

GENERAL INDUSTRY

S SERIES BALL VALVE
Ball valves of various elastomers and Teflon provide superior suction lift for solids-free fluid, low to high viscosity. No bypasses or complex speed controls required. Flow is infinitely adjustable to capacity. Flows to 984 l/min (260 gal.).

S SERIES FLAP VALVE
Recommended for solids-laden fluids, low to high viscosity. Passes line-size solids to 7.6cm (3"). Does not damage or heat pumped product. Adjustable flows to 984 l/min (260 gal.).

P SERIES BALL VALVE
This line of non-metallic pumps are chemically resistant to most acids, alkalis, latexes and other corrosive fluids. Models available in polypropylene, PVDF and PFA.
Adjustable flows to 435 l/min (115 gal.).

E SERIES BALL VALVE
These lighter weight, general duty models have midsections of aluminum or cast iron. Exceptional lift for solids-free fluids from low to high viscosity. Adjustable flows to 984 l/min (260 gal.).

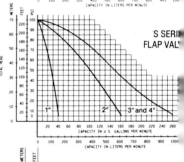

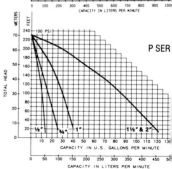

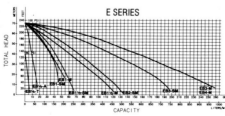

DISTRIBUTORS WORLDWIDE

WARREN RUPP, INC. · A Unit of IDEX Corporation
P.O. Box 1568 · Mansfield, Ohio 44901-1568 U.S.A.
Telephone 419 524 8388

Fax 419 522 7867
FAX FOR A FREE ENGINEER'S CATALOG!

WARREN RUPP — ISO9001 CERTIFIED

SandPIPER

SIMPLE DESIGN. SUPERIOR PERFORMANCE & INNOVATION.

Sealless. No-Lube Air Valve.
Self-Priming. Explosion-Proof.
Variable Flow. No Motors.
Can Run Dry Without Damage. Full Line of Accessories & Controls.

SEALLESS PUMPS

SPILL CONTAINMENT
The ET and ST Series for hazardous fluid features a fluid-filled Teflon or aluminum spill chamber, and additional diaphragms between the air valve and each outer pumping chamber. Option leak detections devices can provide visual, audible or shut-down response. Spill containment protects personnel, equipment and the environment. Flows to 984 l/min (260 gal.).

FOOD & PHARM
The SS Series offers USDA-Accepted and 3-A Standard pumps in stainless steel. Special clamps allow quick disassembly for mechanical cleaning. Pumping chambers feature a smooth, electropolished finish. Gentle pumping action is ideal for shear-sensitive materials. Pumpage is not heated or degraded. Solids to 5cm (2"). Flows to 570 l/min (150 gal.).

POWERUPP®
PoweRupp®, a motor-driven double diaphragm pump, offers all the advantages of progressing cavity, lobe, piston, gear and other positive displacement pumps. Elastomeric replacement parts are far less costly, meaning much lower maintenance costs. In-line serviceable for fewer manhours. Flows to 220 l/min (58 gal.).

WASTE TREATMENT
The W Series is specially-designed for easy, safe handling of municipal and industrial waste. Full opening swing-type check valves pass solids to 75mm (3"). Quick-open clamps provide quick in-line access to all check valve clean-outs. Bottom discharge porting uses gravity to eliminate solids build-up. Flows to 988 l/min (260 gal.).

UTILITY
· PortaPump® submersible, battery-powered pump operates off any 12-volt truck or car battery. Cables and battery clips included. Portable. Weighs only 33 pounds. Fits through a 25cm (10") opening. Model SPA1½-A.
Flows to 162 l/min (43 gal.).
· SludgeMaster® air-powered submersible pump handles mud, leaves, twigs, sand, sludge and trash-laden water. Handles soft solids to 3.8cm (1½").Portable.Weighs 59 pounds. Fits through 35.5cm (14") openings. Designed for rough duty. Optional rock screen. Model SMA3-A.
Flows to 1135 l/min (300 gal.).

ACCESSORIES
Performance-enhancing accessories include:
· Filter/Regulators · Surge Suppressors
· Speed Controllers · Air Dryers

©1994 All rights reserved. ®SandPIPER and PoweRupp are registered tradenames of Warren Rupp, Inc. Teflon is a registered tradename of E.I. duPont.

WARREN RUPP, INC. · A Unit of IDEX Corporation
P.O. Box 1568 · Mansfield, Ohio 44901-1568 U.S.A.
Telephone 419 524 8388
Fax 419 522 7867
FAX FOR A FREE ENGINEER'S CATALOG!

DISTRIBUTORS WORLDWIDE

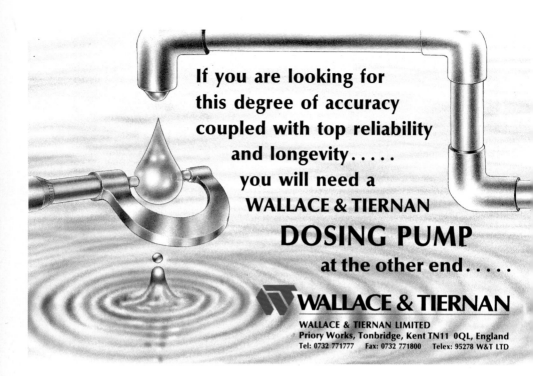

RECIPROCATING PUMPS: NOMENCLATURE, ETC.

consuming system throttling devices and they are self-priming, thereby able to handle some significant amount of gas or vapour. There are three types of mechanism defined reciprocating pumps: piston, plunger and diaphragm or bellows. Valves at each end of the volume alternately close and open to isolate the higher pressure volume from the suction on the compression stroke and the higher pressure discharge from the volume on the return stroke. This description covers the simplex, single acting pump and as can be seen in Fig. 8.2, there is a periodic sine wave pressure condition[1] in the discharge that results in significant pressure pulsations. A single acting duplex pump has a mitigating effect on this because a second control volume is added and its compression stroke is taking place at the same time as the first's return or intake stroke. As cylinders are added[2], see Fig. 8.2, the pulsation levels diminish in an asymptotic fashion. How important these pulsation levels are is strictly determined by the needs of the particular installation. Fig. 8.3 shows theoretical and actual flow characteristics for a triplex pump. Pulsation dampeners act like a flywheel in reducing pulsation levels but they are expensive. A properly installed pulsation dampener in the suction line can absorb the cyclical flow variation and reduce

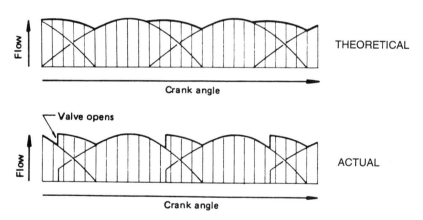

FIGURE 8.3 – Triplex pumps. Theoretical and actual flow characterics.

the pressure fluctuation to that corresponding to a length of 5 to 15 pipe diameters, if kept fully charged. There is a similar benefit to one installed in the discharge line, if pulsation or piping vibration is a problem. However, the latter cannot be analyzed as readily because of the increased range of the variables[3]. A *single_acting pump* is one which has one compression and one intake stroke per cycle. A *double acting pump* is one which has two compression and two intake strokes per cycle. The number of cylinders determines another terminology classification of reciprocating pumps. A pump with one cylinder is called a simplex, one with two a duplex, three a triplex pump etc. There is no sucking action during the intake cycle on a reciprocating pump as there is with rotary pumps. The cylinder must actually see a reduction in pressure such that suction pressure, be it atmospheric or otherwise, can force fluid into it. A typical characteristic curve is shown in Fig. 8.4. The power end of controlled volume and power pumps use pressure, splash or gravity oil lubrication. Steam ends should be lubricated prior to startup in accordance with the pump vendor's instructions, and then will require a continuous feed of several drops per minute from the lubricator while operating. Plungers and piston rods are sealed with packing.

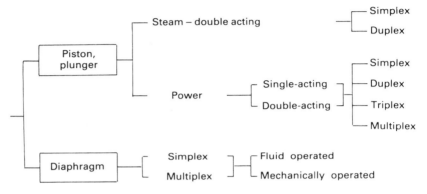

FIGURE 8.4 – Typical characteristic curve of reciprocating pumps.

Definitions

Steam pumps are direct acting pumps that have a steam cylinder that acts in the same way as the pump cylinder. The steam drives the piston or plunger which is locked to the same shaft as the pump piston or plunger. Thus the steam directly drives the pump. They consist of a steam end and a liquid end.

Power pumps are driven through a crankshaft from an outside power source. They consist of a power end and a liquid end. Figs. 8.5 through 8.8 show various types of power ends.

Piston pumps can be either of the above two pumps where the piston driving member is coaxial to and of a larger diameter than the rod it is attached to. The piston normally is circumferentially grooved and has rings in these grooves which seal against the cylinder or cylinder liner i.d.

Plunger Pumps can be either steam or power driven but the rod is the plunger. The sealing rings for a plunger are stationary and are mounted in the cylinder wall, with the plunger sliding axially within these sealing rings.

Controlled volume pumps also called *metering, dosing or proportioning pumps* accurately displace a fixed volume of fluid in a specific time.

Controlled volume piston or plunger pumps have their pistons or plungers in direct contact with the fluid and use packing or sealing rings to restrict leakage.

RELATED TITLES IN THE HANDBOOK SERIES

Condition Monitoring Handbook
First edition to be published in 1995

Filters and Filtration Handbook 3rd Edition

Handbook of Hose, Pipes, Couplings and Fittings

Handbook of Industrial Materials 2nd Edition

Handbook of Power Cylinders, Valves and Controls

Handbook of Noise and Vibration Control

Handbook of Reinforced Plastics
First edition published September 1994

Handbook of SCADA Systems

Handbook of Valves, Piping and Pipelines 2nd Edition

Industrial Fasteners Handbook

Industrial Membranes Handbook
First edition to be published in 1995

Pneumatic Handbook 7th Edition

For further details send for our Industrial Engineering and Materials catalogue [PBG05+A1].

Simply fax or post a copy of this advert complete with your address details to:
Elsevier Advanced Technology
PO Box 150, Kidlington, Oxford OX5 1AS, UK
Tel: +44 (0) 1865 843848 Fax: +44 (0) 1865 843971

ELSEVIER
ADVANCED
TECHNOLOGY

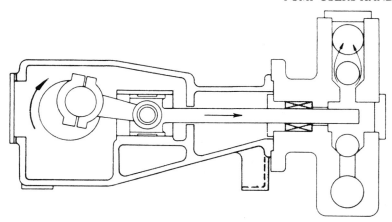

FIGURE 8.5 – Horizontal single-acting plunger power pump.

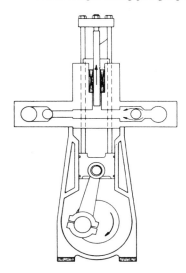

FIGURE 8.6 – Vertical single-acting plunger power pump.

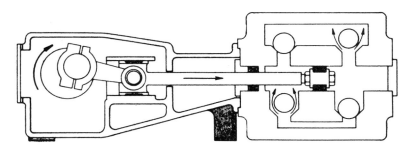

FIGURE 8.7 – Horizontal double-acting piston power pump.

s.a.s.
di BOAGLIO B. & C.

C.V.B.'s ATEMA plunger pumps resolve the most different problems in industrial high pressure pumping. The W series pumps, with horizontal plunging pistons, 3 or 6 (boxer type), with modern and technical design are particularly strong and exceptionally reliable.
Manufactured with contact parts in Brass, Aluminium Bronze, AISI 316 L S.S.: and now also in DUPLEX, have been successfully employed for years in liquid handling, spraying, pressurizing, hydraulic testing, process plants in chemical, pharmaceutical, food industry: particularly for reverse osmosis.
Most of C.V. B.'s pumps are sold as complete pumping unit, ready to work.

A special recovery system consisting in an hydraulic turbine "TURBOMASTER" fitted on the electrical motor, realizes large energy savings in R.O. plants, particularly for desalting.

C.V.B s.a.s.
di Boaglio B&C
Via Passo Buole 151/A
10135 TORINO.
Tel: +39 11317 2058
Fax: +39 11317 2059.

tapflo Air-Operated Diaphragm Pumps

Tapflo corrosion resistant air-operated double diaphragm pumps in Polyethylene or PTFE are manufactured in *6 different sizes* with variable capacities up to 400 l/min. The delivery head is up to 80 mwc and the pumps are selfpriming. They are sealless, have no rotating parts - idelistic for handling agressive liquids and can run dry without damage.

TAPFLO AB
Filaregatan 4, S-442 34 KUNGÄLV, SWEDEN
Tel. +46 303-633 90 Fax. +46 303-199 16

PERFECT PUMPING
IT'S OUR TOTAL COMMITMENT

- High pressure triplex pumps with a choice of materials
- Comprehensive range from 1kw to 50kw
  Capacities up to 260 l/min, pressures up to 500 bar
- Understanding of cavitation and related problems
- Captive acceleration tubes and pulsation dampeners
- Easy maintenance. Reduced down time.
- Prompt personal attention

Cat Pumps (UK) Limited
1 Fleet Business Park, Sandy Lane, Church Crookham, Fleet, Hampshire GU13 OBF. Tel: 01252 622031 Fax: 01252 626655

Steam pumps.

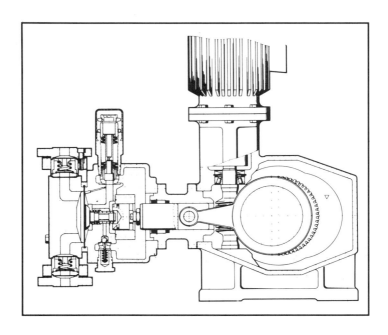

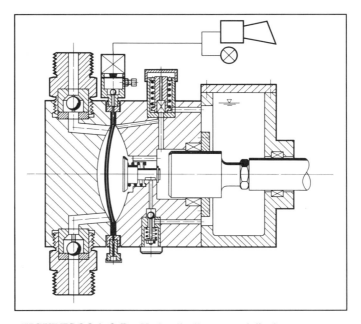

FIGURES 8.8 A & B – Hydraulically actuated diaphragm pumps.

RECIPROCATING PUMPS: NOMENCLATURE, ETC.

Controlled volume diaphragm pumps have flexible diaphragms in contact with the fluid. Figs. *These pumps are true hermetic (sealless) pumps.*

Controlled volume bellows pumps are essentially the same as diaphragm pumps and operation is the same. There is a limited presence of these pumps in the small sizes.

Capacity, Q m³/h, cc/min (gpm) is the total volume output per unit of time of the fluid at suction conditions.

Pump displacement is the volume swept by all pistons, plungers or diaphragms. It is the cross-sectional area, A of the piston, plunger or diaphragm times the stroke, s times rpm times the number of pistons, plungers or diaphragms. A deduction must be made for the volume of the rod on double acting piston pumps.

Slip is the loss of capacity due to leaks past the valves including back flow during delayed closing and past double acting pistons. Slip does not include fluid compressibility or leaks from the wet end.

Pump power input, Pi is the mechanical power delivered to the pump input shaft.

SI $\quad\quad$ Pi = Q x ptd / 1279 h_p, kw

where $\quad$ Q = Capacity, m³/hr
$\quad\quad\quad$ p_{td} = Pressure Difference, discharge minus suction,
$\quad\quad\quad$ η_p = Pump Mechanical Efficiency, the ratio of power

output to power input

USCU $\quad$ Pi = Q x p_{td} / 1714 η_p, hp

Pump power output, P_o is the power imparted to the fluid by the pump.

SI $\quad\quad$ P_o = Q x p_{td} /1279, kW

USCU $\quad$ P_o = Q x p_{td} / 1714, hp

Volumetric efficiency, $\quad$ η_p = P_o/P_i

NPSH of reciprocating pumps is very similar to that of centrifugals. *Npsha* = Total suction pressure minus vapour pressure (for reciprocating pumps this is normally expressed in pressure units as opposed to head units). However, total suction pressure with the higher speed pumps or long suction lines must have the equivalent pressure of the *acceleration head* deducted from it.

Acceleration head, h_a. The fluid in suction and discharge lines of controlled volume pumps is constantly undergoing acceleration and deceleration because of the sinusoidal flow pattern. (Multiple cylinders reduces the inertia in these accelerations and deceleration's). To compensate for this there are instantaneous pressure drops and increases. The instantaneous drop or rise is referred to as acceleration head. This value must be taken into consideration to ensure that the pump will fill properly without cavitation (this is not a loss as what is lost on one cycle is recovered in the next). The acceleration head is a function of the length of the suction line, the average velocity in this line, the rotative speed, the type of pump and the relative elasticity of the fluid and the pipe. For steam and power pumps the formula is[3]:

$$h_a = \frac{LVnC}{Kg} \text{, m (ft)} \qquad \text{Ref.}^3$$

		SI	USCS
where h_a =	Acceleration head	m	ft
L =	Length of suction line	m	ft
V =	Velocity in suction line	m/s	ft/s
n =	Pump speed	rpm	rpm
C =	A factor representing the type of pump*		
K =	A factor representing the relative compressibility of the fluid. 1.4 for hot water, 2.5 for hot oil		
g =	gravitational constant	m/sec²	ft/sec²

*C = 0.2 simplex double acting
0.2 duplex single acting
0.115 duplex double acting
0.066 triplex single acting
0.066 triplex double acting
0.040 Quintuplex single acting
0.040 Quintuplex double acting
0.028 septuplex single acting
0.028 septuplex double acting
0.022 nonuplex single acting
0.022 nonuplex single acting

*Note: C will vary from these values for unusual ratios of connecting rod length to crank radius.
When calculating h_a, sections of the suction line of varying diameters must be calculated for L and V separately and added.
For controlled volume diaphragm pumps the formula is the same but
C = 0.628 for simplex single acting
K = 1.4.

Ref. 3 emphasizes the need for careful design of the inlet system, pointing out that the system must provide relatively constant pressure flow into the pump at a pressure

RECIPROCATING PUMPS: NOMENCLATURE, ETC.

sufficiently higher than the vapour pressure to prevent flashing as it enters the pump chambers. Also that any entrained gas in the fluid, or flashing in the pump can cause damaging vibrations, reduced volumetric efficiency, component failures, reduced life of packing, springs, valves, seats and gaskets. In the extremes, pitting or catastrophic failure of liquid end components can result.

Tanks, as pointed out previously, should have: sufficient retention time and surface area to allow entrained gas to surface; feed lines that terminate below the fluid surface, separation of entering and leaving areas by baffling and vortex breakers at the outlet. See this subject in Chapter 5 under sumps and tanks. Use separate suction lines to each pump of diameter at least equal to the pump suction flange and as short and elbow-free as possible. There should be no high points in the suction piping where gas can collect. The NPSHA with acceleration head considered, should preferably have at least 2 m (7 ft.) safety margin added. Strainers and valves should have port areas at least equal to the suction pipe diameter and a suction pressure gauge should be located 2 diameters upstream of the suction flange.

The NPSHR at the inlet to the pump is determined by the manufacturer from test results. It includes velocity head, acceleration head, suction valve losses, pump internal friction, and (in the case of controlled volume pumps) the losses required to move diaphragms and hydraulic oil, the pressure required to prevent the release of dissolved air or gases in the hydraulic oil or pumpage (note: this is not the same as the allowance for vapour pressure of the pumpage). If vacuum compensation is used, the pressure requirement of the valves must be included. In the case of metering pumps the metering accuracy is dependent on adequate margin of NPSHA over NPSHR. In all pumps this margin is crucial to the well being of the pump.

Discharge piping

Discharge piping for reciprocating pumps should contain a relief valve, shut-off valve and check valve. A by-pass line with shut-off valve should be installed between the discharge and suction lines, close to the pump. Any increasers should be between the check valve and the pump. The discharge line should be well anchored. Dead ends and short radius elbows should be avoided. Ref.[4] provides good advice on anchoring piping near pumps and on the installation of expansion joints.

Steam pumps

These pumps are a combination of a steam engine and a pump. Actually compressed gases could and have been used in place of steam. This pump class has infinite variability available in control of head or capacity. Slip averages around 4% new. Any of these gas or vapour driven pumps are limited to the maximum pressure limitations of the driving medium. Once this pressure is reached they will no longer respond as positive displacement pumps, but will actually stop. If the pressure of the driving medium rises or the pump discharge system pressure requirement drops the pump will restart.

Power pumps

These pumps; driven by motors, engines or turbines are used in ammonia, chemical, oil

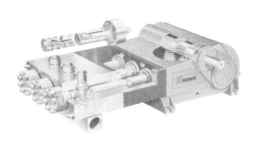 

FIGURE 8.9 – Triplex plunger pump cutaway.   **FIGURE 8.10** – Triplex plunger pump.

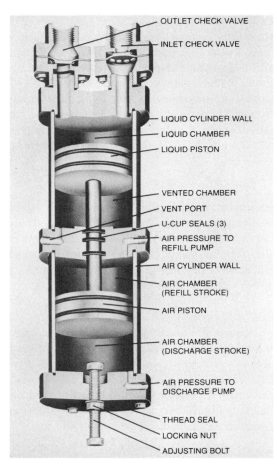

FIGURE 8.11 – Single acting power pump used for metering chemicals.

RECIPROCATING PUMPS: NOMENCLATURE, ETC.

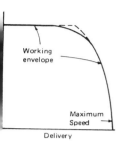

FIGURE 8.12 – Typical diaphragm pump characteristic curves.

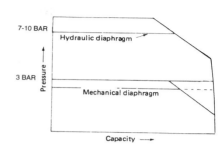

FIGURE 8.13 – Typical diaphragm pump speed characteristics.

FIGURE 8.14 – Dosing pump.

FIGURE 8.15 – Dosing pump.

FIGURE 8.16 – 17 head modular dosing pump.

FIGURE 8.17 – Hydraulically activated tubular diaphragm metering pump.

FIGURE 8.18 – Hydraulically activated diaphragm metering pump.

FIGURE 8.19 – Tubular diaphragm metering pump.

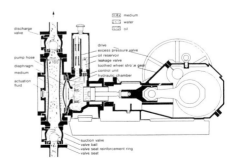

FIGURE 8.20 – Hose, diaphragm, piston pump.

RECIPROCATING PUMPS: NOMENCLATURE, ETC.

and gas pipelines, fertilizer, ash sluicing, steel mill descaling and industrial processes such as high pressure cleaning and cutting, hydro forming and hydrotesting. Figs. 8.9 and 8.10 show triplex plunger pumps the former with a cut-away view. Figs. 8.6-8.17 shows a radial piston pump which was classified as a rotary pump. Fig. 8.11 shows a single acting piston power pump used for metering chemicals. A double acting option is available.

Controlled volume (metering) pumps

Diaphragm; These pumps are positive displacement (except when driven by a gas or vapour medium whose maximum pressure is exceeded by the system requirements) reciprocating controlled volume pumps. They are self-priming, can be run dry without damage and are inherently explosion proof. Packing or mechanical seals are not required. They can be given the safety of a second diaphragm where leakage or damage to the main one could be of significant danger or expense. This serves the same purpose as the secondary containment shell of a canned pump or magnetic-drive pump. Typical characteristic curves are shown in Fig. 8.12 for hydraulic and mechanical diaphragms. The effect of speed is shown in Fig. 8.13. They are generally furnished with either ball or flap type check valves. The former being supplied for applications where relatively large particles exist in the fluid. They are available up to approximately 660 kP_a (4000 psi) per stage. One mechanical type, often used in the construction trade has a spring in the back of it. This helps to prevent damage to the diaphragm, since large objects will cause the spring to compress and the diaphragm to become stationary. Under these conditions the pump is no longer acting as a positive displacement pump. *Dosing pumps* is a term that is applied to double diaphragm metering pumps with a magnetic solenoid type of plunger actuation. They are often tied into electronic control panels that are capable of setting the length of the plunger stroke and the rate. See Figs. 8.14 and 8.15. Diaphragm pumps are used for automotive fuel pumps, abrasive slurries, beer, bilge pumps, treatment of boiler, food products, water and waste water as well as swimming pool fluids, chemicals, clean and dirty coolants, coatings, paint, resins, adhesives, inks, latex, mill scale, mud, sewage, oils, solvents, varnishes, yeast, wine, etc. Many diaphragm pumps have the diaphragm actuated by hydraulic fluid which is pressurized by the piston or plunger and then released. Figs. 8.15 through 8.20 are of this type. Fig. 8.16 is a 17 headed modular pump unit ready to be shipped for blending /proportioning detergent recipes. Each head can meter one ingredient, and all can operate simultaneously. Fig. 8.17 is a hydraulically actuated diaphragm metering pump with a simplex or duplex option. Double diaphragms are available for feeding of difficult and dangerous liquids such as: hazardous, toxic, noxious, and corrosive fluids; abrasive suspensions such as detergent slurry and slip; solutions and meltings tending to effloresce on air contact.

Fig. 8.19 shows a tubular diaphragm metering pump. Fig. 8.20 shows a rather unique pump, which is really three pumps in one, a piston pump, diaphragm pump and hose pump. As can be seen in the figure the only product wetted surface is the inside diameter of the hose itself. The hose is only squeezed by 30% of its volume. This results in a gentle action on the product and hose resulting in longer life and ability to pump highly viscous, aggressive and abrasive products.

Piston and plunger These pumps have a reciprocating piston or plunger in direct contact with the fluid sealed by packing or a mechanical seal.

References
[1] Hicks, T., "Practical Power Service Library" Power Magazine p53
[2] Dickenson, C., "Pumping Manual", 8th Edition, Elsevier Advanced Technology, pp169-174 (1992).
[3] Hydraulic Institute Standards, 14th Ed., The Hydraulic Institute (1983) p252.
[4] "Rubber Expansion Joints and Flexible Connectors", Technical Handbook, 5th Ed. (1979), Fluid Sealing Assn. Philadelphia, Pa..

SECTION 4

Pump Construction

MATERIALS AND CORROSION

SEALS AND PACKING

MATERIALS AND CORROSION

The most common material used in pumps is cast-iron, Class 35 and 40. Often bronze is used for the impellers and impeller wear- rings on centrifugals. This liquid end combination is known as cast-iron bronze fitted. Cast-iron stainless fitted is the next most frequent combination followed by ductile cast iron-bronze fitted, ductile cast-iron stainless fitted, all stainless, steel and all bronze. Titanium is used in many corrosive applications where the life of the previously mentioned materials would not be adequate. Fibreglass and plastics are seeing more and more use in corrosive applications. In erosive applications, hard materials are utilized such as the ni-hards. In erosion - corrosion applications the high chrome steels are utilized. The use of non - proprietary duplex stainless, such as 2205, ASTM A890-Grade 3A and other proprietary stainless steels is increasing rapidly, in place of the austenitics 316. 316L, 317, and 317L. However, there are a vast number of other materials that serve best in certain applications, some of which will be touched on in this chapter. Tables 1-6 cover physicals and chemical make up of the basic materials mentioned above. Table 7 shows the ASTM designation of various materials.

The factors that determine the selection of a material are physical properties, corrosion resistance, erosion resistance and cost. Cost includes material, machining and protective costs such as finishes. The physical properties required could be as simple as the strength required to withstand the internal pressure or that plus wear resistance, corrosion

TABLE 1 – Cast Iron

Class	Tensile Strength min. psi	
20	20,000	For additiodnal requirements applicable to these materials, refer to ASTM specification A48.
25	25,000	
30	30,000	
35	35,000	
40	40,000	
50	50,000	

TABLE 2 – Ductile Iron

ASTM	Grade	Tensile Strength min. psi	Yield Strength min. psi	Elongation in 2 inches min. %
A-536	60-40-18	60,000	40,000	18
	65-45-12	65,000	45,000	12
	80-55-06	80,000	55,000	6
	100-70-03	100,000	70,000	3
	120-90-02	120,000	90,000	2
A-395	.....	60,000	40,000	18

TABLE 3 – Nominal Composition and Minimum Mechanical Properties of Bronze

ASTM ALLOY	B30 C90700	B584 C90500	B30 C92700	B584 C90300	B584 C92200	B584 C83600	B584 C93700
Copper	89.00	88.00	88	88.00	88.00	85.00	80.00
Tin	11.00	10.00	10	8.00	6.00	5.00	10.00
Lead	-	-	1.5	-	1.50	5.00	10.00
Zinc	-	2.00	-	4.00	4.50	5.00	-
Phosphorus	.20	-	.25	-	-	-	-
Tensile, psi	35,000	40,000	35,000	40,000	34,000	30,000	25,000
Yield, psi	-	18,000	-	18,000	16,000	14,000	12,000
Elongation %	10	20	10	20	22	20	8

TABLE 4 – Nominal Composition and Minimum Mechanical Properties of Carbon Steel

Carbon, max.	0.30
Manganese, max.	1.00
Phosphorus, max.	0.50
Sulphur, max.	0.06
Silicon, max.	0.60
Tensile, min. psi	70,000
Yield, min. psi	36,000
Elongation in 2 in., min., per cent	22
Reduction of area, min., per cent	35
ASTM Spec.	A216, WCB

TABLE 5 – Nominal Composition and Minimum Mechanical Properties of Chromium Steels

	Selection 4	Selection 5	Selection 6	Selection 7
Carbon	0.20 max.	0.15 max.	0.30 max.	0.50 max.
Chromium	5.00	12.00	20.00	28,00
Molybdenum	0.50	0.50 max.	-	-
Nickel	-	1.00 max.	2.00 max.	4.00 max.
Silicon	0.75	1.50 max.	1.50 max.	1.50 max.
Tensile, psi	90,000	90,000	65,000	55,000
Yield, psi	60,000	65,000	30,000	-
Elongation, %	18	18	-	-
ASTM Spec.	A217-C5	A743-CA-15	A743-CB-30	A743-CC-50

MATERIALS AND CORROSION

TABLE 6 – Mechanical Properties of Austenitic Steels

	Selection 8 19-9	Selection 9 19-10/Mo	Selection 10 Highly Alloyed Stainless
Carbon, max.	0.08	0.08	0.07
Manganese, max.	1.50	1.50	1.50
Phosphorus, max.	0.04	0.04	0.04
Sulphur, max.	0.04	0.04	0.04
Silicon, max.	2.00	2.00	1.50
Nickel	8.0-11.0	9.0-12.0	27.5-30.5
Chromium	18.0-21.0	18.0-21.0	19.0-22.0
Molybdenum	-	2.0-3.0	2.0-3.0
Copper	-	-	3.0-4.0
Tensile, psi	65,000	70,000	62,500
Yield, psi	28,000	30,000	25,000
Elongation %	35	30	35
ASTM Spec.	A743, CF-8	A743, CF-8M	A743, CN-7M

TABLE 7 – Summary of Material Selections and ASTM Standards Designations

Material Selection	Corresponding National Society Standards Designation ASTM	Remarks
1	A48, Classes 20, 25, 30, 35, 40 & 50	Gray Iron - Six Grades
1(a)	A536 & A395	Ductile Cast Iron - Six Grades
2	B584	Tin Bronze & Leaded Tin Bronze - seven alloys
3	A216-WCB	Carbon Steel
4	A217-C5	5% Chromium Steel
5	A743-CA15	12% Chromium Steel
6	A743-CB30	20% Chromium Steel
7	A743-CC50	28% Chromium Steel
8	A743-CF-8	19-9 Austenitic Steel
9	A743-CF-8M	19-10 Molybdenum Austenitic Steel
10	A743-CN-7M	20-29 Chromium Nickel Austenitic Steel with Copper & Molybdenum
11		A series of Nickel-base alloys
12	A518	Corrosion Resistant High-Silicon Cast Iron
13	A436	Austenitic Cast Iron - 2 types
13(a)	A439	Ductile Austenitic Cast Iron
14		Nickel-Copper alloy
15		Nickel

resistance and anti-galling attributes. Erosion resistance is dependent on the shape, size, hardness, velocity and flow angle of the particles as well as the properties of the material. Corrosion resistance is a function of: the type of fluid being pumped i.e. highly corrosive acid or alkali or non-corrosive solvent, electrolyte or non-electrolyte, impurity laden or pure liquid; the geometry of the space i.e. tight, close or loose clearances; flow velocity, stress in the material etc. The close clearances of rotary pumps as opposed to centrifugals requires different materials for certain applications. A flow velocity of 6-12 m/s (20-40 fps), in a centrifugal pump, could result in no corrosion because of the sweeping action preventing localized build up of salts in one example. In another with abrasive particles, this could cause cause a problem of erosion-corrosion[1]. Some of these occur because the protective oxide film is eroded away and time for it to rebuild is inadequate due to the erosion rate. In the cases of tight clearances, crevice and pitting corrosion rather than broad corrosion attack may take place especially when theses tight clearances allow stagnant conditions that result in a build up of impurities that can result in aggressive local attack. On the other hand we have the case of pure water which is very aggressive as opposed to harder water with common impurities. What works in one process may not work in another plant where the pH is not controlled as closely. Corrosion, as can be seen, is a very complex subject. The proper solution is very product and application oriented. Even corrosion tests on the materials under laboratory conditions can be very misleading at times due to deviation from actual conditions in a specific process. Experience is the best selection tool available for materials, the more specific this is to the specific process in your plant the better. For new processes or new materials, there are indexes available that give relative resistances of available materials that can be the basis of a first try. For general corrosion resistance of various materials with various substances, extensive charts are available[2]. Condensed charts are available from materials, pump suppliers and are part of the Hydraulic Institute Standards. If any known deviations in the process or fluid exist from that used in the tests on which the index was built, then expert assistance should be used to evaluate the effects of these deviant conditions.

Tables 8-12 show relationships of various austenitic, ferritic and duplex stainless steel materials for different indices. The pitting index in Table 8 is defined by a formula, $PI = Cr + 3.3 Mo + 13 N$ ($FeCl2$). This formula adds the relative effect on the pitting resistance of the composition percentage of 3 materials. This relationship is as found from tests on a certain concentration of ferric-chloride under certain conditions of time, temperature etc in the laboratory. Pitting is a localized mechanism, usually caused by chlorides or other halides. It requires more severe conditions for it to occur than crevice corrosion, therefore selection of a material to avoid crevice corrosion, should also avoid pitting corrosion[3]. Crevice corrosion is also a localized condition which occurs in crevices or confined spaces. Gasketed areas are an example of such a confined space that the fluid can get to but once there becomes stagnant. The presence of chlorides or other halides is a factor in most crevice corrosion problems along with a lack of oxygen or reduction in pH[4]. Table 8 shows pitting index in the right-hand column for various materials. The higher the pitting index, the more resistant a material is expected to be. Duplex stainless steels are alloys that are approximately 50% austenite and 50% ferrite. Some corrosion rates are shown for various stainless steels against different acids in Table 9.

MATERIALS AND CORROSION

TABLE 8 – Nominal Compositions (wt%) and Pitting Index (PI)

UNS Number	Alloy Name	Cr	Mo	Ni	Cu	C[A]	N[A]	Other	PI
First-Generation Duplex Stainless Steels									
S32900	Type 329	26	1.5	4.5	—	0.08	—	—	24
J93370	CD-4MCu	25	2	5	3	0.04	—	—	32
Second-Generation Duplex Stainless Steels									
S32304	SAF 2304	23	—	4	—	0.030	0.05–0.20	—	25
S31500	3RE60	18.5	2.7	4.9	—	0.030	0.05–0.1	1.7Si	29
S31803	2205	22	3	5	—	0.030	0.08–0.20	—	34
S31200	44LN	25	1.7	5	—	0.030	0.14–0.20	—	32
S32950	7-Mo PLUS	28.5	1.5	4.8	—	0.03	0.15–0.35	—	35
S32550	Ferralium 255	25	3	5	2	0.04	0.1–0.25	—	37
S31260	OP-3	25	3	7	0.5	0.030	0.10–0.30	0.3W	37
S32750	SAF 2507	25	4	7	—	0.030	0.24–0.32	—	38
J93404	Atlas 958	25	4.5	7		0.030	0.10–.30	—	42
J93345	Escalloy 450	24	3.7	9.5		0.040	0.20	1.25Si	39
Austenitic Stainless Steels									
S30403	Type 304L	18	—	8	—	0.30	0.10	—	18
S31603	Type 316L	16	2	11	—	0.030	0.10	—	24
S31703	Type 317L	18	3	11	—	0.030	0.10	—	29
N08020	20Cb-3	20	2.2	34	3.5	0.07	—	CS	27
N08904	904L	20	4.3	25	1	0.020	—	—	35
SA N08367	AL-SXN	20	6	25	—	0.030	0.18–0.25	—	40
SA S31254	254 SMO	20	6	18	0.8	0.020	0.18–0.22	—	42
Ferritic Stainless Steels									
SF S44660	Sea Cure	0.02	27	3.5	3	.5Ti		—	37
SF S44635	Monit	0.03	25	4	4	.5Ti		—	38
SF S44800	29-4-2	0.003	29	2.1	4	—		—	42

[A]ASTM Specification range or maximum if a single number
SA = Super Austenitic
SF = Super Ferritic
PI = Cr+3.3Mo+1.3N(FeCl$_2$)

Table 9 – Corrosion Rate in Selected Chemical Environments[A]

Solution Temperature	Corrosion Rate (moy)					
	Type 304	Type 316	Type 317L	Alloy 20	Alloy 2205	Ferralium 253
1% Hydrochloric. boiling	—[B]	—	0.1	—	0.1	0.1
10% Sulfuric. 150°C (88°C)	—	—	8.9	—	1.2	0.2
10% Sulfuric. boiling	16420	855	490	43	206	40
30% Phosphoric. boiling	—	—	6.7	—	1.6	0.2
85% Phosphoric. 150°F (66°C)	—	—	0.2	—	0.4	0.1
85% Nitric. boiling	8	11	21	8	21	5
10% Acetic. boiling	—	—	0.2	—	0.1	0.2
20% Acetic. boiling	300	2	—	2	0.1	—
20% Formic. boiling	—	—	8.5	—	1.3	0.4
45% Formic. boiling	1715	520	—	7	4.9	—
3% Sodium Chloride. boiling	—	—	1	—	0.1	0.4

[A]R.M. Davison, H.E. Devereil, J.O. Redmond. "Ferritic and Duplex Stainless Steels" Process Industries Corrosion, ed. B.J. Moniz. W.J. Pollock (Houston. TX, NACE. 1988). p.427.
[B]Not tested.

TABLE 10 – Critical Crevice Corrosion Temperatures[A]

UNS Number	Alloy Name	Critical Crevice Temperature in 10% $FeCL\ 6H_2O$ pH = 1. 24-hour Exposure	
		°F	°C
S32900	Type 329	41	5
S31200	44LN	41	5
S31250	DP-3	50	10
S32950	7-Mo PLUS	80	15
S31803	2205	83.5	17.5
S32250	Ferralium 253	72.5	22.5
S30400	Type 304	<27.5	<–2.5
S31600	Type 316	27.5	–2.5
S31703	Type 317L	32	0
NO8020	20Cb–3 (Alloy 20)	32	0
NO8904	904L	32	0
NO8367	AL–6XN	90.5	32.5
S31254	254 SMO	90.5	32.5

(A)J.O. Redmond. "Selecting Second generation Duplex Stainless Steels.

"Chemical Engineering, vol. 93, no. 20. Oct. 27, 1986. pp. 153 155 and no. 22. Nov.24, 1986. pp. 103-105.

Table 10 shows the Critical Corrosion Temperature for various alloys[5]. Table 11 shows stress cracking resistance.

Hardness is considered to be a measure of a material's resistance to erosion. Erosion can result from damage from particulate in the fluid stream or from cavitation. Forces involved

MATERIALS AND CORROSION

TABLE 11 – Stress Cracking Resistance
(P=Pass, F=Fail)

	Wick Test	Boiling 25% NaCl
Type 316L	F	F
Type 317L	F	F
Alloy 904L	P or F	P or F
Alloy 20	P	P
Avesta 2205	P	P

TABLE 12 – Resistance to Abrasive Wear

Alloy	Volume Loss mm^3
Ferralium 255	97.5
22% Cr-Duplex	110.9
317L	123.3
316L	127.0

Note: (2205 is a 22% Cr-Duplex alloy)

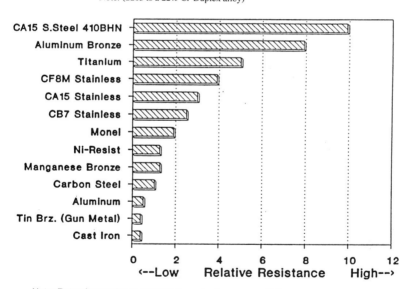

Note: Rate of wear due to cavitation erosion increases with increased temperature.

FIGURE 9.1 – General ranking of cavitation erosion resistance of common cast metals when pumping clear water at ambient temperature

in the implosion of bubbles of vapour can be tremendous and as one would expect material selection can be even more critical than with particulate damage. Table 12 shows the relative resistance to abrasive wear of 4 materials. Figure 1 shows the relative cavitation

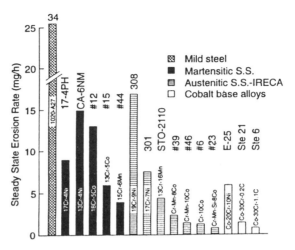

FIGURE 9.2 – Comparison of cavitation resistance of ferretric martensitic and austenitic stainless steels and some cobalt base alloys.

resistance of various common materials. Fig. 9.2[6] shows a comparison of cavitation resistance of some less common materials. These materials are used for the most part in more severe applications of cavitation erosion than those in Fig. 9.1.

Other properties which must be considered are: *strength*, which is a measure of the ability of a material to contain pressurized fluids; *toughness or elongation* which are measures of a materials ductility and the ability to absorb energy after reaching the yield point without breaking; *stress reduction with temperature,* a measure of a materials ability to hold its properties with a temperature increase; and *coefficient of thermal expansion* which is a measure of a material's linear expansion with temperature. How that coefficient compares with other mating materials is the concern; and *galling* which is a measure of how well a material will slide on another without picking up fibres from the other.

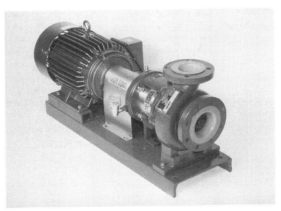

FIGURE 9.3 – Chemical pump with plastic components and liner.

MATERIALS AND CORROSION

TABLE 13 – Properties of rigid plastics suitable for pump casings, body blocks and imellers.

Name (Common or trade) Chemical name	PVC Polyvinylchloride (normal impact)	CPVC Chlorinated polyvinyl-chloride	PE Polyethylene (high density)	PP Poly-propylene (unmodified)	PTFE Polytetra-fluoro-ethylene	RULON All-polymeric reinforced PTFE	PVDF Poly-vinylidene fluoride	ETFE Modified co-polymer of ethylene & TFE	PPS poly-phenylene sulphide
Specific gravity (density)	1.30-1.58	1.49-1.58	0.926-0.940	0.902-0.910	2.14-2.20	2.24 min	1.75-1.78	1.7	1.34
Tensile strength lb/in	6000-7500	7500-11000	3500-4500	4300-5500	2000-5000	1200 min	5500-7400	5000-7500	10000
Compressive strength lb/in^2	10000	9000-16000	—	5500-8000	1700	1200	8680	7100	—
Impact strength ft lb/in notch (Izod test)	0.4-2.0	0.61	1.5-12	0.5-2.2 at 73°F (0.125 x 0.5 in bar)	3.0	6.0	3.6-4.0	No break	0.3 at 75°F (0.5 x 0.25 in bar)
Hardness Rockwell	R113	R12.1	D65 Shore R35-40	R80-110	D50-55 (Shore)	D60-75	D80 (Shore)	R50, D75	R124
Thermal expansion 10^{-5} in/in/°F	2.8	4.4	6	5.8-10.2	10.0	3.3 (°F)	8.5	5-9	5.5
Heat Resistance °F (continuous) °C	130-140 55-60	230 110	200 93	225-260 105-125	500 260	550 290	300 150	300-360 150-180	400-500 200-260
Chemical resistance: effect of weak acids Effect of strong acids	None None to slight	None None	Resistant Attacked by oxidizing acids	None Attacked slowly by oxidizing acids. Avoid chromic acid	None None	None None	None Attacked by hot conc. sulphuric acid	None None	None Attacked slowly by oxidizing acids
Effect of weak alkalis Effect of strong alkalis Effect of organic solvents	None None Resists alcohols, aliphatic hydro-carbons, oils; swells in ketones, esters, aromatic hydrocarbons	None None	Resistant Resistant Resistant below 60°C except to chlorinated solvents	None Very resistant Resistant below 80°C	None None None	None None None	None None Resists most solvents	None None None	None None Resistant below 375-400°F (190-204°C)
Pump components	Casings, impellers, bolts, nuts, shaft sleeves	Impellers, bolts, nuts, Shaft sleeves	Body blocks	Body blocks, bolts, nuts, shaft sleeves	Body blocks	Sleeve bearings	Casings, impellers, bolts, nuts, shaft sleeves	Impellers	Impellers, sleeve bearings

TABLE 14 – Pump body blocks in plastics compared with elastomeric liners.

	Material	Operating temperature range	Applications
BODY BLOCK	Polyethylene	Up to 185°C (365°F)	Excellent for weak and strong acids and weak and strong alkalis; attacked by strong oxidizing acids and aromatic solvents.
	Polypropylene	Up to 185°C (365°F)	Excellent for weak and strong acids and alkalis. Excellent for many solvents.
	PTFE	Up to 265°C (510°F) but may be limited by flexible liner	Excellent for weak and strong acids, weak and strong alkalis, and organis solvents; generally inert to chemical attack
	PVDF	−20°C to 130°C (−4°F to 265°F)	Excellent resistance to solvents
	Epoxy resin	−20°C to 140°C (−4°F to 284°F)	Normally used only when reinforced, eg as GRP
	Polyester resin	−20°C to 140°C (−4°F to 284°F)	Normally used only when reinforced, eg as GRP
	Nylon 6/6	Up to 105°C (221°C)	Good resistance to alkalis, dilute acids, solvents, Glass-filled nylon preformed for strength and rigidity
LINERS	Flexible liners Natural rubber	Up to 165°C (329°F)	Good resistance to weak and strong acids and alkalis; attacked by oxidizing agents; good resistance to oxygenated solvents and alcohols; swells in vegetable, mineral and animal oil. Excellent abrasion resistance
	Buna N	Up to 185°C (365°F) intermittent to 200°C (329°F)	Good resistance to weak and strong acids and alkalis; excellent resistance to aliphatic hydrocarbons; excellent resistance to petroleum oil, petrol, mineral and vegetable oils. Excellent water swell resistance
	Neoprene	Up to 225°C (437°F)	Excellent resistance to dilute acids, weak and strong alkalis; good resistance to concentrated acids; good resistance to oil and petrol
	Hypalon (chlorosulfonated polyethylene)	Up to 210°C (410°F)	Excellent resistance to dilute and concentrated acids, weak and strong alkalis. Exceptional resistance to strong oxidizating acids. Good resistance to concentrated mineral acids.
	Butyl	Up to 225°C (437°F)	Good resistance to corrosive chemicals. Outstanding resistance to dilute mineral acids. Excellent resistance to vegetable and mineral oils and solvents. Excellent heat resistance and low gas permeation
	Viton	Up to 250°C (482°F)	Excellent resistance to oils, solvents and missile fuels and most chemicals at elevated temperatures.
	Compar (polyvinyl alcohol)	Up to 150°C (302°F)	Extremely resistant to organic solvents; attacked by water, weak acids and alkalis
	Nordel	Up to 225°C (437°F)	Resistant to weak acids, most alkalis and ketones

MATERIALS AND CORROSION

Plastic pumps

Plastics are being used for pump casings, impellers and other pump parts on an ever increasing basis. Where volume is high and sizes are small there could be a distinct cost advantage. Aside from that many plastics provide good corrosion resistance to chemicals in general or to specific corrosives, they are generally lighter than metals and they may not contaminate the process liquid like metallic materials do when they wear. Tables 13 and 14 provide application information on certain plastic and elastomers as well as physical properties. Fibreglass reinforced epoxies are being marketed for base plates on chemical service pumps. These base plates are considerable more expensive than the steel fabricated base plates but they provide the corrosion resistance necessary that the steel does not. Pumps with plastic liners and other parts have had good experience in corrosive environments. Fig. 9.3 shows a hermetic magnetic-drive pump that has 7 of its 10 main parts made of plastic including the impeller and the liner. Tefzel and Halar fluoroplastics, CFR, and Teflon make up the plastics.

References

[1] Rayner, R., "New Metallurgy for Process Pumps", World Pumps, (Oct. 1992).
[2] Pruitt,K.,"Compass Corrosion Guide", Compass Publications (1978).
[3] Kovach, C and L. Redmerski, "Corrosion Resistance of High Performance Stainless Steels in Cooling Water and Other Refinery Environments" Corrosion 84, Paper 130 (4/2-4/6,1984).
[4] Dickenson, C,. "Pumping Manual" 8th Edition, Elsevier Advanced Technology (1992) p34
[5] Redmond, J., "Selecting Second Generation Duplex Stainless Steels", Chemical Engineering (10/27/86 and 11/24/86.
[6] Schiavello, Bruno, "Cavitation and Recirculation Troubleshooting Methodology" Proceeding of the 10th Annual Pump Users Symposium, Texas A & M, Houston Texas (March 9-11, 1993) Dr. Jean C. Bailey, Editor

- Extensive range of mechanical seals & flexible couplings to suit varying applications and duties
- Designs with proven technology
- 19 companies worldwide with international & local support
- 50 years experience
- Flexiservice network for 24 hour reconditioning & back up service

**First and Foremost
for al
seals and
couplings
for 50 years**

First & Foremost for
1945 **50** 1995
SEALS & COUPLINGS

Head office: *Nash Rd, Trafford Park, Manchester. M17 1SS*
Flexibox Limited: Tel 0161 872 2484 Fax: 0161 872 1654
Mechanical Seals and Flexible Couplings

pm9thed.3d

Durametallic

GF-200
GAS BARRIER SEAL

A technological breakthrough combining zero emission performance with longer life, energy saving and negligible seal face wear.

Suitable for pumps, blowers & fans in chemical & hydrocarbon processing, pulp & paper manufacture, together with other processing applications involving toxic chemicals or gases.

SEALING SYSTEMS FOR CHEMICAL AND PROCESS INDUSTRIES.

GB-200

*ZERO
EMISSIONS
PERFORMANCE
COUPLED WITH HIGH
QUALITY MATERIALS AND MANUFACTURE.*

DURAMETALLIC UK
Unit 13B, United Trading Estate,
Old Trafford, Manchester M16 0RJ

Tel: 0161 848 7061
Fax: 0161 872 6772

SEALS AND PACKING

Packing

Packing of the compression type which is used in pumps can be described as a deformable material used with a gland to prevent or reduce the amount of fluid leakage between surfaces that move in relation to each other. Figure. 10.1[1] shows a packing gland assembly. As the gland is tightened down, the spiral rings of the square cross-sectioned packing are forced against the back of the bore and this forces a deformation in the radial direction. The gland is continually tightened until there is only a few drops of fluid leaking out along the shaft. Packing must have some leakage for cooling reasons and to prevent scoring of the shaft or shaft sleeve. Four to five rings of packing, as shown, is common. When lubrication or cooling of the packing is inadequate, or the pumped fluid contains particles that can be abrasive, a lantern ring is added and a compatible flush is forced through it. There is a diverse offering of packing materials available from seal and packing vendors. Some vendors in this area have many years of experience in providing the right packing for a specific application and their advice should be sought. Packing use has been dropping in favour of mechanical seals. This is to a great extent, the result of the reduction in user maintenance staffs and the loss of experience from retirement. Conservation efforts and environmental regulations have also been a big factor.

Mechanical seals

Mechanical seals, see Figs 10.2 and 10.3 a-e[2], have a pair of mating faces perpendicular to the shaft as the primary seal. One is generally flexibly mounted in a seal head and rotates with the shaft, the other in the gland where it rotates with the shaft. Two secondary seals are commonly present; one between the gland and the stationary seal mating ring and the other between the rotating seal assembly and the shaft or shaft sleeve. In the latter case there will also be a static sealing element between the shaft and the sleeve. The two mating surfaces are generally lapped flat and are of two dissimilar materials. Carbon is a common material for one of the faces, the other is a metal or ceramic. The choice of materials is a matter of economics, corrosion considerations, temperature and the pressure velocity limits of the materials. Preload, holding the seals together is normally provided by springs

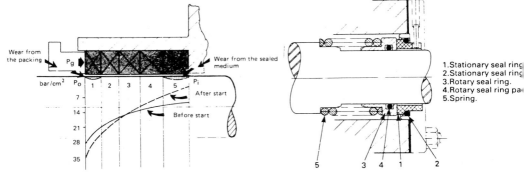

FIGURE 10.1 – Typical radial pressure in a gland before and after start.

FIGURE 10.2 – Typical simple mechanical seal.

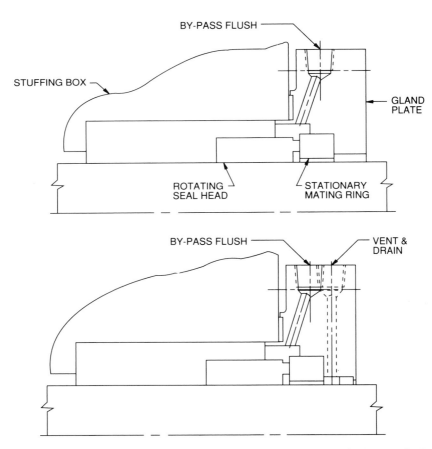

FIGURES 10.3A & B – Two typical mechanical seal arrangements with by-pass flush and vent port.

Precision Mechanical Seals
Custom Engineered or Standard Replacement

- Extensive component stock inventory ensures prompt delivery
- Exclusive design features offer superior performance
- Standard Materials: Stainless Steel; Carbon or Coolcarb®; Buna; Viton; EPT; Neoprene; Ceramic or Ni Resist
- Custom designs, special materials and engineering assistance available
- Designs conform to ANSI and DIN 24960 requirements
- Interchangeable with Types 1, 2, 6, 6A, 7, 21, 43, 47 and other common types worldwide
- Distributor inquiries welcome worldwide

PAC-SEAL INC.
INTERNATIONAL
211 Frontage Rd., Burr Ridge, IL 60521 USA • Phone: 708/986-0430 • Fax: 708/986-1033

European Offices
- PO Box 77326, P. Faliro 17510 • Phone: (30) 1- 981-7893 • Fax: (30) 1- 981-2017
- Unit 21 Venture 20, Lynx West Trading Estate, Yeovil Somerset BA20 2HP
  Phone (44) 01935 411599 • Fax: (44) 01935 411715

Send for Seal Specification Guide

MEET THE FAMILY

SPRING ENERGISED SEALS

One of our family of seals is eager to solve your sealing and delivery problems

METAL 'C' RING

- HIGH & LOW PRESSURES
- HIGH RESISTANCE TO RADIATION, CORROSION & WEAR

METAL CASED ptfe LIP SEALS

METAL 'O' RING

Send for design manual now

SLIPPER SEALS

- HIGH RESILIENCE & SPEEDS (25,000 RPM)
- HIGH & CRYOGENIC TEMPERATURES

More features
Bigger range
Better value

SEAL **LTD.**

Tel: 01922 692447
01922 685159
Fax: 01922 685168

66 LICHFIELD ROAD, PELSALL, WALSALL, WEST MIDLANDS WS3 4HL

LIQUIDYNE®
THE RELIABLE WATER LUBRICATED SHAFT SEAL

IHC Lagersmit has been supplying shaft seals for more than 30 years. Development of the water lubricated LIQUIDYNE shaft seal for the (dredging) industry began about 10 years ago. The worldwide patented water lubricated LIQUIDYNE, has already for years been successfully used in many types of shaft sealing applications such as centrifugal-, dredge-, pulp-, sewage-, cooling-, jet-, process pumps, etc.

The LIQUIDYNE shaft seal is extremely suitable for:
- High pressure-velocity values
- Abrasive polluted mixtures
- Adaption to a variety of conditions
- Inspection of the lip-seals and sealing surface
- All existing shaft diameters

Further major advantages are:
- Reliable, durable and predictable long lifetime
- Virtually maintenance free
- Environmental friendly
- Compact rugged construction with proven design
- Approved by classification authorities

IHC Lagersmit is backed by highly trained specialists with a lot of know-how and experience in shaft sealing and bearing applications, which all stand for rapid reaction and optimal service to complex situations. Our large stock of standard- and spare parts gives a speedy access to competent deliveries and services.

IHC LAGERSMIT
PO Box 5 - 2960 AA Kinderdijk - Holland
Tel. + 31 (0)78 6910472 - Fax + 31 (0)78 6910477

LONG LIFE MAINTENANCE FREE MECHANICAL SEALS

- Wide range of standard seals to DIN 24960 available — common sizes ex-stock
- Customised designs a speciality
- Will handle practically any fluid, sludge or slurry
- All material combinations
- Speedy and economical repair of most makes of seal
- Team of experienced engineers available for advice on applications
- Sales offices/distributors conveniently located throughout the United Kingdom and Western Europe.

JAMES WALKER BRITCO LTD
Britco Works, Orgreave Crescent
Dore House Industrial Estate
Handsworth, Sheffield, S13 9NQ, UK

Tel: 0114 269 0776
Fax: 0114 254 0455

SEALS AND PACKING

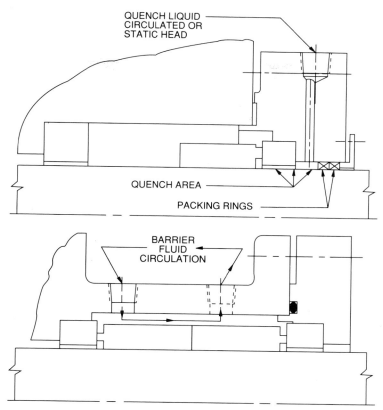

FIGURES 10.3C & D – Single seal with quench and double seal.

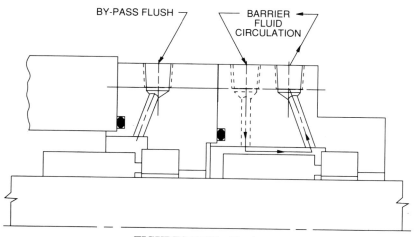

FIGURES 10.3E – Tandem seal.

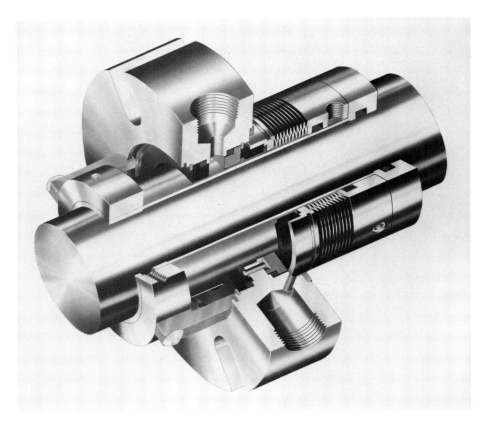

FIGURE 10.4 – Bellows cartridge seal.

or a bellows. The seal faces are generally lubricated and cooled by the material being pumped, clean water or in some cases a buffer fluid. Some form of clamping to the shaft is present to provide torque to rotate the non-stationary parts with the shaft. A big advantage of a mechanical seal is that, unlike packing, it does nothing to score or damage the shaft or shaft sleeve. Fig. 10.3a shows a typical seal with by-pass flush where liquid at the pump discharge is piped directly to the flush connection. When the liquid being pumped is dirty or abrasive a cyclonic separator is generally inserted in this piping. This same arrangement is also used when injection of clean liquid from an outside source is used to keep solids away from the seal faces. Fig. 10.3b shows a vent port which is recommended for collection of hazardous liquids and vapours. Fig. 10.3c shows a quench arrangement. This is recommended as a buffer zone to the atmosphere when the fluids have solids in suspension which crystallize upon contact with the atmosphere. A double seal is shown in Fig. 10.3d. Double seals which are two seals in series, are recommended when highly corrosive fluids are being pumped. Clean liquid circulation into the seal chamber is recommended along with the double seal. A tandem seal, Fig. 10.3e, is

SEALS AND PACKING

recommended when the fluid properties make a single seal impractical, or the fluid is hazardous. A by-pass from discharge is recommended to the inboard seal (inboard refers to the side of a pump towards the driver, the outboard side is away from the driver) along with circulation of a buffer fluid to the outboard seal with tandem seals. Fig. 10.4[3] shows a bellows cartridge seal. Bellows seals are utilized for additional misalignment capability. Elastomeric bellows seals provide the maximum misalignment capability. Cartridge seals

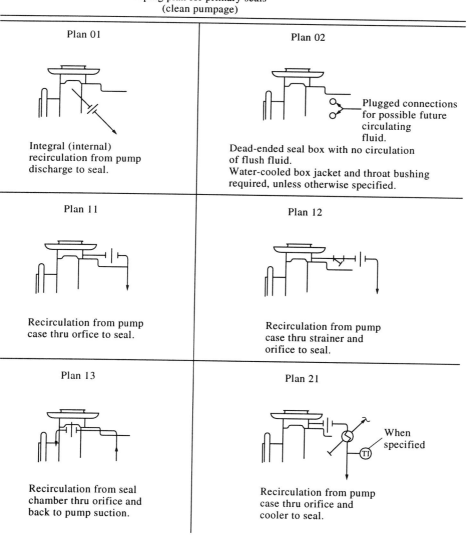

FIGURE 10.5 – API piping schematics.

(Clean pumpage continued)

Plan 22	Plan 23
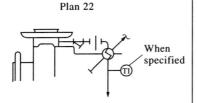 Recirculation from pump case thru strainer, orifice and cooler to seal.	Recirculation from seal with pumping ring thru cooler and back to seal

Dirty or special pumpage piping plans

Plan 31	Plan 32
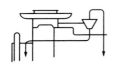 Recirculation from pump case thru cyclone separator delivering clean fluid to seal and fluid with solids back to pump	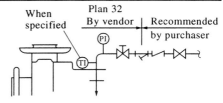 When specified / By vendor | Recommended by purchaser. Injection to seal from external source of clean cool fluid. (See note #2)
Plan 33	Plan 41
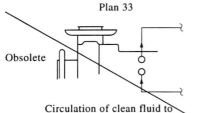 Obsolete. Circulation of clean fluid to double seal from external circulation system. (See note #3)	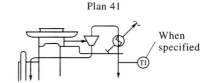 When specified. Recirculation from pump case thru cyclone separator delivering clean fluid thru cooler to seal and fluid with solids back to pump suction.

Notes

1. These plans are representative of commonly used systems. Other variations and systems are available, and should be specified in detail by purchaser or as mutually agreed between purchaser and vendor.
2. Plans 32 and 33 purchaser shall specify the fluid characteristics, and vendor shall specify the required GPM and PSIG.

FIGURE 10.5 – API piping schematics.

SEALS AND PACKING

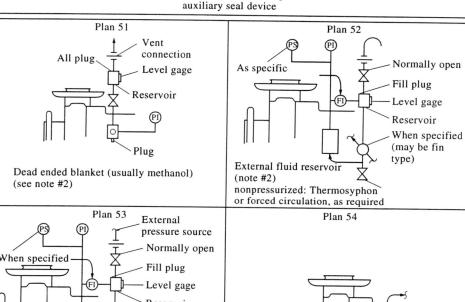

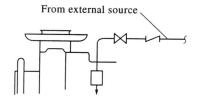

FIGURE 10.5 – API piping schematics.

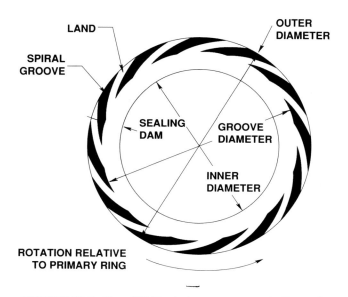

FIGURE 10.6 – Type 28LD spiral groove non-contacting seal.

as shown in Fig. 10.4, are becoming more and more popular. All dimensions are preset at the factory and the measurement of dimensions and setting of the seal, activities with chances of error, are eliminated. Fig. 10.5[4] shows piping plans for seals for various application conditions. Much has been done in the last decade in the way of advancing seal art especially in the area of non-contacting seal faces. Fig. 10.6[2] shows one such seal. Pressure is built up in the spiral grooves during operation which causes the faces to separate.

Seal and packing failures constitute one of the three largest failure areas in centrifugal pumps. These failures are due to misalignment, piping strains, excessive radial and axial loads, improper seal settings, high temperatures, unclean lubricant and coolant and presence of air or gases in the lubricant and other less frequent causes.

References
[1] The Pumping Manual, 8th Edition, Elsevier Advanced Technology (1992).
[2] International Sealing Systems, Mechanical Shaft Seals Recommendation Guide, John Crane, Bulletin S- 2018-2
[3] The X-Series Cartridge Mounted Dura Seal, Form # 566, The Durmetallic Corp.
[4] API Piping Schematics, American Petroleum Institute

SECTION 5

Testing, Installation and Start Up, Vibration and Balancing

TESTING

INSTALLATION AND START UP

VIBRATION

BALANCING

TESTING

Pump testing

Pump testing is generally one of two types: *performance testing* to ascertain the performance capabilities of the pump and *hydrostatic testing* to ascertain the leak tightness and pressure integrity of the pump. Performance testing is usually carried out in the pump vendor's plant, on his test rig, prior to shipment. The tests can be witnessed by the purchaser or his representative or a non witnessed certified test can be provided by the vendor. On very large engineered type pumps model tests are often run on smaller scale models, this is referred to as *model testing*. Testing on-site is sometimes carried out, but this is in installations where the capital costs of instrumentation required can be justified. It is the author's experience, that on-site testing is frequently inaccurate because the instrumentation is not well understood by the operating or maintenance personnel who are called to use it on an infrequent basis, and is not kept in calibration. This is not to say that a minimal amount of instrumentation on-site for monitoring changes in performance is not worthwhile. As a matter of fact one of the biggest problems encountered in field troubleshooting pumps is the lack of simple discharge and suction gauges, that are very important in revealing a pump or systems condition.

Manufacturer's test stands can be classified as open or closed. The open test stand is one that will be found where a large amount of pumps are tested. The open tank allows the discharge piping to move three dimensionally to accommodate the different discharge connection locations of the various pumps being tested. This movement is generally accommodated by a gantry. The suction piping with flexible connections also has some limited three dimensional movement capability. In many cases, the pumps are mounted on bases that can be moved three dimensionally also. The disadvantages of the open test stand are firstly, its flexibility and looseness does not allow meaningful vibration readings to be taken on the pump, and secondly, the pressure cannot be reduced on the tank for NPSH testing. The latter is not serious since other means of testing for NPSHR such as sump level reduction, or varying the capacity while holding speed and suction head constant. Even suction throttling performs the same function adequately when set up properly. The closed tank is used where the set up time is not as critical to the cost of the order and the rate of set-ups is less. These set-ups are relatively rigid and vibration testing is meaningful. Even here, however, the vibration levels will be higher than those expected on-site. Europump[1] found in the survey leading up to the referenced document that vibration readings on the rigid manufacturers test stands were 2 mm/s higher than when installed. Closed tanks are generally utilized for R & D Testing,

Certified performance tests are highly recommended on any pump. Unfortunately, in the case of smaller pumps the costs are not proportional to size. Hence the cost of a performance test can be a relatively high percentage of the pump cost. The opposite being the case for the larger pumps. Since there are so many manufacturing variables that can affect pump performance, especially with castings, there is always a small percentage of pumps tested that do not meet the ISO[2] or HI[3] acceptable tolerance band of head or capacity on the first test. These pumps are then corrected with an impeller trim if the head is high or a chip if the head is low or some other less common correction such as a inlet blade modification to improve NPSHR for example. Naturally, orders that do not call for testing have this same low inherent failure percentage but are understandably not corrected before shipment. The certified test also provides an index that can be used for monitoring once the pumps are in operation. As an example; the certified head capacity curve will provide a flow measurement that can be correlated to the pressure difference between suction and discharge gauge readings, especially if they are located in the same places as they were on the certification test. Wear, rubbing and other problems causing a degradation of performance can be monitored by running a power consumption vs pressure difference test on-site and monitoring any changes over time. As one author[4] put it "... field testing does not require maximum accuracy to be of value. The important thing is to make the initial test and subsequent tests using the same instrumentation and procedure each time in order to determine when conditions change." So whether you have a full shop or field

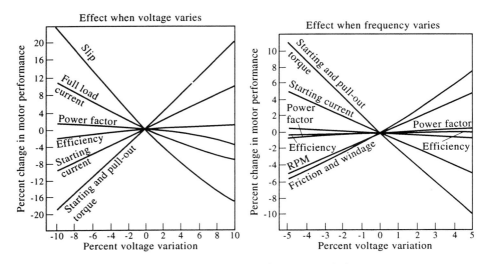

FIGURE 11.1 – Effect of voltage and frequency variation on motors.

performance test or not, do not lose the opportunity to perform index tests that can be used for comparison later. There are many pitfalls that can cause differences between the factory and field tests. For example sump conditions can affect the suction conditions detrimentally; a centrifugal pump head is controlled by the system resistance. If this system resistance is different from the factory test then the two will not agree, electrical power differences between the site and the factory test rig can cause performance differences (see Fig. 11.1). If you have a 2950 rpm motor (50 rpm slip) and the voltage at the site is 10% higher than what was present at the factory test, then according to Fig. 11.1 you have 17% less slip or a speed increase of 8 to 9 rpm. ; pressure taps can be inaccurate if not properly installed. Differences between the ISO and HI test codes or standards are slight and a facility that can conform to one could conform to the other without major effort and expense. The testing requirements of positive displacement units are very similar to centrifugal units. These codes standardize the requirements for calibration frequency and methods of instrumentation; the location and installation of instrumentation and instrumentation taps, the test procedure including the number of points to be taken and the data acquisition and reporting. A copy of one or both is recommended from a standpoint of test witnessing and on-site index testing because of their educational and reference value.

The reduction in costs and increase in reliability of data acquisition systems has resulted in the increased use of these systems on production test rigs. The result is that dual speed or variable speed testing can be run at the factory test if the pump will be run at either of those two conditions on-site. The data acquisition systems allow data to be presented almost instantaneously on PC's in the form of head capacity, NPSH, and power characteristic curves. With the ability of these systems to average many more test points per unit of time than manual methods, the scatter of data points on these curves is materially reduced.

Normal hydrostatic pressure tests are at one and a half times the design working pressure of the pump. However, some pumps have a different design working pressure on the suction side than the discharge side. Double suction, split case pumps are an example. These pumps are hydrotested in the factory with special fixtures. Seals and packing are also not chosen to operate at one and a half times the rated condition. Often these pumps are hydrotested with packing that is removed after the test and replaced with the job packing or mechanical seal. Hence, when piping is hydrotested in the field, it is important that the safe pressure not be exceeded. Some manufacturers provide the maximum allowable field test pressure on the nameplate. If the manufacturer is required to perform a leak test, at shut-off conditions, this should be called for on the order.

References

[1] Guide to Forecasting the Vibrations of Centrifugal Pumps, Europump (1990)
[2] Centrifugal Mixed Flow and Axial Pumps – Code for Hydraulic Performance Tests for Acceptance – Classes I and II, ISO 9906 (199). "Positive Displacement Pumps – Code for Hydraulic Performance Tests for Acceptance, Classes I and II, ISO" (199).
[3] "Centrifugal Pump Test Standards", Hydraulic Institute HI 1.6 (1993).
[4] Luley, R., "Pump Testing: Factory vs Field, Water & Sewage Works" (July, 1974)

INSTALLATION AND START-UP

Storage

When a pump arrives at the job site it should be checked for shipping damage, for lifting instructions and for the installation, operating and maintenance manual. The instruction manual should be given credence over the generalized instructions here. The next task is to prepare it for storage unless it is going to be moved directly into place. The preparation can vary with the length of time it is expected to be in storage and the storage conditions; e.g. inside or outside, heated or not, winter or summer etc.. It is important that the pump be installed in a dry place. If freezing conditions are possible, some heat should be supplied even if it is only a low wattage heater and any drain plugs should be removed to let any retained water from testing out. The pump should be covered with plastic after the flanges have been covered along with the ends of the shaft up to and including the shaft openings to the pump, stuffing box and housings. Generally the pump or pump unit as shipped has adequate protection for short term storage in a dry, ventilated space that is maintained above freezing. If a complete motor/pump unit has been delivered then the same treatment should be given the motor and the coupling should be completely covered or wrapped. Once a month the shaft should be turned. If the storage time is going to be lengthy, the pump should be thoroughly dried, by blowing hot air through it or other means. All internal surfaces that can be reached should be coated with a film of light oil to avoid rusting. Preferably, the vendor should be advised ahead of time and charged with the responsibility for long term storage preparation.

Installation

The pump *location* should be where it is accessible for maintenance. This means that enough room must be provided for the removal of the rotating assembly and any covering parts that must be removed first. Enough space should be allowed for operating personnel to adjust packing, view oil sight glasses and gauges, add oil, grease etc.. It should be as close horizontally as possible to the tank, sump or other source of fluid that it will be pumping and within reason as far below the connection to such source as possible.

The *foundation* should be of adequate mass to absorb vibration e.g. five times the weight of the pumping unit. In the case of reciprocating units, a mass of ten times the weight of the pumping unit is recommended. Generally the fabricated bases require that their structure be completely filled with grout up to the sub-base top plate and provisions must be made to allow this. The baseplate should be raised from the floor so as to keep water from the base plate, bolts and base of the pump. It should also contain hold down bolts that have been hooked around a rebar and encircled by sleeves 3 to 4 times the bolt o.d., that will allow some accommodation of the bolt to the bolt-holes in the pump base. The resistance of these bolts to being pulled out of the foundation is most critical in the case of vertical units where forces can act at relatively large distances from the baseplate.

Once the foundation is in and cured the pump unit should be lifted into place with the foundation bolts in the respective bolt holes and the bottom of the base sitting on pairs of wedges that are at $^3/_4$ in. – 1 $^1/_2$ ft. high. See Fig. 12.1. These wedges should have a slight taper, be located approximately every 0.6 m (2 ft.) and especially next to each foundation bolt. Once the unit is sitting on all the wedges, move the suction flange up to its mate. If they don't match up squarely without bolt binding of the foundation bolts, piping adjustments must be made. Disconnect the coupling halves. The unit is now ready for *leveling*. If a complete pump and drive unit was purchased then it was aligned at the factory. However, shipment often causes the alignment to shift somewhat. With a level on the horizontal shafts and then the vertical coupling faces, adjust the level with the tapered wedges as shown in Fig. 12.2. If a bare pump on a base was purchased from the pump vendor then the coupling halves must next be installed on the pump and motor shafts in accordance with the coupling manufacturer's instructions first. Then the base should have 1/4 in. to 3/8 in. of shims added over each bolt hole between the bottom of the motor feet and the top of the base. This shim pack should be made up of no more than three shims of different thicknesses. It will facilitate present and future alignments.

With the pump level, the next step is a rough *parallel and angular alignment*. Parallel alignment means that the shaft axis of the pump and motor are in parallel planes, but are not necessarily concentric with each other. Angular alignment means the shafts are concentric with each other, but not necessarily parallel to each other. A straight edge and a taper gauge or a set of feeler gauges will suffice for the rough alignment. See Figs 12.3 and 12.4. Angular alignment can be checked by inserting the taper or feeler gauge between

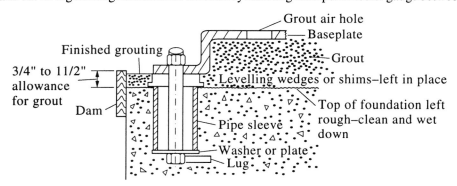

FIGURE 12.1 – Typical foundation bolt design.

INSTALLATION AND START-UP

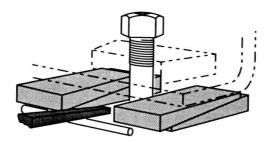

FIGURE 12.2 – Method of levelling.

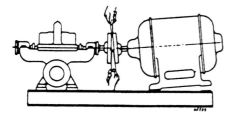

FIGURE 12.3a – Measuring vertical angular misalignment.

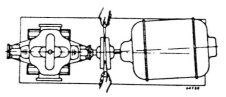

FIGURE 12.3b – Measuring horizontal angular misalignment.

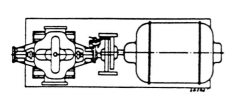

FIGURE 12.4a – Measuring horizontal alignment.

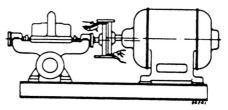

FIGURE 12.4b – Measuring vertical alignment.

the coupling halves at the top bottom and both sides and, on a complete unit, adjusting the wedges only. In the case of the bare pump and base, after the motor is bolted down onto the spacer pack and base an angular alignment in the vertical plane should first be attempted, with the wedges, followed by side to side adjustment of the back of the motor. If the base does not have holes for the motor feet bolts, then it is necessary first to get the motor into rough alignment and then scribe the i.d. of the holes in the motor feet onto the base. The motor should be removed, holes drilled and motor realigned. Parallel alignment can be checked laying the straight edge on the outside diameter of the coupling hubs at the top, bottom and both sides. Necessary adjustments should be made to bring the straight edge into contact with both coupling OD's at all four points around the periphery. Each time the wedges are adjusted, (or the motor shims) and the bolts taken up further, adjustments may be required. Make sure the gap between shafts is in accordance with the instruction manual or coupling instructions. Now the flanges can be bolted up, one at a time. If the flange faces are not in parallel alignment and easily brought together then adjustments may have to be made to the piping system.

A more precise alignment should now be completed using *dial_indicators*, or one of the newer alignment techniques such as with lasers. Dial indicators, as shown in Fig. 12.5 give a more precise reading of misalignment than the straight-edge and taper gauge method we just used. One of the first things that should be done with a dial indicator arrangement is to duplicate it as simply as possible on a piece of pipe with the indicator assembly on top in the vertical position and the indicator set to zero.. This assembly should be rotated 90° to the side and the reading taken, then rotated another 90° to the bottom vertical position for another reading and then another 90° to the opposite side for another reading. These readings measure the *sag* in the indicator assembly at each position and should be added algebraically to the corresponding readings taken on the unit. Now fasten the dial indicator to the top of the coupling hub on the pump shaft and set the indicator to zero. Mark the motor coupling hub around the button and rotate both shafts 90°, take a reading and repeat three more times for a complete turn making sure each time that the indicator button is in the circle. These readings with the sag correction indicate the side to side and/or up or down corrections that have to be made to the motor and its shims for parallel alignment. The corrected readings at this point should not exceed 0.125 mm (0.0005 in.). For angular alignment move the indicator so that the button is pressing against the face of the same coupling half it was on previously. See Fig. 12.5. If the extension of the assembly has increased or decreased then the sag should again be measured, duplicating the new geometry. Set the indicator to zero and measure the four quadrant misalignments as before. Make the necessary adjustments in the motor shims to bring into angular alignment to the same tolerance as above. Note that no adjustments to the plate or shims between the pump and the base has been called for. Adjustments here should not have to be made and are not recommended unless in some specific instance the instructions call for it.

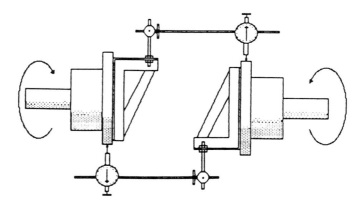

FIGURE 12.5 – Alternate method of alignment.

FIGURE 12.6 – Laser alignment.

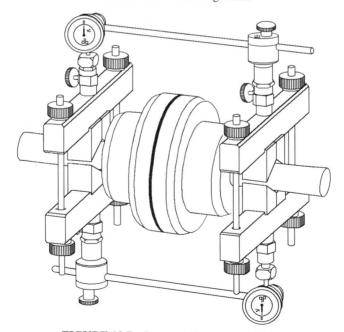

FIGURE 12.7 – Reverse alignment method

There are more *advanced alignment systems* available today that can halve the time it takes for alignment. The laser alignment system[1,2,3], see Fig. 12.6, is very effective but relatively expensive. It should be explored for installations with many pumps. The reverse dial indicator method[4], see Fig. 12.7, can be more precise and quicker than the conventional dial indicator methods covered here at little extra expense. It inherently corrects for the sag.

The unit should now be *grouted* in. First snug up the foundation bolts evenly but not too tightly and make alignment adjustments if necessary. Fill the baseplate up completely until grout is coming out of the holes provided in the top of the base fabrication. The leveling pads should be left in place. After the grout has hardened (two days approximately) tighten the foundation bolts down firmly. Grout prevents the unit from shifting, provides a clean smooth surface, and provides additional mass and rigidity to the base. The mass reduces the vibration from the pump and the rigidity resists deflection. The mixture is generally 2:1 sand over cement with enough water added to let it flow freely

Pre-start actions

The motor, with the coupling halves still unconnected, should be bumped over to ascertain direction of rotation. In the case of three phase power, it is only necessary to switch two of the three phases to obtain the opposite rotation. Check the packing to see if it is in place,. If not, follow the instructions in the instruction manual and the packing box for installation. Packing rings should have their joints staggered. If there are five rings, then each joint would be about 72° from the next. If a lantern ring is furnished again follow the instructions. Generally two rings of packing would be inserted before the lantern ring. Do not take up too tightly on the gland. If a mechanical seal is furnished the instruction on that type of seal should be followed, since there is such a variety. Make sure all auxiliary connections such as packing or seal water connections are hooked up and instrumentation is installed. Check the pump, coupling and motor and their instructions to see if there is adequate lubricant available. Caution! – do not over lubricate. Bolt the discharge flanges together, assuming they are in good alignment. Fill the piping to the top of the pump with liquid. vent the top of the pump and check for piping leaks. This is especially critical on the suction side where suction pressures will be below atmospheric pressure, and air will be sucked in. Rotate the pump shaft by hand or strap to make sure there is no binding. Now fill the piping completely and recheck the alignment. If it has moved then there is probably some pipe strain on the unit due to the pipe hangers not supporting the weight of the water. This should be corrected before starting the unit. A final alignment should now be made and the maximum misalignment brought to within 0.075 mm (0.003 in.). If the pump has an open impeller, follow the instruction manual in setting the proper clearance between the wear plate and the impeller.

Start up

The unit can now be started and carefully monitored. If speed control is available, bring it up to speed slowly making any adjustments in the system that are necessary. After 3 or 4 hours of operation clean out any strainers. After the system has been completely adjusted and all filters cleaned again, the time is proper to run index tests duplicating the design

INSTALLATION AND START-UP

conditions and factory test as closely as possible. If the process fluid is hot, a hot alignment should be made. After the unit has been running for some time and full process temperatures have been reached shut the unit down and immediately do an alignment. This will cover the thermal expansion of the pump relative to the driver.

Fault finding or trouble shooting

Tables 11A 1- 7 cover fault finding for the various types of pumps. Because of the big differences between pumps of a given type and their applications, it is highly recommended that the manufacturers instruction manuals be consulted first and these generalized charts be used for supplemental assistance.

References

[1] Cavanaugh, J., "Care and Feeding of Alignment Problems", Pumps and Systems (Jan. 1994)
[2] Weiss, W., "Laser Alignment Saves Amps, Dollars" Plant Services (April 1991).
[3] Bloch. H., "Pumps, Compressors, Valves and Piping Get Faster Machinery Alignment With Laser-optic Upgrade", Power (Oct. 1987).
[4] Evans, G., "Shaft Alignment Tools", Pumps and Systems (Jan. 1994)

VIBRATION

Vibration is defined[1] as The oscillating, reciprocating, or other periodic motion of a rigid or elastic body or medium forced from a position or state of equilibrium. Measurement devices are available that can measure the displacement, velocity or acceleration of this periodic motion. In the case of pumps vibration can be caused by unbalance, misalignment, bent shaft, hydraulic forces (especially at off design flows), anti-friction bearing problems, oil whip in hydrodynamic bearings, looseness, resonance and electrical unbalance of the motor. One consultant[2] lists the top nine vibration – related pumping problems as follows:

1. Improper suction conditions.
2. Recirculation caused by pump not properly designed for the application.
3. Vane pass frequency vibrations caused by improper gap between impeller discharge and diffuser overlap.
4. Pump not accurately aligned for actual heat rise of pump and driver and pipe strain.
5. No attention to coupling fitting, balance or keys.
6. Improper selection of seals, fitting and flush/cooling.
7. Improper lubrication (e.g. too cold, water ingestion and condensation).
8. Improper selection of base-plate and grouting procedures.
9. Improper maintenance procedures in shop; bearing selection/fitting wrong.

The problem with vibration starts when we try to put limits on it. What is acceptable vibration and what is not is not an easy determination. The more rigid a pump is the less vibration it should see. The same pump, however, can be very rigid in the horizontal position, where gravity is holding it against its supports and flexible like a reed when it is in the vertical configuration. What is the criteria for determining limits. One logical criteria is that we do not want rotor lateral (perpendicular to the shaft axis) vibration amplitude to exceed the clearance in the wear ring area.

We do not want packing pounded out or mechanical seals to leak and their faces to wear due to this relative motion caused by the rotor vibration, nor do we want bearings, shafts etc to fatigue and fail. However, everyone of these will give different limits and how we

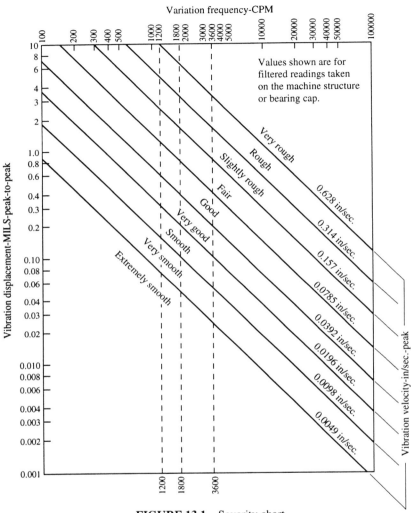

FIGURE 13.1 – Severity chart.

correlate them to externally measured vibration readings or even proximity probes raises another question. One of the first set of curves (Rathbone Chart) that was put together was by an insurance company[3] based on subjective opinions as to whether the machines in question were smooth or rough running. It also did not differentiate between the type of machinery. A later significant work was that of Blake[4], who added differentiation of the type of machine and also actual maintenance data to his input. Next came the IRD General Machinery Vibration Severity Chart which was based on the Rathbone chart but with allowable limits adjusted downward based on discussions with machine operators. The differentiation for various types of machinery was absent. However, a recommendation

VIBRATION

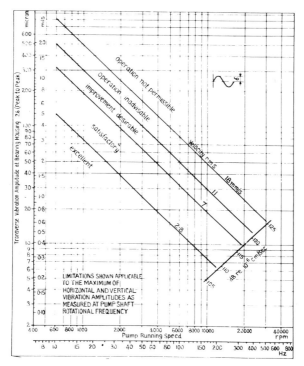

FIGURE 13.2 – Vibration severity chart showing vibration limits.

was made as follows: If no history of vibration measurement is available to you, You may begin with the General Machinery Vibration Severity Chart. See Fig. 13.1. The work further points out that an "understandable and useful definition of vibration tolerance can be reached" and proposes the following: *"Vibration Tolerance* is the allowable deviation in vibration amount from what has proven by experience to be satisfactory. There are two pump organizations Europump[5] and HI[6] that have put together vibration charts based on the combined experience of a majority of the worlds pump manufacturers. The Europump guidelines are more recent and show some correlations with their experience that are being correlated with HI experience and will probably be adopted in part at least by the Hydraulic Institute in the future. The continuing cooperation of these two organizations will hasten the time when there is no significant difference between their recommendations. Fig. 13.2 shows the acceptable field vibration limits (feed pumps for 500-600 MW units). Fig. 13.3 shows the vibration limits for a horizontal clear water pump running at 1500 rpm by Europump. While there are differences, the use of either of Hydraulic Institute or Europump references which are both based on vast experience is recommended over general severity charts. The limits are being reduced based on the tighter quality standards being imposed on balancing and tolerance control in the pump industry today and the next edition of the HI Standard will reflect this trend. Note in Fig. 13.3, the correction factors for the number of vanes and off-bep operation.

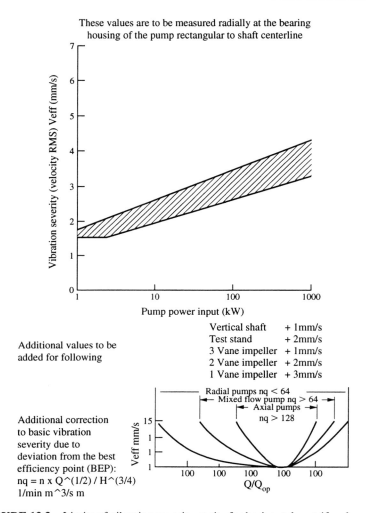

FIGURE 13.3 – Limits of vibration severity at site for horizontal centrifugal pumps pumping clear liquids (multivane imp.) pump speed = 1500 rpm

The next area of contention amongst authors is whether the limits should be presented in displacement, velocity, acceleration or some combination of them. The weight seems to be in favour of velocity limits today. Maxwell's paper[7] points out that the reason velocity is used in classifying machinery vibration in the 10 to 1000 cps range is that it is independent of frequency in this range. But velocity limits can be overly loose at the low frequencies and overly tight at the high frequencies. Looking at Fig. 13.2 for example, the constant velocity line of 0.2 in./s shows a limit of of 8 mils displacement at 500 rpm (or cpm) compared with 5 mils in the 14th edition and lower in the 15th. At the other end of the scale, the same velocity line at approximately 13,000 cpm allows an amplitude of only 0.3 mils/s which manufacturers of pumps with high numbers of vanes point out is impossible to attain and not in line with experience. A disadvantage of velocity is that it is not related to either stresses or forces[8]. Marscher[9] gives detail in support of this and goes on to argue persuasively for interim limits based on displacement until an adequate data base is built up with spectral (filtered) data. HI's curves have used displacement and velocity in the past. The point that the author wishes to make from all this background material is that if you stay with limits based on empirical experience, then the theoretical or practical flaws in presentation are not serious. The last conflict of presentation is the use of filtered vs. unfiltered readings. Here we have those who argue that users do not have the more sophisticated and expensive spectral equipment that can take filtered vibration readings and therefore unfiltered readings should be used to present limits. On the other side, there are those that argue that limits do not mean anything if they are not addressing the frequencies that can be associated with causes of the high levels of vibration. Table 1 shows the multiples of running speed cps that are associated with common causes of high non-resonant vibration in the vast majority of cases.

TABLE 1 – Frequency of Common Vibration Excitations, in multiples of running speed

0.3N	Diffuser Stall
0.5N	Oil Whirl
0.8N	Impeller Stall
1N	Unbalance
	Bent Shaft
	Misalignment
	Looseness (lesser magnitude)
	Eccentric Journals
	Motor Electrical
	Offset casing halves, double suction
	Uneven flow entering impeller eye
2N	Cracked shaft
	Misalignment (lesser magnitude)
	Looseness
3N	Misalignment (less magnitude than 2N)
xN	Vane pass frequency, x = # vanes
yN	Gear tooth frequency, y = # gear teeth
1-4N	Drive belts
Range of several N	Antifriction bearings
2N	Reciprocating units

Here again the presentations of Europump and HI are still going to give you the best guidelines available and except in rare cases they should keep you out of trouble, if followed. You should, however, be aware of the limitations of unfiltered readings.

An *unfiltered reading* is an *overall reading* that includes the weighted effect of all vibration amplitudes for every frequency included in the reading. For instance, Fig. 13.4

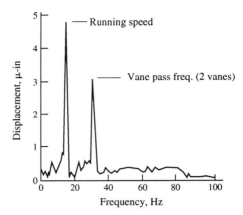

FIGURE 13.4 – Frequency spectrum.

which is a complete spectra of readings that one would obtain by scanning every frequency in that band with filtered readings. An unfiltered reading would give one reading for this entire band which might be anywhere from 1.1 to 1.5 times the peak vibration that is the biggest contributor depending on the make-up of the rest of the vibration energy represented in the reading. The more high peaks, the closer to 1.5. The less high peaks, the closer to 1.1. So you can see that with a given unfiltered reading you can have a variation of almost 40% in the vibration amplitude of the filtered vibration that may be of concern. With the tighter limits today this is probably safe in most cases. And certainly no worse than the bulk of experience which is based on unfiltered readings.

Natural frequency

Natural frequency is defined[10] as the frequency of free vibration of a system. Every part in a pump has a natural frequency and if some exciting force acts on it while it is standing alone at that frequency, the part will start vibrating. There is likewise natural frequencies of the pump assembly that we are concerned with here. The natural frequency of a pump is found by attaching eccentric weights to an exciter motor and running it at various speeds. When the natural frequency is closely approached there will be a high jump in the vibration level. When the frequency of an exciting force matches that of a natural frequency, they are said to be in *resonance* with each other. Resonance problems form the bulk of vibration problems with vertical pumps. They are extremely rare in horizontal pumps. This is, as mentioned before, due to the big difference in stiffness between the two even with the same model and size pump. Pump manufacturers can determine the natural frequencies of their bare pumps but other variables come into play. Foundation bolting causes a wide range of variability. The natural frequency is lowered by an increase in mass and increased by an increase in stiffness. Increased damping will normally give only a slight downward shift to the natural and a rounding off of the peak at a slightly lower magnitude. For this reason it is not generally pursued. However, a tunable damped dynamic vibration absorber has been highly successful in reducing amplitudes by a factor of 6, but unfortunately cost is

VIBRATION

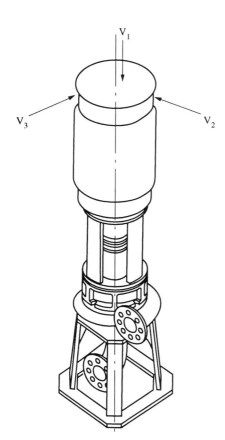

FIGURE 5 – Vertical seperately coupled clear liquid or non-clog pump.

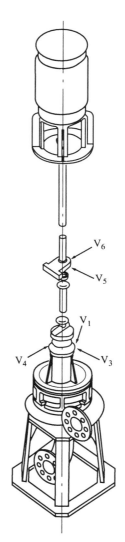

NOTE: V5 and V6 should be checked to insure bearing support is rigid. Vibrations at this point could be transmitted to the pump or the motor.

FIGURE 6 – Vertical clear liquid or non-clog, flexible shafting driven pump.

of the same magnitude as the pump. Mass changes come with the variety of motors that may be applied to a given pump with their different weights. Stiffness of a vertical unit can vary due to the different lengths of the motors but the biggest variation comes from the differences between the the systems that the given pump becomes part of. Two of the styles of vertical units are shown in Figs. 13.5 and 13.6, along with the locations that vibration readings should be taken. Some manufacturers have made up vibration avoidance charts for the separately coupled type shown in Fig. 13.5. These charts show the location of the natural frequency based on a certain weight motor and a band of 15% of this frequency on either side. This avoidance area is adjusted for motors of different weight. Should the sales department find that a selection falls in the avoidance band they can talk to the motor vendor about a different weight motor or they can substitute a much stiffer pedestal than the one shown. This has resulted, in the case of one manufacturer, in almost a total elimination of the problem on this type of pump. The few cases that have occurred have either been due to oversight or a abnormal amount of flexibility in the system in the case of improperly anchored foundation bolts or abnormally stiff system. The flexible line shaft pump assembly shown in Fig. 13.6 has its motor mounted on the floor above, so the motor mass is not a factor on the unit below. However, every job has a different line shaft length, support system etc. A good percentage of problems come from included angles of motor to line shaft not being equal to the included angle between the line shaft and the pump shaft. (This gives a 2N vibration), bent shafts; overly tight Universals (1N); inadequate support beams for the floor above[11], under the motor; and beams to which the line shaft bearing supports are mounted. With the large and increasing percentage of these pumps that have adjustable speed control, avoiding a natural in the operating range is sometimes not possible, even with avoidance charts. In these cases the speed control must be set up such that it will pass through the natural frequency but not dwell there. Note that nowhere here is re-balancing recommended as a correction for the high vibration caused by the resonance. While that could be done and may even make the vibration tolerable, it is not permanent and increased vibration from uneven wear, etc. will bring it back. The natural frequency should be shifted in one direction or the other by at least 15%. The choice of direction is often obvious if the forcing frequency is already to one side or the other of the natural. Vertical sewage pump naturals are often in the vicinity of the vane pass frequency. To check and see if it is the hydraulics that is exciting the natural the pump can be run dry and readings taken. *Torsional vibration* occurs in these line shafting systems on an occasional basis. There is a tendency in these type of drive shafting systems for the shafts to actually wind (twist) in one direction and then unwind in a periodic fashion to some extent. This is not a problem unless a torsional critical (natural) vibration frequency is reached and the amplitudes shoot up. This type of problem is more common when gears or reciprocating units are involved. It can be corrected by finding the source of excitation or changing some component shaft to move the natural away from the excitation, e.g. a larger diameter line shaft. Torsional dampeners are effective for correction of gear and reciprocating torsional resonance problems.

Rotor Critical Speed

Rotor Critical Speed in centrifugal pumps has a rare occurrence in catalogue type units but does occur occasionally in engineered type pumps. It is then up to the pump manufacture to correct the situation by changing the rotor mass, stiffness or wear ring stiffness[12] or if the exciting frequency is that of vane pass then he could change the number of vanes or add a volute splitter for corrective action.

Hydraulic Resonance

Hydraulic Resonance occurs when there is an amplification of the pulses that occur in a pump because of resonance with the natural frequency of a section of piping. This can occur with centrifugal pumps and occasionally does, but much less frequently than with positive displacement pumps with their higher pulse levels. This build up of dynamic pressures can be very serious. It can result in high pressures, fatigue failures, and high noise and vibration levels. The incompressible medium being pumped carries the pulsation into the piping and even if it is low as is the case with centrifugal pumps it is magnified many times due to the resonant condition. Altering the natural frequency of the piping section, or adding a pulsation dampener will resolve the problem . Lowering the speed and increasing the impeller diameter accordingly or raising the pump speed and trimming the diameter in the case of a maximum diameter impeller could resolve the problem if the changes would be significant, and is the most economical solution.

Vibration Monitoring or Condition Monitoring

Vibration Monitoring or Condition Monitoring is becoming more and more common. In all our previous discussion of the allowable limits for vibration, nothing was as important as noticing a change in vibration, assessing its cause and correcting it. Whether you go to a full blown vibration monitoring system or just take vibration readings once a week and watch the trends, you are taking a big step forward toward preventative maintenance.

References

[1] Webster's New Universal Unabridged Dictionary, Barnes and Noble Publishers, (1992)
[2] Nelson, W., "Addressing Pump Vibrations Parts I and II, Pumps and Systems" (Mar. and Apr, 1993.
[3] Rathbone, T., Vibration Tolerance, Power Plant Engineering, (Nov. 1939)
[4] Blake, M., New Vibration Standards for Maintenance, Hydrocarbon Processing and Petroleum Refiner (Jan. 1964) Vol. 43, No.1.
[5] "Guide to Forecasting the Vibrations of Centrifugal Pumps", Europump (May 1991)
[6] Hydraulic Institute Standards, 14th Edition (1983).
[7] Maxwell, H., "Absolute Vibration Velocity Limit Needed For Pump Test Standard", Power Engineering (June 1986).
[8] Ludwig, G and O Erdmann, "Gas Turbine Vibration Limits – A Fundamental View", ASME (April !973).
[9] Marscher, W., Talk on Vibration at Hydraulic Institute, May 25, 1993.
[10] Harris, C.,"Shock & Vibration Handbook", 3rd Edition 1988
[11] Meyer, R., "Solve Vertical Pump Vibration Problems", Hydrocarbon Processing (Aug. 1977)
[12] Gopalakrishnan,S., "Critical Speed in Centrifugal Pumps", ASME 82-GT-277

BALANCING

Balancing

Balancing is the process of bringing the mass center of a body near to or coincident with its axis of rotation, such that the residual unbalanced force and couple are acceptable. The most perfectly machined solid disc could exhibit rotating unbalance due to the non-homogeneity of material density and/or eccentricity in the clearance between the shaft and the disc. The first shifts the centre of gravity away from the geometric centre while the second shifts the centre of gravity and the geometric centre away from the rotational axis. Castings are prone to non-homogeneities and complex shapes such as impellers, are affected adversely by casting and machining tolerances or casting deviations such as those due to pattern wear. Vane spacing and vane i.d. variations are examples. A 45.4 kg (100 lb) impeller shifted in its clearance by just 0.0254 mm (1/1000 in.) would result in 115.3 g/cm (1/10 lb/in. or 1.6 oz/in.) of unbalance. To offset this unbalance requires any combination of weight addition and the radius of weight addition equalling the 115.3 g/cm (1.6 oz/in.) (e.g. 1.6 oz at 1 in. or 0.16 oz at 10 in., placed on the opposite side of the impeller, 180° from the unbalance). So far we have covered the implicit situation of a thin or relatively narrow impeller. However, it is possible for a relatively wide impeller to have its centre of gravity on the rotating axis and still have problems. Although it is in perfect single plane (static) balance, it is not in two plane (dynamic) balance. Consider a thin racing bike tyre and rim. If their centre of gravity is on the axis of rotation, there is no unbalance. Now consider a wide automobile tyre and rim (Fig. 14.1) and assuming the tread is in balance, we are left with the two side walls and the inside and outside discs of the rim. Take the classic case where the center of gravity exists at the axis of rotation, and half way between these discs and sidewalls. The assembly is in single plane (static) balance at this centre plane. However, the two sidewalls could have an equal unbalance, 180° opposite each other leaving a couple unbalance. A dynamic balance can alleviate the results of a couple unbalance by eliminating the residual unbalance in the planes of the sidewalls. In the more general case, the c.g. does not coincide with the axis of rotation and the unbalances in the two planes are not 180° apart. The two plane (dynamic) balance will correct the single plane (static) unbalanced force and couple. If you recall, the location of

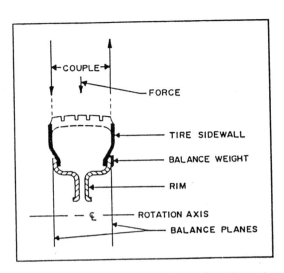

FIGURE 14.1 – Auto tyre two balance plane illustration.

balance weights on your car's tyres are at the outside diameter of the rims on the inside and outside of the tyre, and at some random phase angle to each other rather than exactly 180° apart. Actually the impeller is a complex body of many radial planes. Nevertheless, corrections of a rigid body such as an impeller in just two of those planes can result in the c.g. being brought into proximity to the rotational axis, thus substantially reducing the unbalanced force and couple.

Effects of unbalance

Effects of unbalance are lower seal life: higher vibration and to some extent; lower bearing life, and increased shaft stress. In pumps the first two are the most important. Bearings are sized based on the radial loads which can be of a magnitude of ten or more times the effect of vibration loads. Unbalance shows up at the frequency corresponding to rotative speed. (Unfortunately, misalignment can manifest itself at this same frequency, causing unbalance to be labelled the culprit when more often than not it is not). Fatigue of impeller blades, increased looseness of foundation bolts, brinelling, rubbing of rotating parts against stationary parts and resonant frequency problems are some of the effects of extreme unbalance that occur.

Balance requirements

The amount of unbalance allowable varies with the relative rigidity of the stator or frame, the capability of the bearings, the shaft speed and rotor mass. In the case of pumps, radial loads near shut off are such a large factor in bearing sizing that relative rigidity is the main factor in determining balance requirements. ISO 1940[1] provides a method for experimentally determining the maximum allowable unbalance as measured on the bearing housing and in the absence of such experimental data provides recommended maximum unbalance

BALANCING

grades. See Table 1. The required balance quality for each size and type of pump is determined experimentally in accordance with ISO 1940 in the following manner:

1. The impeller is balanced statically or dynamically, as required, to the minimum residual unbalance attainable. The pump is then assembled and aligned to the low end of the alignment tolerance range to keep the noise to signal ratio as low as possible
2. The unit is started and run at full speed. Vibration readings are taken on the bearing housing, (filtered readings at rotative frequency are preferable)
3. A known unbalance is then added and the readings are taken again. This is repeated with increasing unbalance weights until the vibration reading exceeds the background levels by less than 1 mil displacement. ISO 1940 describes this point as one where the unbalance noticeably affects the vibration, the running smoothness or the functioning of the machine. The value of residual unbalance at this threshold establishes the maximum G-grade quality level for that pump. Arbitrarily reducing the maximum allowable residual unbalance determined in this manner by 25% for an additional factor of safety is not required but should be considered. It may drop the G-grade quality level. Note in the case of pumps with ball-bearings there is little attenuation in the vibration from the shaft through the bearing and casing. When journal bearings exist the attenuation is much greater and should be taken into account.

Balance quality classification

ISO 1940 has established balance quality (G grades) which permit a classification system. Balance quality may be defined as the product of the specific unbalance and the maximum service angular speed of the rotor[2].

$$G = e\omega$$
where:
e = specific unbalance g/mm / kg [(lb/in.) /lb]
ω = rotative speed, rad/min

$$e = U/m$$
where:
U = permissible residual unbalance, g/mm (lb/in.)
m = rotor mass, kg (lb)

Experience has shown that the permissible specific unbalance varies inversely as the speed, and rotors of the same type share the same maximum permissible G-grade quality levels. The statistical empirical data for rotors of the same type point to the above relationship $e\omega$ = constant. it can also be argued on the basis of mechanical similarity that geometrically similar rotors running at equal peripheral speeds, have the same stresses in their rigid rotors and bearings. The balance quality grades are based on this relationship. Figs 14.2 and 14.3 are from ISO 1940. Fig. 14.2 is the basic figure in metric units and Fig. 14.3, the derived figure in USCU. Each balance quality G grade comprises all the

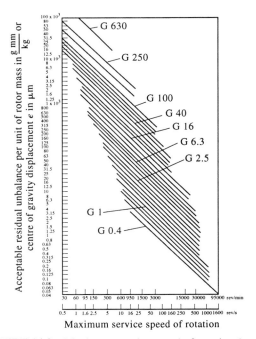

FIGURE 14.2 – Maximum service speed of rotation (metric).

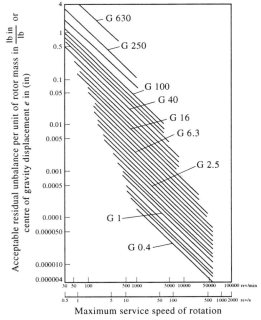

FIGURE 14.3 – Maximum service speed of rotation (USCU).

BALANCING

permissible residual unbalances from zero to the maximum at the left of the G curve. These are plotted against maximum operating service speed N and allowable residual unbalance U. The G-grades shown are separated from each other by a factor of 2.5. Actually any G-grade in between can be meaningfully specified. (Note! The G-grade absolute value is meaningful only in Fig. 14.2 and is the metric value in mm/s. It is not converted to U S units in Fig. 14.3).

Balance quality G-Grade selection

If a line of pumps has had the maximum allowable unbalance determined experimentally in accordance with ISO 1940, then these G-grades could be specified. A reputable manufacturer would be balancing to these grades from internal specifications. In the absence of experimental determination, ISO gives recommended G-grades for different types of machinery, Table 1. It recommends G6.3 for pumps. NFPA-20 (National Fire Protection Association) also calls for G6.3 and HI will be recommending it. However, as with all general types of recommendations this default table should be weighed carefully with the type of pump, its relative rigidity and the application considered. It is the writer's experience with experimentally determined limits of several lines of pumps that G6.3 would only be required in 15% of the cases. In the rest the experimentally determined grades are higher. This means that more correction time and cost is involved than necessary in 85% of the cases using G6.3. Even so, in most cases this is what the user's are calling for and what they are getting. This should result in a reduction in vibration due to unbalance. However, consider the case of a hard-metal slurry pump. These units are designed with very rugged frames and bearings because the application will cause wear to progress very rapidly. The cost to balance to G6.3 compared to G25 would, because of the material hardness, be very expensive, completely unjustified and unnecessary. There has been a steady trend of improved sensitivity capability of balance machines, both in the single plane balancing of thin rotors and the plane separation of two plane (wider) rotors. As this trend continues, either the range of the rotor weights handled on the same balancer

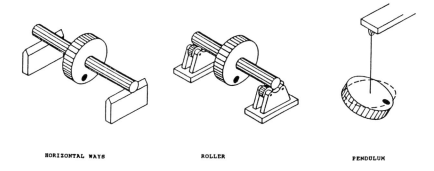

HORIZONTAL WAYS   ROLLER   PENDULUM

FIGURE 14.4 – Three methods of static balance

TABLE 14.1

Balance quality grade G	$e\omega^{1,2}$ mm/s	Rotor types — *general examples*
G 4000	4000	Crankshaft drives[3] of rididgly mounted, slow marine diesel engines with uneven numbers of cylinders[4]
G 1600	1600	Crankshaft drives of rigidly mounted, large, two cycle engines
G 630	630	Crankshaft drives of rigidly mounted, large four-cycle engines Crankshaft drives of elastically mounted, marine diesel engines with uneven numbers of cylinders[4]
G 1600	1600	Crankshaft drives of rigidly mounted, large, two cycle engines
G 630	630	Crankshaft drives of rigidly mounted, large four-cycle engines. Crankshaft drives of elastically mounted, marine diesel engines
G 250	250	Crankshaft drives of rigidly mounted, fast four-cylinder diesel engines[4]
G 100	100	Crankshaft drives of fast diesel engines with six or more cylinders[4] Complete engines (gasoline or diesel) for cars, trucks and locomotives[5]
G 40	40	Car wheels, wheel rims, wheel sets, drive shafts Crankshaft drives of elastically mounted, fast, four-cycle engines (gasoline or diesel) with six or more cylinders[4] Crankshaft drives for engines of cars, trucks and locomotives
G 16	16	Drive shafts (propellor shafts, cardan shafts) with special requirements Parts of crushing machinery Parts of agricultural machinery Individual components of engines (gasoline or diesel) for cars, trucks and locomotives Crankshaft drives of engines with six or more cylinders under special requirements

BALANCING

Balance quality grade G	$e\omega$ [1,2] mm/s	TABLE 14.1 – continued Rotor types — general examples
G 6.3	6.3	Parts of process plant machines Marine main turbine gears (merchant service) Centrifuge drums Fans Assembled aircraft gas turbine rotors Flywheels Pump Impellers Machine-tool and general machinery parts Normal electrical armatures Individual components of engines under special requirements
G 2.5	2.5	Gas and steam turbines, including marine main turbines (merchant service) Rigid turbo-generator rotors. Rotors Turbo-compressors Machine-tool drives Medium and large electrical armatures with special requirements Small electrical armatures Turbine-driven pumps
G 1	1	Tape recorder and gramophone drives Grinding machine drives Small electrical armatures with special requirements
G 0.4	0.4	Spindles, disks and armatures of precision grinders Gyroscopes

[1] $w = 2pn/60^a$ $n/10$, if n is measured in rev/min and w in rad/s.

[2] In general, for rigid rotors with two correction planes, one-half of the recommended residual unbalance is to be taken for each plane; these values apply usually for any two arbitrarily chosen planes, but the state of unbalance may be improved upon at the bearings. For disk-shaped rotors, the full recommended value holds for one plane.

[3] A crankshaft drive is an assembly that includes the crankshaft, flywheel, clutch, pulley, vibration damper, rotating portion of connecting rod, etc.

[4] For the purposes of this International Standard, slow diesel engines are those with a piston velocity of less than 9 m/s; fast diesel engines are those with a piston velocity of greater than 9 m/s.

[5] In complete engines, the rotor mass comprises the sum of all masses belonging to the crankshaft drive described in note 3 above.

can be increased with the same G-level capability, or the G-level capability of the balancer over the initial range of rotor weights can be improved. The G-grade specification of 2.5 for an impeller, is now attainable with most new balancers but may be of little value if the shaft shift in the impeller bore causes a significant relative unbalance. HI has determined that practical looseness limits for clearances between arbors and impeller bores are as follows:

These limits while a practical limit for arbors will not assure a G6.3 residual unbalance for a impeller to be run at 3600 rpm when it is removed from the balance arbor and mounted on the pump shaft. In this case, a G10 specification or rotating assembly balance is recommended. Threaded impellers or tapered shafts should meet the G6.3 specification and possibly a G2.3. If that is required. It is the writer's recommendation that balance grades below G6.3 not be specified without good reason, and if necessary that a rotating element balance be called out.

Maximum looseness (Diametral)

Rotor Journal Dia. in.	≤ 1800 rpm	1800-3600 rpm
0.1.499 in.	0.0015 in.	0.0015 in.
1.5 – 1.999 in.	0.0020 in.	0.0015 in.
2.0 and over	0.0025 in.	0.0015 in.

Types of Balance

Single plane

A static or force balance for relatively thin, lower specific speed impellers can be done by spin balancing or on static rails or disc. Fig. 14.4 shows three methods of static balance. Fig. 14.5 shows a thin, low specific speed impeller being spin balanced.

FIGURE 14.5 – Thin impeller on balance machine.

BALANCING

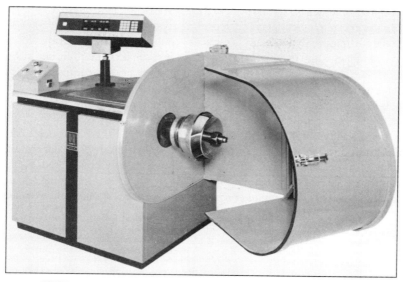

FIGURE 14.6 – Wide, high specific speed impeller on balancer.

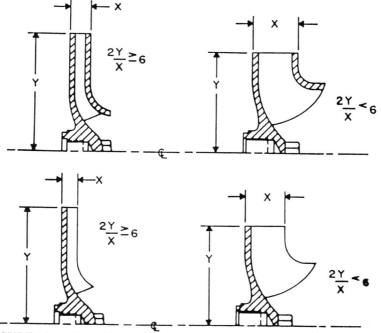

FIGURE 14.7 – Rule of thumb for balancing thin, open or closed impellers.

FIGURE 14.8 – Rule of thumb for balancing wide, open or closed impellers.

Two plane

A balance process which brings the principal axis of the rotor, that may or may not intersect the rotational axis into essential coincidence[2]. If a rotor is balanced in two planes then by definition it is balanced dynamically. Practically, however, if the planes are close enough to be considered one, the residual unbalanced couple can be negligible, such as with the impeller shown in Fig. 14.5. Fig. 14.6 shows a higher specific speed impeller undergoing a two plane balance on a balancer. A common rule of thumb for determining whether a two plane balance is required is to call for single plane balance when the outside diameter of the impeller is six or more times the width (including the shrouds) at the o.d., and a two plane balance if the ratio is less. This rule is followed by ANSI B73 for chemical service pumps. As the ability of balance equipment to work with closer plane separation over a wide range of rotor sizes improves the ratio will undoubtedly be lowered. Figs 7 and 8 show these rules applied to both open and closed impellers. Any rotational balance must be done at a speed that is sufficiently high to reach the sensitive range of the balancer and low enough to avoid windage effects[2].

Rotor assembly balance

It is possible to balance a complete rotor between bearings. This is generally not necessary on pumps that operate at /or below 3600 rpm unless G-grades below 6.3 are called for. The impeller is generally the significant mass and the significant contributor to centre of gravity shifts from the rotational axis. Rotor assembly balance should call for half height keys where necessary. The corrections for the assembly balance should be done at the largest diameter practical to minimize the weight corrections. All parts should then be match marked so that they can be reassembled in the same way after the next disassembly. A rebalance of the assembly will still be required if the impeller has a clearance fit with the shaft.

The vast majority of the pumps built have rigid rotors so the complex balancing process for flexible rotors will not be covered here. A flexible rotor is defined as one whose first critical speed is below the design rotative speed.

References

[1] "The Balance of Rotating", Rigid Bodies, ISO (1973).
[2] Muster and Stadelbauer, "Balance of Rotating Machinery" Shock and Vibration Handbook, 3rd Edition, McGraw Hill.

SECTION 6

Drives

ELECTRIC MOTORS

TURBINES, ENGINES, GEARS, V-BELT DRIVES
AND COUPLINGS

VARIABLE SPEED DRIVES AND SPEED CONTROL

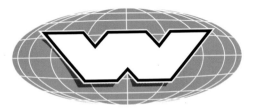

HIGHLY EFFICIENT

Brook Crompton's W range has been designed to ensure high efficiencies across all loads to reduce running costs.

- **Manufactured to BS EN ISO 9001.**
- **Low Noise.**
- **Good Starting Performance.**
- **High Efficiency at all loads.**
- **Same motor suitable for 380 or 415 volts.**
- **Multi-Mount Adaptability.**
- **Suitable for Hostile Environments.**
- **Weatherproof/Hoseproof/Dustproof to IP55.**
- **Mechanical Strength.**
- **Ease of Installation.**
- **High Specification, Great Value.**

POWERFUL CONNECTIONS

St Thomas' Road, Huddersfield HD1 3LJ, England. Tel: (0) 1484 422150. Fax: (0) 1484 516873.

ELECTRIC MOTORS

AC motors

AC motors are either *Synchronous* or *Induction* units. The Synchronous motor operates in synchronism with the power frequency. It is basically a constant speed device. There are two types: Self-excited and d.c.-excited. The Self-excited units are made up of three different constructions: *Hysteresis* units are fractional kw and hp units used for servo motors, *Permanent Magnet* units are available up to 4 kW (5 hp) and Reluctance units are available up to approximately 20 kW (30 hp). Reluctance and permanent magnet motors are used on adjustable speed drives where exact speed matching is required. The d.c.-excited motor requires slip rings for excitation. A starter with full motor protection and ability to apply the d.c. excitation at the proper moment is required, in addition to the d.c. supply. These units are 20 kW (30 hp) and above. They are not generally used in VFD applications. The most distinct advantages of synchronous motors are their higher efficiency (1% higher than a comparable induction motor) and the ability to provide power factor correction. The Induction motor is the most robust and simple of the two and is extensively used throughout all industries. It, by the nature of its design, develops torque through slip, the difference between the synchronous frequency and the actual rpm.

$$Ns = 120 \ f/p$$
where N = synchronous rpm
f = frequency
p = poles

$$Na = N_s - N_{sl}$$
where Na = actual rpm
Nsl = slip rpm

Slip rpm varies from approximately 3%-4% in the fractional sizes to less than 1% above 50 kW (75 hp). The poles can be 2, 4, 6, 8, 10, 12 etc. Speeds VS poles for 50 Hz and 60 Hz power are shown in Table 1.

The most common type of induction motor is the one with the squirrel cage winding. So called because the combination of bars at the o. d. of the rotor and the end rings they

TABLE 1 – Synchronous speeds vs poles

Poles	50Hz	60Hz
2	3000	3600
4	1500	1800
6	1000	1200
8	750	900
10	600	720
12	500	600
14	416	514
16	375	450
18	335	400
20	300	360

are connected to look like a squirrel cage. As the rotating magnetic field moves past the rotor bars, it induces current into the bars. The shape of these bars has a direct influence on the shape of the motors starting torque curve. There are four NEMA classifications of motors based on starting torque characteristics: NEMA Designs A,B,C and D. Design A, Fig. 15.1 units are generally engineered for specific applications. The breakdown torque is higher than the more common design B. Design B motors, Fig. 15.2 are the *de facto* standard units. Design C motors are the next most popular units, Fig. 15.3. Design D units, Fig. 15.4, have high starting torque and are available in two higher slip ranges, 5-8% and 8-13%. A typical shape of the rotor bars is shown next to the respective curves. These shapes are adjusted in each specific case by the manufacturer. A typical motor starting torque curve is shown in Fig. 15.5 along with a pump system head curve and definitions of the various torques involved. *Locked rotor torque* is the minimum torque a motor develops at rest for rated frequency and voltage. *Pull-up torque* is the minimum

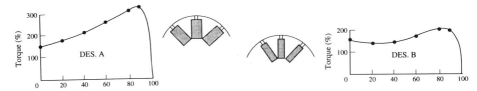

FIGURE 15.1 – Design A floating torque and motor bars. **FIGURE 15.2** – Design B floating torque and motor bars.

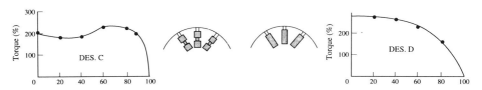

FIGURE 15.3 – Design C floating torque and motor bars. **FIGURE 15.4** – Design D floating torque and motor bars.

EURODRIVE...
removing the gamble

It looks as though our competitors have had their chips in the race for U.K. market leadership.

The combination of a product range whose reliability is legendary, the ability to offer genuine delivery within 24 hours of receiving customers' instructions and impressive after sales back-up, has taken the gamble out of buying power transmission equipment for over 9,000 U.K. customers.

**FOR FAST SERVICE AND A SAFE BET
CALL (01924) 893855**

S.E.W.-EURODRIVE
INTERNATIONAL POWER TRANSMISSION ENGINEERS

Beckbridge Industrial Estate, Normanton, Yorkshire WF6 1QR.
Telephone: 01924 893855 Fax: 01924 893702
24 HOUR EMERGENCY SERVICE HOTLINE 01924 896911
SOUTHERN OFFICE TEL: 0181 458 8949. MIDLAND OFFICE TEL: 01527 877319.
SCOTTISH OFFICE TEL: 01786 478730.

GROW WITH EURODRIVE – SUCCESS BREEDS SUCCESS – WORLDWIDE

WORLD PUMPS
INCORPORATING PUMPS POMPES PUMPEN

ISSN 0262-1762

The official magazine of *EUROPUMP*, the European Association of Pump Manufacturers

Providing a decisive source of practical and technical information on pumps and seals to thousands of

- specifiers
- designers
- users of pumps, seals, valves and motors
- engineers
- consultants

Every month, *World Pumps* brings you

- top quality feature articles
- the latest business and financial news
- product news
- coverage of major exhibitions
- reviews of relevant literature

and the unique Product Finder, to help you select the products and services that suit you exact requirements, and much more ···

For industries including

- building services & construction
- offshore oil & gas production
- corrosion protection
- water & sewage treatment
- irrigation & drainage
- chemicals & biotechnology
- food & beverages

Your subscription will also include

- The International Directory of Pumps and
- The International Directory of Seals and Sealing Products

For Your FREE sample copy of **World Pumps** please complete the following

Name .... Position .... [JT]
Organization ....
Address ....
State .... Post/Zip Code ....
Country ....
Tel .... Fax ....
Nature of business .... [SIC:]

WORLD 3D   [PIP10 + A3]

and post or fax this page to
Elsevier Advanced Technology
P O Box 150, Kidlington,
Oxon OX5 1AS, UK
Tel: +44 (0)865 843848,
Fax: +44 (0)865 843971

[AD94]

ELSEVIER ADVANCED TECHNOLOGY

"The heavyweight journal of the industry"
THE DAILY TELEGRAPH

ELECTRIC MOTORS

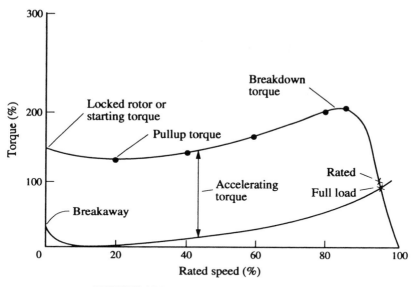

FIGURE 15.5 – Motor torque and system curves.

torque developed from locked rotor condition to breakdown torque condition during start up. *Breakdown torque* is the maximum torque that the motor develops.

The other type of AC Induction Motor is the *Wound rotor motor.* These units are variable speed and variable torque units. They are started with a secondary resistance in the rotor. As this resistance is reduced the speed increases and the unit comes up to full speed. The secondary resistance can be designed to dissipate the heat resulting from operation at reduced speeds down to 50% of full load speed.

Squirrel cage induction motors can be wound to give two, three or four different speeds. They can be started in several different ways, as listed in Table 2 along with the corresponding voltages and torques.

TABLE 2 – Squirrel cage induction motor starting methods

Starter Type	Percent Full Voltage		
	Motor Voltage	Line Current	Motor Torque
Full Voltage(x-line)	100	100	100
Autotransformer .8 tap	80	64	64
.65 tap	65	42	42
.50 tap	50	25	25
Primary Reactor .80 tap	80	80	64
.65 tap	65	65	42
.50 tap	50	50	25
Primary Resist. (typ.)	80	80	64
Part Winding (1/2-1/2)	100	70	50
Wye Start – Delta-Run	100	33	33

Motors may also be started by variable frequency as covered in the chapter on VFD's.

Common insulation systems today are classes A, B, F and H having temperature limits of 90, 130, 155 and 180°C respectively. Motor insulation life is generally considered to be halved for every 10°C rise up to these limits. In addition to thermal device protection provided in the starting equipment motors may also have thermistors embedded in their windings in the areas of anticipated hot spots. A partial list of motor enclosures follows:

1. *Open.* Permits air circulation over the windings without any special restrictions.
2. *Open drip-proof.* An open machine with air openings designed in such a way that any liquids or solids falling on the motor at a 15° angle from the vertical will not pass through the openings.
3. *Splash proof.* An open motor designed to prevent liquids or solids from impinging directly on the stator and rotor if they come toward the unit at any angle up to 100° from the vertical.
4. *Open, externally ventilated.* A separate motor driven blower, piggy-back mounted on the machine is used for air circulation.
5. *Weather Protected-Type I.* An open motor with air circulation openings designed to minimize the entrance of rain, snow and other airborne particles to the electrical parts. The openings are arranged to prevent the entry of a 19 mm (3/4 in.) rod.
6. *Weather Protected-Type II.* Same as Type I except that the intake and discharge passages are designed to discharge high velocity particles ingested without them entering the electrical parts area.
7. *Totally Enclosed.* This motor is designed without openings, but is not air or liquid tight.
8. *Totally Enclosed, Fan Cooled.* This machine has a fan, connected to the motor shaft, that circulates air over the outside of the motor casing.
9. *Explosion proof.* This unit is designed to withstand an internal explosion of gas or vapour and to prevent ignition by that internal explosion of gases and vapours outside of the motor.
10. *Submersible.* A totally enclosed motor equipped with sealing features to permit operation while submerged in a specified medium at a specified depth.

TURBINES, ENGINES, GEARS, V-BELT DRIVES AND COUPLINGS

Turbines

Turbines are often used as pump drivers where steam is readily available. Even gas turbines are occasionally used to drive larger pumps where a fuel supply is available. Steam is often available in utilities and in industrial applications where steam is being generated on premises. In such applications steam turbines should be considered as one of the alternatives. Speed control is inherently available and the cost savings of using it instead of a discharge throttling valve can be very significant. Power outages need not affect the operation of the steam turbine, controls being the only probable tie in to the electric power system. Steam turbines are non-sparking in nature and as long as the steam temperatures are below the gas or vapour flash point they would be safe in such applications that require explosion proof considerations. They have adequate starting and running torque to drive all types of pumps.

IC engines

IC engines are being used in large numbers fire pump applications. The stand-by availability of diesel engines for fire pump service in the event of power failures is common for that purpose and thousands are being furnished each year. However, their use is not limited to that application. Lower speed diesel engines are being furnished for continuous operation in many applications. Their efficiency (35 to 40%) brings them close to being cost competitive from a operating standpoint with electric power drives. Maintenance costs have been a problem area, but increased maintenance contract availability from the manufacturers and competent repair shops is changing that. The diesel engine lends itself to speed control. So cost savings generated can make a difference. In the absence of adequate or reliable electric power in some areas, they are the logical choice. Gas engines are available for stand-by, light duty and portable installations but the fuel consumption is much greater and operating costs can be 3 to 4 times that of the diesels. The same engine can be sold with intermittent or continuous ratings by the use of a de rating factor. Intermittent engines have a 90% factor and continuous an 80% factor by which the basic rating is multiplied. There are other site related deration factors that must be considered also. Elevation considerations call for a deration of 3.5% for every 300 m (1000 ft.) above sea-level and temperature considerations call for a deration of 1% for every 5.6°C (10°F) above 15°C (60°F). This may vary globally. The standard rotation direction of an IC. engine is counter-clockwise when looking at the flywheel end of the engine. Special care must be taken in the foundation design for IC. Engines to avoid resonance and excessive vibration conditions from the dynamic loads and elastic soils[1].

V-belt drives

V-belt drives are extensively used in pump applications. They are probably the lowest cost of all the drive alternatives. They involve dual sheaves, generally of adjustable pitch which are selected based on the centre distance between the two shafts. 120 kW (150 hp) drives are common. The drives may be mounted horizontally or vertically and at a 45° angle. Speed ranges of 8.5-1 are available. For speed increasing services the belt drives are often combined with gear sets. Allowance must be made in the driver size for the losses in the belt drive.

Gears

Gears are generally used together with constant speed motors or engines to step the driver speed down to the pump speed requirement and less frequently to step it up. They are often the best choice for ratios above 6:1. Spur, helical and herringbone gearing are used for parallel shafts. The helical is preferred. On a first cost basis, it is more expensive but its performance characteristics which include having more than one tooth in engagement at a time and significant noise reduction over the spur can justify it. The spur is generally seen in the low power applications. Bevel and hypoid gearing are used for right angle drives. One caution on gear losses that must be considered is that a gear loss of 1.5-2% at rated load may only drop 0.5% at minimum load. This means that the gear losses can become a significant percentage of the pump load at the minimum operating conditions.

Couplings

Couplings are essential elements of pump drive systems. Their main function is to transmit power from the output shaft of the driving machine to the input shaft of the driven machine. Secondary functions are to accommodate thermal growth of one or both of the shafts and axial shaft motion from other origins and misalignment between the two shafts. There are two broad categories, solid (rigid) and flexible. The *solid coupling* is used when precise alignment of shafts must be maintained and where the rotating assembly of one machine is used to support the rotating assembly of the other machine. The coupling requires that the shafts be in perfect alignment before the coupling is installed and that they are not prone to go out of alignment. It is used for example between sections of line shafting. The *flexible coupling* makes up the preponderance of couplings used and comes in many forms and shapes. Figs 16.1-16.4 show four of the most popular types of couplings. Table 1 is a comparison chart showing the relative merit of each of the types for various features that are listed and an explanation in the remarks column. *Spacer couplings*, Fig. 16.5 are used

TURBINES, ENGINES, GEARS, V-BELT DRIVES AND COUPLINGS

FIGURE 16.1 – Steelflex coupling.

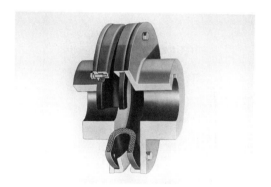

FIGURE 16.2 – Torus coupling.

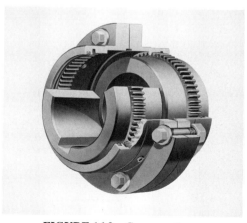

FIGURE 16.3 – Gear coupling.

230 PUMP USERS HANDBOOK

FIGURE 16.4 – Disc coupling.

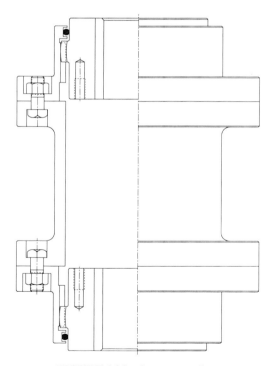

FIGURE 16.5 – Spacer coupling.

TURBINES, ENGINES, GEARS, V-BELT DRIVES AND COUPLINGS

TABLE 1 – Coupling comparison chart

Key: empty box = least favourable; full box (XXXX) = most favourable. Ratings below are shown as number of X's (0 = least favourable, 4 = most favourable).

Feature	Steelflex – Fig.1	Torus – Fig.2	Gear – Fig.3	Disc – Fig.4	Remarks
Torque range	XXX	X	XXXX	XX	Gear cplgs. pack a lot of punch in a small package while elastomer cplgs. get prohibitively expensive in large sizes
Bore range	XXX	X	XXXX	XX	An indication of the breadth of the line
Speed capability	XX	X	XXX	XXXX	Because of closely machined & fitted components disc & gear couplings have little inherent vibration
Misalignment capacity	X	XXXX	XX	XX	Steelflex is rated for 1/4°, gear & disc about 3/4°, and Torus is still higher
Temperature range	XXX	X	XX	XXXX	Torus is limited to 150°F. While a disc coupling can handle 450°F
Torsional softness	XX	XXXX			Allows windup to dampen shock loads and vibrations
Torsional stiffness	XX		XXXX	XXXX	Disc couplings allow for exact synchronization of equipment and can raise natural frequencies
Ease of installation	XX	XX	X	X	Includes types of fits typically utilized, number of components, lubrication and alignment
Ease of service	XXX	XXX		X	Degree of difficulty in disassembly and replacement of wearing component
Minimum maintenance	XXX	XXX		XXXX	Frequency of routine service, parts replacement and relubrication
Initial price < 4" shafts	XXXX	XX	X		Steelflex is usually the most economical thru 200 HP motor applications
Initial price > 4" shafts	XX		XXXX	X	Gear couplings are frequently used on L.S. shaft connections due to their low price
Spare part costs	XXXX	XX		X	Spare part cost for wearing equipment
Lifetime operating costs	XXXX	XX		XXX	Considers initial price & labor, spare part cost & labor plus frequency of routine maintenance
Life expectancy	XX	X	X	XX	A measure of durability & rugged performance over the long haul
Stock bore availability	XXXX	X	XXXX	X	Number of stock bores carried in inventory
General availability	XXXX	X	XXX	X	General marketplace availability thru distribution channels and speed of response
Minimum radial loads	XX	XX		XXXX	Forces exerted on bearings due to misalignment
Minimum bending moment	XX	XX	XX	XXXX	Forces exerted on bearings due to misalignment
Generated axial forces	XX	X	XXX	XXXX	Forces produced by the coupling which act on connected shaft bearings
Transmitted axial forces	X	XX		XXXX	Force transmitted thru the coupling

where space is needed for disassembly or where a significant improvement in misalignment capability is desired. Couplings for internal combustion engine driven pumps should be selected after a torsional vibration analysis has been performed due to the pulsating torque that exists with these drivers. This analysis can determine just how torsionally stiff the coupling should be.

References

[1] Newcomb, K., "Principles of Foundation Design for Engines and Compressors", Transactions of ASME (April 1951).

VARIABLE SPEED DRIVES AND SPEED CONTROL

Speed control

The difference between throttling and variable speed is akin to having a high wall of fixed or ever increasing height with a pile of coal as high on one side and having to shovel the pile over to the other side of the wall, in the case of throttling to one of a variable height wall that stays the same height as the top of the coal in the case of the variable speed. In the first case there is a lot of energy wasted and the lower the pile of coal the greater the wasted effort. In the second case the wasted effort is minimized because the wall is never in the way. For example if you take a constant speed HQ characteristic with the unthrottled system curve as shown in Fig. 17.1, the pump is operating at point 1. Now if decide you want to drop the head and capacity to point 2 you must throttle back on the discharge valve until the pump moves back on the HQ curve and gets to point 3. The head at point 2 is the height of the coal but h_2-h_2 is the height of the wall above the coal that represents wasted energy or head. Now look at Fig. 17.2 and you can see that with speed control N_2 gives you

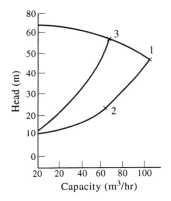
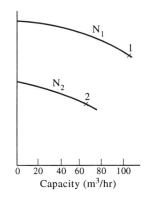

FIGURE 17.1 – Throttling capacity control.   **FIGURE 17.2** – Variable speed capacity control.

a pump operating point directly at point 2. Not only have you saved on energy but consider the benefit in terms of the reduction in recirculation forces on the pump's noise and vibration and potential reduction in damage. The throttling valve is no longer required.

Variable speed drives

In addition to the variable speed capabilities of turbines and engines covered in the last chapter, there are other types such as fluid drives, eddy current drives, adjustable voltage a.c. drives, wound rotor motors, variable frequency drives, and d.c. motors. All of these are in use today on pump speed control.

Fluid drives

Fluid drives offer a smooth speed reduction to as low as 15% of the rated speed of the centrifugal pump, (in the case of positive displacement pumps the low limit may be 30%). They provide soft start acceleration. The main drawbacks are cost, large footprint, and the losses. Losses can be in the nature of 3-4% at rated speeds but as speed decreases the slip losses are proportional to the slip/rated speed ratio or stated another way, efficiency = output speed/input speed. This is akin to throttling.

Eddy current drives

Eddy current drives rely on the generation and decay of an electromagnetic field between components in the driver and driven rotors. Torque is proportional to slip, so the driven unit comes up to a speed where the torque matches the load. Losses are close to that of the fluid drives above with efficiency = output speed/input speed.

Adjustable voltage a.c. drives

Adjustable voltage a.c. drives consist of a high-slip motor (10%), constant frequency adjustable voltage control system. They are available in capacities up to 60 kW (80 hp) but the variable frequency drive is emerging as the most preferred method.

Wound rotor induction motors

Wound rotor induction motors together with a rheostat or other solid state secondary resistance control provide a means for soft start variable speed control that has found frequent use with pumps, especially in waste water applications. They have been available up to 2250 kw (3000 hp) in all standard voltages. Except in the small sizes, this drive has been the most popular in the past with good functionality and a relatively low first cost.

D.C. motors with SCR's

D.C. motors with SCR's have seen use for speed control of pumps. The power supply provides an adjustable D.C. voltage that flows to the armature and a constant D.C. voltage that flows to the shunt field. The motor speed is a function of the voltage that is impressed across the armature. These drives are relatively stiff compared with other electric drives in that resistance to speed change due to torque is high.

Variable frequency drives

Variable frequency drives have been in use for over ten years. They provide soft starts and very good efficiencies at reduced speeds. Of late, advancements in the technology of these

VARIABLE SPEED DRIVES AND SPEED CONTROL

drives and substantial reduction in manufacturing cost have made them the preferred choice. Standard pre-engineered units are now available up to the 450 kw (600 hp) sizes, with custom engineered sizes available up to three times that high. One source[1] reports that payback periods are now in the eight month to two year range. Keeping a constant voltage to frequency ratio when changing frequency maintains a constant flux in the motor and full load torque capability.

$$V/Hz = constant$$

Holding this ratio, results in the motors power capability changing directly with speed. Since:

kW = T x N/7113 where T = torque, Nm; N = rpm   or
hp = T x N/5250 where T = torque, lb/ft; N = rpm

The current will remain constant as the speed is changed, under these conditions.

VFD controls convert a.c. power to d.c. and then back to a square pulse form of sinusoidal wave through the inverter, which may be Pulse Width Modulation (PWM), Constant voltage Source, 6 Step Variable Voltage (VVI) or Current Source Variable Voltage Technology (CSI). The type of load is an important factor in selecting the VFD, Pumps impose variable torque loads (T α N^2) with torque varying approximately as horsepower as shown in Fig. 17.3 for a system with no static head. Typically, variable torque VFD's will have efficiency in the 96-97% range at full load and drop down to only 86-87% at 15% load, which is a very substantial improvement over throttling and slip drives. They can be overloaded by 110% for one minute. The head savings with speed control by VFD following the system curve VS a throttling valve following the HQ characteristic curve for any given capacity can be seen in Fig. 17.3. Multiplying this by

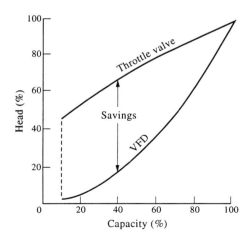

FIGURE 17.3 – VFD head savings vs. capacity.

the VFD efficiency and the time spent at each capacity over a period of time such as a day, will give the power savings over that period.

The displacement power factor of a PWM drive is approximately 0.95 at all points of operation. It is not affected by motor speed, load or power factor. The displacement power factor of a VVI drive is given in Table 1. PWM drives are the most commonly used.

The following should be evaluated when selecting a VFD so that the requirements match the drives capabilities:
- What type of load, i.e. Constant or variable torque?
- Any shock components present?
- Are heavy inertial loads present?
- Motor considerations?
- Speed range for heavy loads?
- Does the system have a static head component?

The following recommendations are from VFD manufacturers:
- Inverter duty motors are now available for VFD's, Conventional motors can be used but they must be selected with some care. High Efficiency Rated motors are preferable.
- For pumps with their variable torque loads use a motor with a 1,15 service factor, but do not use the service factor.
- The Design B motor is best suited for VFD's.
- Startup torque requirements must be watched closely, i.e. positive displacement pumps are likely candidates for a larger size VFD.
- There are elevation and temperature deration factors for VFD selection.
- Isolation transformers may be necessary due to the harmonics generated.

Vertical pump/motor assemblies must be checked for reed critical natural frequencies when applied to variable speed.

There are distinct advantages of VFD/pump drives:
- Significant power savings.
- Soft start.

TABLE 1 – Power factors for VVI drives

Percent Speed	Power Factor
100	.90
90	.85
80	.76
70	.67
60	.58
50	.50
40	.41
30	.32
20	.23
10	.14

VARIABLE SPEED DRIVES AND SPEED CONTROL

- Starting power penalties eliminated or reduced.
- Possible rebates from utilities for power savings.
- No starter or throttle valve required.
- Increased pump life due to reduced recirculation and soft starts.
- On multiple pump control, the need for pump over-rated capacity when fewer pumps are on can be eliminated. In some cases one less pump may be required.

References

[1] Baljevic, P., "Adjustable Frequency Drives for Pumps and Fans" Proceedings Rocon Conference, New Jersey Institute of Technology (Nov. 10-12, 1993).

SECTION 7

Applications

WATER PUMP APPLICATIONS

FIRE PUMPS

CHEMICAL PROCESS PUMPS

FOOD, BEVERAGE AND PHARMACEUTICAL PUMPS

PETROLEUM PRODUCTION, PIPELINE AND PRODUCT PUMPS

PULP AND PAPER PUMPS

SOLIDS AND SLURRY PUMPS

WASTE WATER / SEWAGE PUMPS

WATER IS THE ONE THING NO LIVING CAN DO WITHOUT...

- GRUNDFOS PUMPS MAKE SURE IT'S THERE!

We all depend on access to fresh water, but not all of us have it flowing right in front of our noses.

Grundfos have developed a whole range of pumps especially for extracting fresh water from wells of any size and any depth. Grundfos pumps are designed for years of troublefree operation in the most difficult circumstances. With our solar powered pump systems we can even provide water where there is no electricity or other sources of power.

Contact us if you want to know more about Grundfos pumps for water supply.

GRUNDFOS®
Leaders in Pump Technology

WATER PUMP APPLICATIONS

Water Applications

Water Applications for pumps are just about infinite. Water pumps come in all sizes from *automobile water pumps* to giant circulating pumps for utilities, Fig. 18.1, and almost every type of pump is applicable for some specific application or applications. *Residential well pumping systems* where utility mains are not available are generally of the centrifugal - eductor type described in Chapter 3 or submersible units (Fig. 18.2). Flexible vane pumps have been used in this application also. The *de facto* standard for water pump materials in the range of 6-8.5 pH is cast iron, bronze fitted. Above this range, the hard water (alkaline) requires all iron or stainless fitted pumps. In the soft water (acidic) range below 6.0 pH bronze is not suitable. What is best, depends on the degree of acidity, nature of the acidity, the nature of impurities present, velocity and temperature as well as the design geometry. The latter aspect being especially critical in the presence of tight clearance, stagnant areas. Today, ferritic or duplex stainless steels, high chrome steels, nickel cast irons, Monel,

FIGURE 18.1 – Very large utility circulating water pump.

FIGURE 18.2 – Residential submersible well water pump.

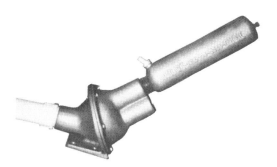

FIGURE 18.3a – Irrigation ram pump.

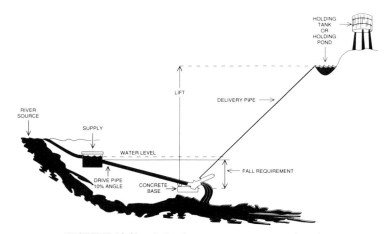

FIGURE 18.3b – Irrigation ram pump system elevation.

VORSPRUNG DURCH IDEEN

Wilo – Pumpen in Perfektion

– der Vorsprung für alle Anwendungen im Gebäude:

- Heizung
- Sanitär
- Schwimmbadtechnik
- Bewässerungssysteme
- Abwassersysteme

Top-E-Wilo mit Control-Modul

Wilo – Komplettleistungen

– der Vorsprung für gemeinsamen Erfolg:

- Qualität und Zuverlässigkeit
- Umweltorientierung
- Innovative Systemlösungen
- Qualifizierte technische Beratung
- Umfassender Service – wie Kundendienst und Verkaufsunterstützung

Wilo-Pumpen für den OEM-Bereich

WILO GmbH · Nortkirchenstraße 100 · D-44263 Dortmund-Hörde
Telefon (02 31) 4 10 20 · Telefax (02 31) 4 10 23 63 · Telex 8 22 697

WILO
Pumpen-Perfektion

36th year of publication: all previous editions sold out

PUMPING MANUAL
9th Edition
By Christopher Dickenson

Section One: Introduction
SI Units
Pump Evolution
Pump Classification
Pump Trends

Section Two: Pump Performance and Characteristics
Fluid Characteristics
Pump Performance
Calculations, Type Number and Efficiency
Area Ratio
Pipework Calculations
Computer Aided Pump Selection

Section Three: Types of Pumps
Centrifugal Pumps
Axial and Mixed Flow Pumps
Submersible Pumps
Seal-less Pumps
Disk Pumps
Positive Displacement Pumps (General)
Rotary Pumps (General)
Rotary Lobe Pumps
Gear Pumps
Screw Pumps
Eccentric Screw Pumps
Peristaltic Pumps
Metering and Proportioning Pumps
Vane Pumps
Flexible Impeller Pumps
Liquid Ring Pumps
Reciprocating Pumps (General)
Diaphragm Pumps
Piston, Plunger Pumps
Self Priming Pumps
Vacuum Pumps

Section Four: Pump Materials and Construction
Metallic Pumps
Non Metallic Pumps
Coatings and Linings

Widely recognised as the first source of reference on all aspects of pump technology and applications the Pumping Manual will enable you to...

Specify the right pump for the task
Design cost-effective pump systems
Understand the terminology
Ensure effective installation, operation and maintenance of all your pumping equipment
and much more!

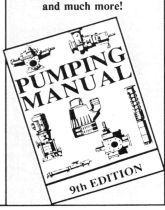

Section Five: Pump Ancillaries
Engines
Electric Motors and Controls
Magnetic Drives
Seals and Packaging
Bearings
Gears and Couplings
Control and Measurement

Section Six: Pump Operation
Pump Installation
Pump Start-up
Cavitation and Recirculation
Pump Noise
Vibration and Critical Speed
Condition Monitoring and Maintenance
Pipework Installation

Section Seven: Pump Applications
Water Pumps
Building Services
Sewage and Sludge
Solids Handling
Irrigation and Drainage
Mine Drainage
Pulp and Paper
Oil and Gas
Refinery and Petrochemical Pumps
Chemical and Process
Dosing Pumps
Power Generation
Food and Beverages
Viscous Products
Fire Pumps
High Pressure Pumps

Section Eight: User Information
Standards and Data
Buyers Guide
Editorial Index
Advertisers Index

800 pages
1500 figures and tables
ISBN: 1 85617 215 5

For further details post or fax a copy of this advert complete with your business card or address details to:
Elsevier Advanced Technology, PO Box 150, Kidlington, Oxford OX5 1AS, UK
Tel: +44 (0) 1865 843848 Fax: +44 (0) 1865 843971

CIP4699

ELSEVIER ADVANCED TECHNOLOGY

WATER PUMP APPLICATIONS

Hastelloy, phosphor bronze and titanium or non-metallics could provide the best specific solution to a given application. The harder the material, the higher the abrasive resistance when abrasive wear or erosion/corrosion is a concern. Sanitary pumps are used for *Clean In Place* applications where water pumps are used for *rinsing*. These pumps can be of centrifugal or any of the rotary designs with the possible exception of the lobe pump. The latter generally depends on higher viscosities to control the leakage past its axial clearances. Small *sampling pumps* in water utility plants are generally small centrifugals, 1-2.5 m^3/h., 5-10 gpm with heads generally less than 10 m (35 ft). Figs 18.3, 18.3a and 18.3b show a ram pump installation for irrigation that will provide approximately 9 m (30 ft.) head for every foot of fall when water enters at a 10° slope.

Non - residential well water pumps used for small utility service and aquifer source agricultural pumps are generally vertical turbine type diffuser pumps as shown in Fig. 18.4. These pumps in the U S come under ANSI-AWWA specification E101-1977, "Vertical Turbine Pumps-Line Shaft and Submersible Types". Submersible volute pumps are also used to some extent in this application where the simplicity of the submersible offsets the higher operating efficiency of the turbine type. On very deep wells, the volute type submersible with its reduced line-shaft lengths becomes the preferred choice. Reciprocating pumps are economically sound choices where relatively high head, low capacity applications are the case. Once the source gets within normal suction lift range of 3-6 m (10-20 ft), the centrifugal volute pump becomes the preferred choice for source or *circulation duty* Fig. 18.5, with the exception of some low capacity applications that

FIGURE 18.4 – Vertical turbine diffuser pump.

FIGURE 18.5 – Centrifugal volute split case circulating water pump

High Pressure Stage Casing Pumps MD
Modern perfected feed pump technology – economic, reliable, efficient

Shaft seals
- Packing or mechanical seal
- Packing and mechanical seal can be interchanged without modification of stuffing box

Efficient axial balance
- By balance piston or balance disc
- Both seals are at suction pressure
- Low residual thrust with balance piston design
- No axial bearing required with balance disc

Safe sealing of casing, even with rapidly changing temperatures
- by confined O-rings
- with operating temperatures above 180 °C by metal – to – metal seals

Several impeller sets per pump size
- Better efficiencies over a wide range
- Lower energy requirement, economical operation

Special suction impeller for 1st stage
- Good NPSH value
- depending on size double-suction impeller 1st stage possible

Large branch sizes
- Ideal flow pattern ensured by positioning tie bolts external to flow chambers
- Low branch velocities, reduced noise levels
- Higher allowable forces and moments

Labyrinth seals
- Contact free sealing
- No entry of spray water into bearing housing

Stuffing box housing
- Coooled or uncooled design
- Easily accessible cooling chamber
- Clearly arranged connections
- Effective, uniform cooling

Extended wear rings
- Assist stabilisation of pump rotor
- Smooth running even on part load and overload

Replaceable wearing parts
- Guarantee constant high efficiency over pump life
- Cost saving on essential repairs and spare parts
- High availability and short down times

Robust shaft
- Critical speed > operating speed
- Small shaft deflection
- Speeds up to 4000 rpm
- Areas subjected to wear are protected

Long-life, robust bearings
Radial bearings:
- splash oil lubricated antifriction bearings
- ring oil lubricated plain bearings
- force feed oil lubricated plain bearings

Thrust bearings:
- force feed oil lubricated plain thrust bearings
- splash oil lubricated antifriction bearings also available

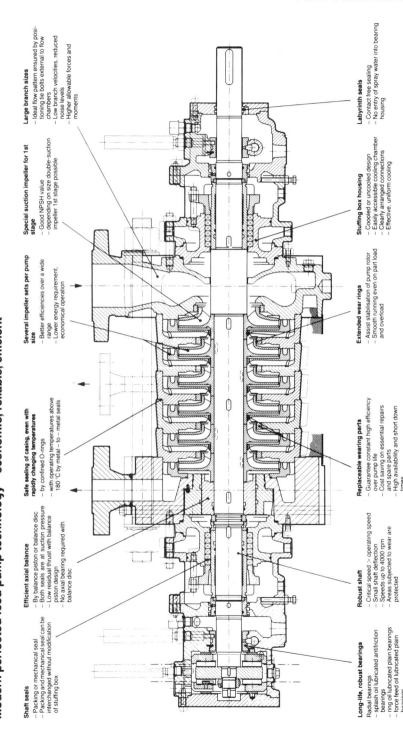

FIGURE 18.6 – Vertically split multistage circulating pump.

WATER PUMP APPLICATIONS 243

favour the reciprocating type. End suction centrifugal and split case double suction, single stage pumps are being used in the lower capacity, low to medium head ranges, up to approximately 3.6 m^3/min (1000 gpm) and 205 m (670 ft) head. Above this range single stage, double suction, split case pumps are used up to 200 m^3/min (60, 000 gpm) and 205 m (670 ft) head. For heads above this, multistage axially split or vertically split, stage Fig. 18.6 or barrel Fig. 18.7 type casings are employed. Fig. 18.8 shows an axially split boiler

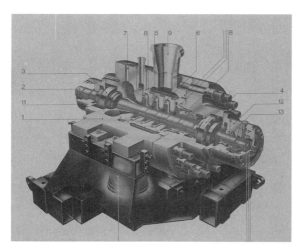

FIGURE 18.7 – Barrel type pump (double casing).

FIGURE 18.8 – Axial split boiler feed pump.

feed pump in a Texas refinery. Fig. 18.9 shows a 14 stage vertically split stage pump in a German industrial power generating plant. Fig. 18.10 shows a 5 stage barrel type boiler feed pump for the 600 MW power station at Maasbracht, Netherlands. Figs 18.11, 18.12 and 18.13 show the main boiler feed pumps at Hong Kong's Castle Peak power station. These pumps are vertically split double casing barrel pumps. Fig. 18.14 shows the central power plant main boiler feed pump in Edmonton, Alberta, Canada. Regenerative turbine pumps can also be used for small relatively low pressure boilers.

FIGURE 18.9 – 14 stage vertically split circulating pump.

FIGURE 18.10 – 5 stage barrel type boiler feed pump.

FIGURE 18.11 – Main boiler feed pump.

FIGURE 18.12 – Main boiler feed pumps.

FIGURE 18.13 – Barrel type boiler feed pump (double casing).

FIGURE 18.14 – Marn boiler feed pump.

For *chilled water and cooling tower (hvac condenser) water circulation* duty in commercial and office buildings, the end suction and double suction units are again the favoured pumps with the end suction being used up to a maximum of about 3.5 m^3/min (92.4 gpm). Even below this range the vertical or horizontally split units are often specified because of their increased ruggedness. Fig. 18.15 shows a two stage vertical turbine pump used on a cooling water circuit. Pumps for *condensate services* are generally centrifugal, Fig. 18.16, but reciprocating units have been used for this service. Condensate and cooling tower pumps have low NPSHR requirements that must be satisfied. *Heating pumps* for

WATER PUMP APPLICATIONS

247

FIGURE 18.15 – 2 stage vertical turbine diffuser pump.

FIGURE 18.16a – Vertical condensate pump.

FIGURE 18.16b – Vertical condensate pump.

industrial, commercial and office buildings run the capacity range of very small to large capacity single stage centrifugal units of the same types covered under chilled water service. Heating pumps for residences are of the mass produced small centrifugal type such as shown in Fig. 18.17. A small capacity, relatively high head pump for *industrial type circulating* applications is the fabricated centrifugal pump shown in Fig 18.18. This pump is ideal for applications where very narrow, radial pump passages (low specific speed) are required. This restricted geometry is a problem with cast impellers, due to the poor surface finish and possibility of fins. These pumps, in the small size ranges are produced on a high quantity basis, generally in stainless construction and provide higher efficiencies than could be attained with the cast units. They should not be applied in

FIGURE 18.17 – Residential centrifugal hot water circulating pump.

FIGURE 18.18 – Fabricated industrial type circulating pump.

WATER PUMP APPLICATIONS

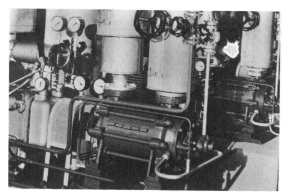

FIGURE 18.19 – 7 stage central heating circulating pump foåçr residences.

FIGURE 18.20 – Mixed flow vertical turbine agricultural sprinkling pump.

FIGURE 18.21 – Axial flow propeller agricultural sprinkling pump.

services where water hammer or severe discharge or suction cavitation could exist. A 7 stage pump for central heating supply of residences in Germany is shown in Fig. 18.19. *Residential and commercial lawns* utilize centrifugal volute or vertical turbine pumps for sprinkling.

Agricultural sprinkling or circulating pumps can be of the centrifugal, mixed flow (Fig. 18.20) and axial (Fig. 18.21) types. Fig. 18.22 shows some vertical dry pit pumps with line

FIGURE 18.22 – Vertically oriented dry pit pumps on irrigation service.

FIGURE 18.23 – Single stage axial flow pump on irrigation service.

WATER PUMP APPLICATIONS

FIGURE 18.24 – Double suction split case pumps for irrigation.

shafts for irrigation supply in Loukkos, Morocco. Fig. 18.23 shows a single stage axial flow pump with adjustable propeller for irrigation project in El Amoun, Egypt. Fig. 18.24 shows double suction split case pumps in the Monduran Dam pumping station in Queensland, Australia supplying irrigation water to the sugar cane industry. *Main utility supply pumps* are shown in Figs 18.1, 18.25, 18.26 and 18.27. Fig. 18.25 shows a single

FIGURE 18.25

FIGURE 18.26 – 3 stage axially split supply pumps.

FIGURE 18.27 – Vertical turbine wet pit pumps.

stage, double suction pump which is one of 35 in the Cutzamala supply system in Mexico City. Fig. 18.26 shows three 3 stage axially split pumps at CEASB, Brazil. Fig. 18.27 shows vertical turbine wet pit pumps for the Alto Tiedte water supply system in Sao Paulo, Brazil. End suction and split case units are used for *fire protection systems* such as

WATER PUMP APPLICATIONS

sprinklers, hydrants and hose lines. See the chapter on fire pumps for further discussion of these units. *Flood control pumps* are large mixed flow or axial flow pumps used for moving large quantities of water over a low static height (levee). See Fig. 18.28, which is a 42 in. axial flow, propeller pump on the Pearl River in Mississippi, USA at the East Jackson Pump Station. COS are 67, 000 gpm, 15 ft. TDH, 600 hp and 435 rpm. A *residential submersible sump pump* is shown in Fig. 18.29. Fig. 30 shows a progressing cavity pump on a clear water application. A primary circulating pump used in pressurized reactors operating at 75, 000 USGPM, 2000 psig and 540°F is shown undergoing its 500 h test in Fig. 18.31. There are many other water applications, but the types of pumps used are the same.

FIGURE 18.28 – 54" axial flow flood control pump.

FIGURE 18.29 – Residential submersible sump pump.

FIGURE 18.30 – Progressing cavity pump on clear water.

WATER PUMP APPLICATIONS

FIGURE 18.31 – Pressurized reactor primary circulating pump.

FIRE PUMPS

Fire pumps

Fire pump is a term applied to pumps that provide the water to fire protection systems in buildings in case of a fire and/or to a fire hose main. The codes that must be met for these pumps are the NFPA-20[1] Code of the National Fire Protective Association in the case of North America which is enforced and interpreted through Factory Mutual, FM and Underwriters Laboratory (ULI[2]). The ULI document is UL 448. Many fire pumps for other countries around the world are ordered to these specifications. Lloyds of London has a fire pump code that is recognized in many parts of the world. Country or local governments may specify the code or listing agency that must be complied with or the insuring company may be the one that specifies the code that must be met and or the listing agency, e.g. ULI or FM. The latter two provide listings of approved fire pumps. They witness compliance tests to their specifications on prototype pumps and do surveillance testing on a periodic, unscheduled basis for compliance.

The main requirements of these specifications are as follows:
- The pump must be rated at one of 21 ratings ranging from 25 to 5000 gpm (95-18925 l/min) in North America and at a pressure not less than 40 psi (276 kp$_a$) at zero suction pressure.
- The pump must produce 150% more than rated capacity at not less than 65% of the rated head with total shut off head not exceeding 140% of the rated head. No droop is allowed in the characteristic curve when descending in capacity to shut off flow.
- A horizontal pump must be capable of a 15 foot (4.5 m) lift at 150% capacity but is not to be applied with a suction lift requirement.
- The Suction pipe for single or multiple pumps shall be sized such that when all pumps are running at 150% capacity, the gauge pressure at the suction flange of the pump shall be no less than 0 bars (0 psig).
- The pump must be hydrostatically tested at 150% of the sum of its shut off head plus its maximum allowable suction pressure, but in no case less than 250 psi (17 bars), for a period of not less than 5 minutes. In the case of vertical turbine pumps the discharge casing and the pump bowl assembly shall be tested.

- Motors shall not be overloaded as follows:
  Those with 1.15 service factor no greater than 115%.
  Those with a 1.0 service factor, no greater than 1.0.
- Motors are not rated for fire pump service.
- Diesel engines on the other hand are rated for fire pump service. In NA, they are rated at the Society of Automotive Engineers standard of 300 ft (91.5 m) above sea level and 77°F (75°C) and 29.61 in. (75.21 cm) Hg. They are derated by 3% for each 1000 ft (305 m) altitude above sea level and 1% for every 10°F (5.5°C) rise in temperature.
- Gasoline engines are not allowed at this time.
- Turbines are acceptable drivers for fire pump service

Fire pumps are required when adequate water supply and pressure is not assured, such as the cases where a high-rise building had too high a static head required for the municipality water main or the water supply pressure was not dependable.

Fire pumps are generally run once a month for check out. During the time in between a jockey pump comes on as required to keep the sprinkler head piping pressure up to rating. A jockey pump is generally a small end suction pump but could be a positive displacement or regenerative turbine type pump.

Fire pumps are mounted on a base with their driver, see Fig. 19.1 for an electric motor driven fire pump and Fig. 19.2 for a diesel driven unit. Fig. 19.3 shows a diesel driven fire pump with additional accessories such as the fuel tank. In some cases these fire pumps are enclosed in a housing, which is complete with louvres for ventilation, all electric wiring and controls as well as the fire pump unit. Such a unit is shipped complete to the job site, where it is only necessary to hook up the connections to the fire protection system mains and the electric power. It is common to furnish an electric motor driven unit with a diesel

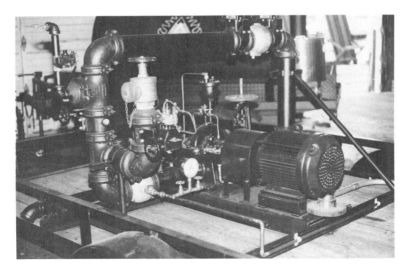

FIGURE 19.1 – Motor driven fire pump installation.

FIRE PUMPS

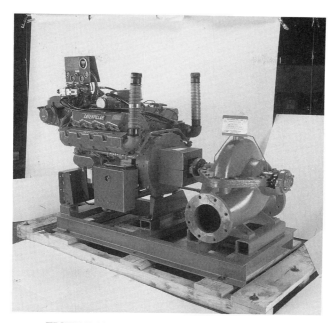

FIGURE 19.2 – Diesel engine driven fire pump.

FIGURE 19.3 – Diesel engine driven fire pump package with accessories.

back in locations where the power supply is not completely dependable. A picture of a vertical turbine fire pump is shown in Fig. 19.4. Vertical turbine pumps are used where the suction of the pump is not flooded from a higher level supply. Horizontal pumps are

FIGURE 19.4 – Vertical turbine fire pump.

FIGURE 19.5 – Portable fire pump.

FIRE PUMPS

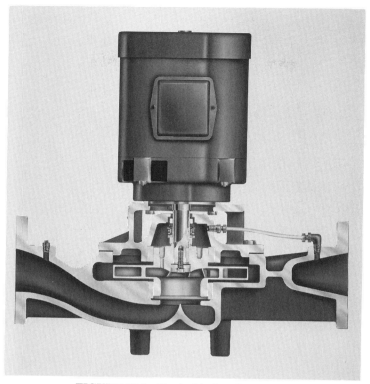

FIGURE 19.6 – Vertical in-line fire pump.

not generally allowed by the codes in such instances even though the same codes or standards require that the horizontal pumps also be able to work against a suction lift of up to 4.5 m (15 ft). Portable fire pumps are often used on site for emergency purposes and by the local fire fighting departments. Fig. 5 shows a portable fire pump of this type. Induction motor driven in-line pumps as shown in Fig. 6 are also used in the lower capacity ranges. This particular in-line unit has a mechanical seal. Mechanical seals are not acceptable to ULI and Factory Mutual at this time. They are, however, accepted in Canada and other countries and many are installed in the US when buiding heights result in discharge pressures that are beyond the capability of packing. Non-listed fire pumps can sometimes be used as specified or approved by the local authority having jurisdiction and this is the route taken when building heights exceed the packing limitations. Many fittings are also listed and approved and can generally be supplied by the fire pump manufacturer.

References

[1] ANSI-NFPA-20. "Centrifugal Fire Pumps", 1993 Edition, National Fire Protection Association Quincy, MA, USA

[2] UL-448, "Standard for Safety – Pumps for Fire Protection Service" (11/27/84), Underwriter's Laboratories, Inc.

THE HEAVY DUTY PROBLEM SOLVERS.

SEAL-LESS • MAG-DRIVE • NON-METALLIC
CORROSION RESISTANT • ANSI-DIMENSIONAL
CENTRIFUGAL PUMPS

Reliable, Field-Tested Performance for the Process Industries' Zero Emissions Requirements

Exceptionally innovative, the ANSIMAG Pump Family features a remarkably simple design. All wetted parts are non-metallic to handle a wide range of corrosives and solvents up to 250°F without corrosion. An innovative rear casing eliminates all unwanted heat, reducing wear and energy costs. And because ANSIMAG pumps are magnetically driven and seal-less, there are no leaks, no emissions, no costly seal maintenance — no problems.

ANSIMAG pumps meet B73.1 ANSI-dimensional standards so all of our pumps can easily replace your existing pumps without expensive pipe changes. ISO 2858 and JIS standards are also available.

K SERIES PUMPS WITH 6" OR 8" IMPELLER
ANSIMAG K series pumps with 6" impeller are available in 1.5" x 1", 3" x 2" and 4" x 3" for flows up to 700 gpm and heads up to 180 ft. with motors up to 20 hp. K Series with 8" impeller is available in 1.5" x 1" for flows up to 180 gpm and heads up to 290 ft. with motors up to 25 hp.

K SERIES SELF-PRIMING PUMPS
Developed especially for extreme heavy duty applications, the K Series Self-Priming pump is available in 3" x 2" for flows up to 280 gpm and heads up to 160 ft. with motors up to 25 hp. The flanged gooseneck is an available option.

F SERIES PUMPS WITH 10" IMPELLER
ANSIMAG F Series pumps are available in 2" x 1" and 3" x 1.5" with a 10" impeller for flows up to 270 gpm and heads up to 440 ft. with motors up to 40 hp.

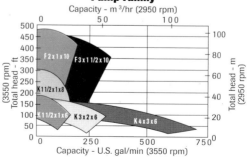

ANSIMAG Pump Family

Made In the U.S.A.

ANSIMAG®

*Simple by Design*SM

DEALERS WANTED WORLDWIDE
ANSIMAG, INC.
1090 Pratt Blvd. • Elk Grove Village, IL 60007 USA • 708/290-0482 • Fax: 708/290-0481

ITT Richter Chemie-Technik

The Answer to Corrosion.
Heavy duty design, long service life.

Standard and close-coupled.

- PFA, PTFE
- PVDF
- UHMW-PE

… with single and double mechanical seal 0,1 - 300 m^3/h (0,4 - 1200 gpm)

… sealless with magnetic drive 0,05 - 200 m^3/h (0,2 - 800 gpm)

ITT RICHTER CHEMIE-TECHNIK GmbH
P.O. Box 10 06 09 · D-47883 Kempen
Phone (0 21 52) 1 46-0 · Fax (0 21 52) 1 46-1 90 · Tx. 853211

CHEMICAL PROCESS PUMPS

Chemical pumps

Chemical pumps essentially include all the pump types. Each of which is best fitted to some specific process application or another. Corrosiveness, toxicity, viscosity, effects of shear, specific gravity, vapour pressure, air entrainment etc., are problems that are wide ranging in this industry. Mr Robert Hart, Senior Engineer of Dupont pointed out in a recent speech to the Hydraulic Institute, 2/94, that pumps in process plants are forced to run in some degree of cavitation, the process fluids are generally at or close to the saturation point and commonly tend to outgas somewhere in the process. Mr Hart pointed out that this gas has more effect on NPSH than head and that dissolved gases coming out of the solution can detrimentally affect a sensitive sealless pump. He also pointed out that because of

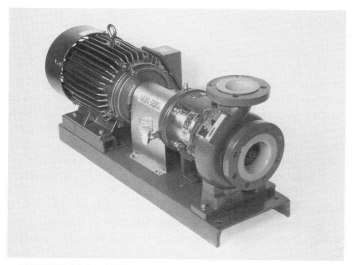

FIGURE 20.1 – ANSI B73 hermetic (sealless) chemical pump.

 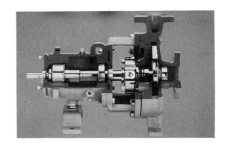

FIGURE 20.2a – ANSI B73 chemical pump.

FIGURE 20.2b – ANSI B73 chemical pump cutaway.

FIGURE 20.3 – Centerline supported chemical pump.

capitalization cost problems, the policy of no installed spares is being followed extensively in new construction. These factors are significant to the chemical pump industry in the selection and application of process pumps and their Mean Time Between Failure (MTBF). Dupont according to Mr Hart tries to hold vibration velocity below 0.2 inches/second (0.5 cm/sec.) and considers maintenance costs to be proportional to vibration velocity level. They also keep a 5' or 35% NPSH margin, whichever is larger, or call for the equipment to be tested and try to maintain a 5°F (2.8°C) temperature sensitivity cushion.

Chemical pumps must have materials that will stand the corrosiveness and abrasiveness of the process fluid, they must have a reliability compatible with the circumstances under which they must be compatible with the application cost, safety and regulatory considerations.

As pointed out in Chapter 6, centrifugal chemical pumps that comply with ANSI B73

CHEMICAL PROCESS PUMPS

or ISO 2858 are the most widely used equipment in the chemical industry. Both of these standards are dimensional standards, the former for North America and the latter for Europe and to a large degree the rest of the world. Fig. 20.1 shows an ANSI B73.1 pump of the hermetic (sealless) type and Fig. 20.2 shows an ANSI B73.1 pump of the seal or packing type. The pump in Fig. 20.1 has a corrosion resistant plastic lining that provides considerable protection against corrosion from specific process fluids. These pumps are available from suppliers of ISO 2858 pumps also. This, in many cases, may allow the use of less expensive metallurgy. Process temperature must also be taken into consideration in the pump selection. Fig. 20.3 shows a high temperature process pump with centreline support (an API feature – see Chapter 22). Figs. 20.4 and 20.5 show a hermetic "can" pump

FIGURE 20.4 – Canned motor hermetic pump (sealless) with external cooler.

FIGURE 20.5 – Bank of canned motor pumps in process plant.

FIGURE 20.6 – Canned motor pumps in nuclear reactor primary circuit.

FIGURE 20.7 – Hermetic high pressure (1200 bar) pump.

in cutaway with external cooler installed in a process plant. Radiation considerations came into play in the selection of the canned pumps shown in Fig. 20.6 for the primary circuit of a nuclear reactor test plant. The pressure requirements of 1200 bar suction pressure resulted in the design shown in Fig. 20.7 and VOC emission considerations played a part in the selection of a mag-drive hermetic pump for a chemical factory as shown in Fig. 20.8.

CHEMICAL PROCESS PUMPS

Fig. 20.31 shows a 12 stage centrifugal unit with pressures generated up to 20,000 ft for injecting solvents into a chemical process.

Positive displacement pumps have their place in this industry. Fig. 20.9 shows a bank

FIGURE 20.8 – Hermetic magnetic drive pump in a chemical plant.

FIGURE 20.9 – Rotary piston pump for chemical service.

FIGURE 20.10 – Peristaltic pump unit.

FIGURE 20.11 – Cutaway of Fig. 20.10 showing rotating mechanism and hose.

CHEMICAL PROCESS PUMPS

of three rotary piston pumps in a chemical production plant. Fig. 20.10 shows a peristaltic pump unit and Fig. 20.11 the rotating mechanism and hose. This pump is rated as high as 15 bar by the manufacturer which indicates a positive displacement solution for low flows and fairly high pressures. This peristaltic line of pumps is used for the transfer of ferric chloride and oxide, the dosing of lime, ammonium and ferric sulphate, flocculants, additive and pigment injection into pressurised lines, sampling, and condensate recovery from vacuum systems. It can handle process fluids containing solids and high viscosity fluids.

Drum or barrel pumps

Drum or barrel pumps are of many types depending on the chemicals being pumped, the safety aspects and power supply. The hand operated dispensing pump line of one vendor offers coices of double acting piston, siphon, rotary, and gear types. Another line of the same vendor is a sealless design with a choice of battery powered electric motor or air motor drives. This line is used for acids, alkalis, inflammable solvents, water, chemical solutions, plating solutions and oils. A third line uses centrifugal and progressive cavity pumps, handling the same broad fluids as mentioned just above plus pastes and food products with viscosities up to 300,000ssu. When pumping flammable or combustible liquids from one container to another both containers should be effectively bonded and grounded to prevent the discharge of sparks or static electricity which could cause an explosion. Figs. 20.12 and 20.13 show barrel mounted drum pumps. To reduce the amount

FIGURE 20.12 – Variable speed drum pump.

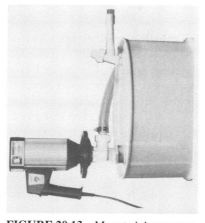

FIGURE 20.13 – Mounted drum pump.

of fluid left in the drum being emptied and to avoid life requirements, the pumps are often mounted at the bottom of the pipe or tube column that is inserted into the drum. These column lengths are optionally selected depending on whether carboys, metal or plastic drums or the deeper vats and reactors are used. The pumps are available in a range of standard materials such as polypropylene, PVDF, 316 stainless steel and aluminium. A double diaphragm pump mounted on a mobile cart is shown in Fig. 20.14. These pumps are moved about the plant for chemical transfer from tank supply lines and drum sources. This is an extremely common arrangement in the chemical industry.

FIGURE 20.14 – Double diaphragm pump – cart mounted.

Diaphragm pumps

Diaphragm pumps are used frequently where their hermetic, self-priming and dry running features are needed. Figure 20.15 shows a standard ball diaphragm pump feeding a filling machine for various industrial cleaning fluids. Fig. 20.16 shows a unique diaphragm pump that features Teflon spill containment chambers, extra barrier diaphragms, and electronic leak detection. This design is becoming increasingly in demand, as a response to tougher environmental regulations on chemical spills. The leak detector is wired to an audible alarm or pump shutdown mechanism. Signal is activated if the primary pumping diaphragm is compromised. The spill is contained in the Teflon chamber until repair can be made. This prevents leaking chemicals from being exhausted into the environment. It also protects the air valve of the pump from damage. The installation is pumping an abrasive, viscous, acid-based product. An oscillating piston/diaphragm pump is shown in Fig. 20.30a and a reverse osmosis installation of that type of pump handling 60 l/min at 70 bar is shown in Fig. 20.30b.

CHEMICAL PROCESS PUMPS

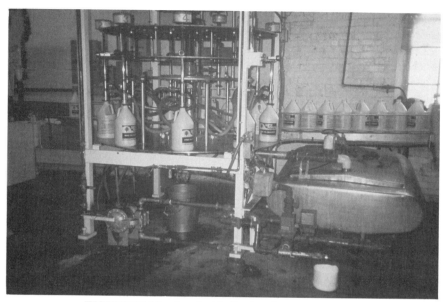

FIGURE 20.15 – Standard ball valve diaphragm pump.

FIGURE 20.16 – Double diaphragm pump with Teflon secondary spill containment chamber.

Metering or dosing pumps

Metering or dosing pumps are used extensively for controlled chemical addition to a process. They can be of almost any type depending on the metering accuracy required. Centrifugal pumps are used where the flows are large, inlet is flooded and the risk of air in-leakage through the seal chamber or stuffing box is low and the accuracy required is attainable. Sliding vane rotaries are used where the accuracy is not over 90%. Other rotary and reciprocating pumps generally will have accuracies above 90% with the exception being large clearance lobe and screw units. The reciprocating pump has the best accuracy. Variable speed drives facilitate accuracy and ease of control. Fig. 20.19 shows the example of a diaphragm metering package for metering catalysts, amines and silicones into a reactor in a synthetic rubber process. Fig. 20.20 shows a skid for metering and mixing of water with highly concentrated sulphuric acid. Fig. 20.21 shows three individual packages of very similar design for metering phosphate, morpholine and hydrazine for water treatment in a large industrial plant. Fig. 24 shows a dosing system for an application system of wire lubricant.

Axial flow propeller pumps

Axial flow propeller pumps are used where high flows and low system head conditions exist. Fig. 20.17a shows one being tested prior to shipment to a large chemical company in the Netherlands. Apparently an empty casing is being used as an elbow. Fig. 20.17b shows the unit itself.

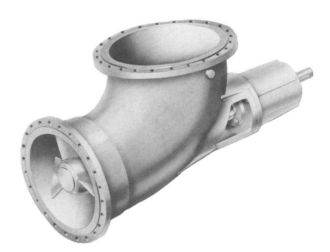

FIGURE 20.17a – Test setup for large axial flow propeller pump.

Vertically oriented volute type sump pumps

Vertical volute type sump pumps are used in chemical and industrial plants for collection of wastes and drainage as well as being incorporated into a process where NPSH is a factor. See Fig. 20.18

CHEMICAL PROCESS PUMPS

FIGURE 20.17b – Axial flow propeller pump.

Progressive cavity pumps

Progressive cavity pumps are often used on viscous products. Fig. 20.22 shows a progressive cavity pump in a calcium carbonate process. The pump was picked for its ability to handle viscous abrasive slurries.

Chopper pump

A chopper pump is shown in Fig. 20.23 in a chemical application.

High pressure plunger pumps

High pressure plunger pumps are used in the chemical industry as process pumps for detergent slurries, liquid ammonia, carbon dioxide liquid and CO_2 extraction plants. The

FIGURE 20.18 – Vertically oriented sump pump.

FIGURE 20.19 – Diaphragm pumps in synthetic rubber process.

CHEMICAL PROCESS PUMPS

FIGURE 20.20 – Diaphragm pump skid for metering and mixing water with highly concentrated sulphuric acid.

FIGURE 20.21 – Diaphragm pump packages metering phosphate morpholine and hydrazine for water treatment in large industrial plant.

FIGURE 20.22 – Progressive cavity pump in calcium carbonate process.

FIGURE 20.23 – Chopper pump in chemical process.

CHEMICAL PROCESS PUMPS

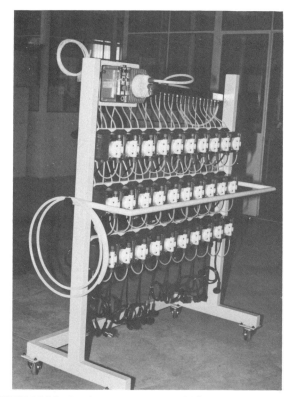

FIGURE 20.24 – Dosing system pump skid for rayon wire lubricant.

FIGURE 20.25 - High pressure plunger pump.

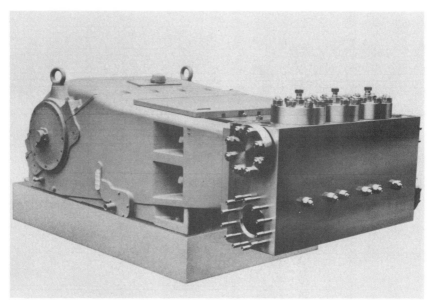

FIGURE 20.26 – High pressure plunger pump.

FIGURE 20.27 – High pressure plunger pump for internal vessel cleaning.

unit shown in Figs. 20.25 and 20.26 can handle capacities up to 2300 l/min at a pressure of 100 bar. All parts in contact with the liquid are of titanium due to the aggressive liquid they will be handling. Suction and discharge pulsation dampeners are furnished to attain reasonably low pulsation levels. This type of unit is also used for large extrusion and

CHEMICAL PROCESS PUMPS

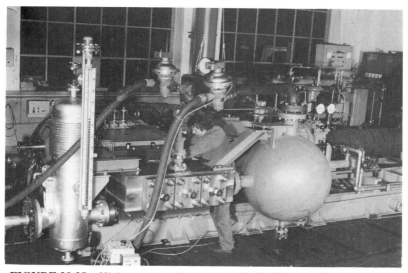

FIGURE 20.28 – High pressure plunger pump for liquid ammonia production.

FIGURE 20.29 – High pressure plunger pump for polymerization process.

FIGURE 20.30a – Wobble plate rotary piston pump with diaphragm on reverse osmosis service.

FIGURE 20.30b – Cutaway of pump in 20.30a.

hydraulic forging presses, rolling mills, and ass either pipeline pumps or salt water injection pumps in connection with crude oil and natural gas wells. The unit in Fig. 20.27 is used in the chemical industry mainly as a stationary high pressure cleaning pump for internal vessel cleaning, soil washing, demolition of concrete, descaling of paint ect. Fig. 20.28 shows a 200 bar pump undergoing test. This pump complete with mounted pulsation dampeners, will be used in the production of liquid ammonia. Fig. 20.29 shows a 200 bar, 25,000 l/hr pump feeding a reactor in a polymerization process.

CHEMICAL PROCESS PUMPS

FIGURE 20.31 – 12 stage – 20,000 ft head pump for solvent injection into chemical process.

FOOD, BEVERAGE AND PHARMACEUTICAL PUMPS

The reader is referred to Chapter 6 for a discussion of Sanitary Construction and the pertinent codes and regulations. These codes and regulations refer to the construction of kinetic or positive displacement type pumps, and apply to most of the pumps used in these industries. The 3A standard available from the International Association of Milk, Food and Environmental Sanitarians is recognized by pump companies and their customers internationally as one of the criteria for sanitary or aseptic hygienic design. Other standards, such as the USFDA (United States Department of Food and Drug Administration) standards and the USDA (United States Department of Agriculture) and regulatory agencies in other countries also play a part. The surface finish, particularly in the pharmaceutical industry, is increasingly being specified with low microfinish values. This polishing is expensive but a factor in bacteria build up.

Pumps in these industries in addition to being designed to meet these standards and resist the entrance of bacteria, germs and the like, are also designed for quick disassembly and reassembly as well as for clean in place (CIP) processing. Viscosity and temperature of the process fluid are considerations, along with emissions on some processes in the pharmaceutical areas. The materials must meet the codes and resist corrosion in the processes they are placed (including the secondary CIP processes).

End suction pumps

End suction pumps are the most widely used pump in this industry.

Rotary pumps

Rotary pumps are the next most commonly used where shear sensitivity, viscosity, low flows, etc., obviate the use of the centrifugal. Just about all types of rotary are used. Mechanical seals and seal-less pumps are used extensively instead of packed pumps in these industries. Double seal usage is increasing rapidly.

FIGURE 21.1 – Progressing cavity pump circulating whole grapes.

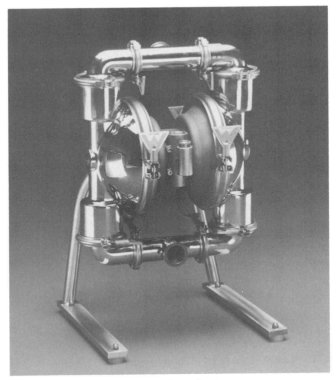

FIGURE 21.2 – Sanitary diaphragm pump.

FOOD, BEVERAGE AND PHARMACEUTICAL PUMPS

Progressing cavity pumps

Progressing cavity pumps are used in the food industry where high viscosities, solids in suspension and/or shear concerns prevail. Typical processes are cottage cheese, peanut butter, corn oil, jelly, gluten, corn syrup, sauces, yeast, tomato puree, apple sauce and salad dressing, yoghurt, tomato soup, animal fat, deviled ham, fish cake, lard, etc. In the beverage industry, they are used for spent grain or hops, vitamin C slurry, starch, grapejuice, grape pulp, citric acid and beer. See Fig. 21.1 of a progressing cavity pump handling whole grapes.

Diaphragm pumps

Diaphragm pumps such as shown in Fig. 21.2 are made with FDA approved materials are used in such applications as tomato paste, meat and poultry, pie fillings, cooking oils, dressings and relishes, jams and jellies, wine and beer processing, juice concentrates, body lotions, toothpaste and yeast slurry as examples. They are made of stainless steel with wetted surfaces polished and all elastomeric materials approved by FDA and able to stand up under frequent CIP cleanings. Figs 21.3 and 21.4 show two detergent blending systems with a 17 headed piston diaphragm metering pump. Fig. 21.5 system is a fully automated, continuous one blending up to 16 components in various recipes to produce liquid detergent products at a rate of 1667 kg/h (20 tons/h). Fig. 21.6 shows a metering pump installation for rum blending. Fig. 21.7 shows a cosmetic lotion metering and mixing installation and Fig. 21.8 is a photo of an installation for the cold sterilization of beverages. Fig. 21.9 is a continuous in-line metering blending system for beverages. .

FIGURE 21.3 – Diaphragm pumps on detergent blending application.

FIGURE 21.4 – Diaphragm metering pumps for detergent blending.

FIGURE 21.5 – Metering system for liquid detergent components.

FOOD, BEVERAGE AND PHARMACEUTICAL PUMPS

FIGURE 21.6 – Metering system for rum blending.

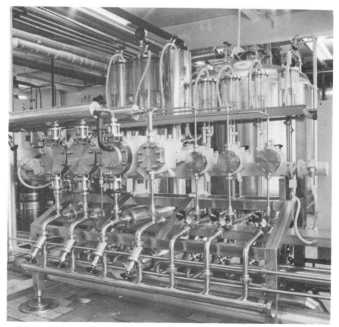

FIGURE 21.7 – Multihead diaphragm pump system for cosmetic blending.

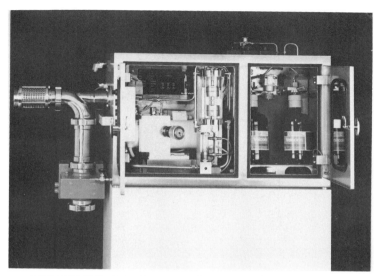

FIGURE 21.8 – Metering system for cold disinfection of beverages.

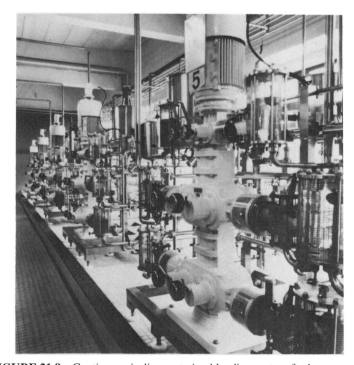

FIGURE 21.9 – Continuous in-line metering blending system for beverages.

FOOD, BEVERAGE AND PHARMACEUTICAL PUMPS

Piston and plunger pumps

Piston and plunger pumps are used where higher pressures are required. Fig. 21.10 shows a high pressure piston pump module. Fig. 21.11 shows a five plunger pump feeding an homogenizing valve, processing emulsion product prior to bulk packaging and Fig. 21.12 shows a high pressure process pump integral to an evaporator system.

FIGURE 21.10 – High pressure piston pump module.

FIGURE 21.11 – Five plunger pump feeding homogenizzing valve on emulsion product.

FIGURE 21.12 – High pressure pump for homogenizer in evaporator system.

Rotary pumps

Fig. 21.13 is a photo of a dairy installation. The pump has resilient rotors installed in a cream pasteurizing system and feeding the plate heat exchanger. Fig. 21.14 is a metal rotor pump in a pharmaceutical installation handling products of medium to high viscosity such as slurries, particulates at pressures of 435 psi.

Centrifugal end suction pumps

Centrifugal end suction pumps such as Fig. 21.15 are used where flows are relatively high and pressures moderately high. They can be furnished with low NPSHR where necessary and efficiency is important. This particular application is on applejuice. Fig. 21.16 is of a centrifugal pump in a pharmaceutical installation in a ceramic membrane filtration package system. Fig. 21.17 shows a sanitary beverage installation and Fig. 21.18 one of sanitary sampling pumps in a brewery.

FOOD, BEVERAGE AND PHARMACEUTICAL PUMPS

FIGURE 21.13 – Resilient rotor pump in cream pasteurization system.

FIGURE 21.14 – Rotary pump in pharmaceutical service for medium to high viscosity slurries or particulates.

FIGURE 21.15 – Sanitary centrifugal end suction pump for CIP on apple juice process.

FIGURE 21.16– Ceramic membrane filtration package system.

FOOD, BEVERAGE AND PHARMACEUTICAL PUMPS

FIGURE 21.17 – Sanitary centrifugal pump on beverage installation.

FIGURE 21.18 – Sanitary sampling pump in brewery.

PETROLEUM PRODUCTION, PIPELINE AND PRODUCT PUMPS

The scope of this chapter is to cover pumps used in the recovery of crude, the refining and the processing and use of petroleum products.

The processes covered use just about every pump type, but again the vast majority of pumps used are centrifugals. Almost all the crude production and refinery plants throughout the world specify pumps to API (American Petroleum Institute Standards). This Standard, tailored directly to the industry, is a candidate for ISO status. The significant considerations in the selection of many of the pumps are viscosity, high temperatures, high pressures, safety (explosion and fire), emissions and reliability singularly or in combination.

Centreline foot support is a characteristic of API pumps where high temperatures are involved see Figs. 22.1 and 22.2. Double casing construction on the very high pressure

FIGURE 22.1 – API heavy duty multistage axially split pump.

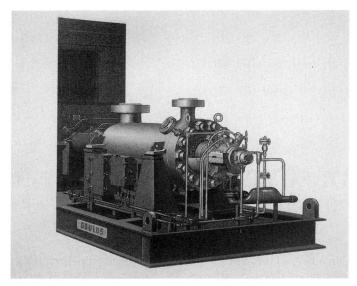

FIGURE 22.2 – API-610 high pressure multistage barrel casing pump

injection units. Fig. 22.3 shows the use of a vertically split barrel casing around an axial split pump for high pressure, speed and temperature services such as water injection. Fig. 22.4 shows a 10 stage pump for refined hydrocarbons at an Alaskan refinery. Operating temperature is 230°F. Higher loadings per blade area are being handled today with higher speeds available from industrial gas turbines. This is requiring higher allowable stress materials and tighter quality standards to prevent stress cracking of the impellers.

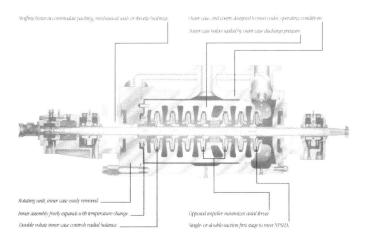

FIGURE 22.3 – Vertically split barrel casing around an axially split pump for high pressure service.

PETROLEUM PRODUCTION, PIPELINE AND PRODUCT PUMPS

FIGURE 22.4 – 10 stage pump for refined hydrocarbons.

Materials such as covered by Fig. 9.2 are utilized. Figs 22.5 and 22.6 show API Vertical In-Line Pump installations of Butane and Raffinat Re-run Overhead Column Service. These pumps offer many advantages in process plant service such as freedom from

FIGURE 22.5 – API 610 vertical in-line pump for butane and raffinet re-run overhead column service.

298   PUMP USERS HANDBOOK

FIGURE 22.6 – API 610 vertical in-line pumps for airport jet fuel transfer.

FIGURE 22.7

alignment problems and expense, temperature effects and flange overloading from piping strains. See Fig. 22.7 showing 6 vertical in-line pumps at the new Pittsburgh, PA, USA airport for jet fuel transfer. Axially split pumps such, as Fig. 22.1, are preferred for ease or speed of disassembly and reassembly in many cases up to the limit of their pressure

PETROLEUM PRODUCTION, PIPELINE AND PRODUCT PUMPS

capability, 165-200 bar (2500-3000 psig). Above this the vertically split barrel type, Fig. 22.2, is utilized.

Injection or waterflood pumps are used to replace the crude oil in the space it occupied in the oil field before being pumped out. The injection is to the bottom of the oil field. Axial-split pumps are utilized normally up to their pressure capability then the double casing or vertically split segmented barrel type take over. See Figs. 22.8, 22.9 and 22.10.

FIGURE 22.8 – 13 stage water injection pump.

FIGURE 22.9 – 12 stage water injection pumps axially split.

FIGURE 22.10 – 13 stage waterflood pumps with gas turbine drive.

Fig. 22.8 shows four 13 stage water injection pumps in a Texas oil field. Fig. 22.9 shows two 12 stage water injection pumps with 3000 psig head capability each in the Alaskan North Slope fields and Fig. 22.10 shows two 13 stage waterflood pumps with gas turbine drive at Lake Maricaibo, Venezuela.

FIGURE 22.11 – 6 stage gas turbine driven pump.

PETROLEUM PRODUCTION, PIPELINE AND PRODUCT PUMPS

Pipeline pumps are also normally of the axial type to 200 bar with the barrel type for pressures to 300-350 bar. Series operation is common for most pipelines with parallel operation generally confined to high elevation changes. Pipeline pumping installations are shown in Figs 22.11 through 22.15. Fig. 22.11 shows a 6 stage gas turbine driven pump

FIGURE 22.12 – 6 stage axial split pump on HC products.

FIGURE 21.13 – 7 stage axial split pipeline pump for light products.

FIGURE 22.14 – 6 stage axial split crude oil pump.

FIGURE 22.15 – Diesel driven axial split pipeline pump.

on H-C products for the mid-America pipeline Herman, Nebraska, USA. Fig. 22.12 shows a 6 stage pump on refined petroleum products through the Southern Pacific pipeline Portland, Oregon, USA. Fig. 22.13 is a 7 stage pump for light products such as propane, midwestern USA. Fig. 22.14 is a 6 stage pipeline pump in midwestern USA and Fig. 22.15

PETROLEUM PRODUCTION, PIPELINE AND PRODUCT PUMPS 303

is a diesel driven pipeline pump with a step up gear in Kansas, USA. Fig. 22.25 is a pipeline pump installation in Saudi Arabia.

Fig. 22.16 shows a multistage pump in a Korean refinery. Fig. 22.17 shows a single stage pump in a Korean refinery and Fig. 22.18 shows a horizontal multistage main oil line shipping pump. Fig. 22.24 is a charge pump for a hydrocracking tower.

FIGURE 22.16 – Multi-stage pumps in Korean refinery.

FIGURE 22.17 – Single stage pump in Korean refinery.

FIGURE 22.18 – Axial split multi-stage main oil line shipping pump.

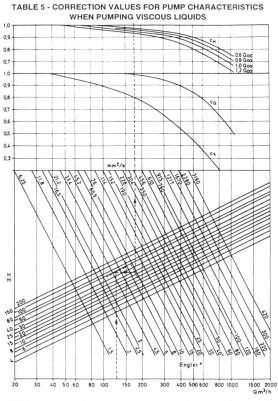

FIGURE 22.19 – Performance correction chart for viscous liquids.

PETROLEUM PRODUCTION, PIPELINE AND PRODUCT PUMPS

Along with boiler feed pumps the high energy pumps used in these services are subject to some of the most extreme conditions imposed on pumps.

Viscosity effects on pump on head and efficiency VS the rated and or tested performance on water can be significant on centrifugal pumps. Viscosities up to 4000 CSt (18,000 SSU) can be seen in these applications. The Hydraulic Institute Standards, 14th Edition provides a Performance Correction Chart for Viscous Liquids, see Fig. 22.19 based on tests of conventional single stage pumps of 2 to 8 inch size, handling petroleum oils, and cautions against its use outside of that range. It recommends performance tests when a particular viscous liquid is to be handled. The performance difference from water shows up as a drop in head or efficiency for a given capacity. An approximation of these changes can be calculated by means of the pump Reynolds Number, but this is best left to the pump manufacturer who can utilize empirical data on a specific pump. Table 1 gives viscosity conversion data from one set of viscometer units to another.

Positive displacement pumps are also affected by viscosity changes, but the problems and their seriousness are different. Reciprocating pumps handling high viscosities are adjusted to a reduced speed in most cases. Rotary pumps actually can show improved efficiencies, due to the reduced slip that accompanies a viscosity increase. Their speeds are also decreased somewhat when viscosity is increased. Fig. 22.20 shows a rotary-vane

FIGURE 22.20 – Rotary vane pumps handling additives and solvents at a oil company blending and packaging terminal in USA.

TABLE 1 – Viscosity Conversion Table

Seconds Sayblot Universal ssu	Kinematic Viscosity Centistokes*	Seconds Sayblot Furol ssf	Second Redwood 1 (Standard)	Second Redwood 2 (Admiralty)	Degrees Engler	Degrees Barbey	Seconds Parlin Cup #7	Seconds Parlin Cup #10	Seconds Parlin Cup #15	Seconds Parlin Cup #20	Seconds Ford Cup #3	Seconds Ford Cup #4
31	1.00	—	29	—	1.00	6200	—	—	—	—	—	—
35	2.56	—	32.1	—	1.16	2420	—	—	—	—	—	—
40	4.30	—	36.2	5.10	1.31	1440	—	—	—	—	—	—
50	7.40	—	44.3	5.83	1.58	838	—	—	—	—	—	—
60	10.3	—	52.3	6.77	1.88	618	—	—	—	—	—	—
70	13.1	12.95	60.9	7.60	2.17	483	—	—	—	—	—	—
80	15.7	13.70	69.2	8.44	2.45	404	—	—	—	—	—	—
90	18.2	14.44	77.6	9.30	2.73	348	—	—	—	—	—	—
100	20.6	15.24	85.6	10.12	3.02	307	—	—	—	—	—	—
150	32.1	19.30	128	14.48	4.48	195	—	—	—	—	—	—
200	43.2	23.5	170	18.90	5.92	144	40	—	—	—	—	—
250	54.0	28.0	212	23.45	7.35	114	46	—	—	—	—	—
300	65.0	32.5	254	28.0	8.79	95	52.5	15	6.0	3.0	30	20
400	87.60	41.9	338	37.1	11.70	70.8	66	21	7.2	3.2	42	28
500	110.0	51.6	423	46.2	14.60	56.4	79	25	7.8	3.4	50	34
600	132	61.4	508	55.4	17.50	47.0	92	30	8.5	3.6	58	40
700	154	71.1	592	64.6	20.45	40.3	106	35	9.0	3.9	67	45
800	176	81.0	677	73.8	23.35	35.2	120	39	9.8	4.1	74	50
900	198	91.0	762	83.0	26.30	31.3	135	41	10.7	4.3	82	57
1000	220	100.7	896	92.1	29.20	28.2	149	43	11.5	4.5	90	62
1500	330	150	1270	138.2	43.80	18.7	—	65	15.2	6.3	132	90
2000	440	200	1690	184.2	58.40	14.1	—	86	19.5	7.5	172	118
2500	550	250	2120	230	73.0	11.3	—	108	24	9	218	147
3000	660	300	2540	276	87.60	9.4	—	129	28.5	11	258	172
4000	880	400	3380	368	117.0	7.05	—	172	37	14	337	230
5000	1100	500	4230	461	146	5.64	—	215	47	18	425	290
6000	1320	600	5080	553	175	4.70	—	258	57	22	520	350
7000	1540	700	5920	645	204.5	4.03	—	300	67	25	600	410
8000	1760	800	6770	737	233.5	3.52	—	344	76	29	680	465
9000	1980	900	7620	829	263	3.13	—	387	86	32	780	520
10000	2200	1000	8460	921	292	2.82	—	430	96	35	850	575
15000	3300	1500	13700	—	438	2.50	—	650	147	53	1280	860
20000	4400	2000	18400	—	584	1.40	—	860	203	70	1715	1150

*Kinematic Viscosity (in centistokes)

$= \dfrac{\text{Absolute Viscosity (in centipoises)}}{\text{Density}}$

When the Metric System terms centistokes and centipoises are used, the density is numerically equal to the specific gravity. Therefore, the following expression can be used which will be sufficiently accurate for most calculations:

Kinematic Viscosity (in centistokes)

$= \dfrac{\text{Absolute Viscosity (in centipoises)}}{\text{Specific Gravity}}$

When the English System units are used, the density must be used rather than the specific gravity.
For values of 70 centistokes and above, use the following conversion:

SSU = centistokes x 4.635
Above the range of this table and within the range of the viscosimeter, multiply the particular value by the following approximate factors to convert to SSU:

Visscosimeter	Factor
Saybolt Furol	10.
Redwood Standard	1.095
Redwood Admiralty	10.87
Engler—Degrees	34.5

Viscosimeter	Factor
Parlin cup #15	98.2
Parlin cup #20	187.0
Ford cup #4	17.4

PETROLEUM PRODUCTION, PIPELINE AND PRODUCT PUMPS 307

pump installation that handles solvents, process oils, lubricants, additives and plasticizers at a blending and packaging installation. Fig. 22.21 shows an offshore skid for condensate injection into natural gas with reciprocating type metering pumps. Fig. 22.22 shows an on-

FIGURE 22.21 – Offshore skid for condensate injection into natural gas.

FIGURE 22.22 – Offshore skid for anticorrosion agent injection into crude oil with reciproacting type metering pumps.

shore skid for injecting anti-corrosion inhibitors into crude oil with reciprocating type metering pumps and Fig. 22.23 shows an off shore skid of reciprocating type metering pumps for the injection of anti-corrosion agents into a natural gas delivery system.

FIGURE 22.23 – Offshore skid for reciprocating type metering pumps injecting anticorrosion agents into natural gas.

FIGURE 22.24 – Charge pumps for hydrocarbon tower.

PETROLEUM PRODUCTION, PIPELINE AND PRODUCT PUMPS

FIGURE 22.25 – Pipeline pump – Saudi Arabia.

The Essential Book for Pump Users/Specifiers

New, Expanded, ANSI-Approved 1994 HI PUMP STANDARDS

New, extremely valuable reference for users and specifiers of pumps and related equipment; it features:

- All new revised information on all major pump types.
- 692 pages of information; replaces all previous editions.
- Dual U.S. and metric units throughout.
- Complete test standards by pump type.
- For full range of pump types: nomenclature; definitions; design and applications; installation; operation and maintenance information.
- Developed through ANSI canvas approval process and endorsed by most major pump manufacturers. For pump users, consultants, contractors, engineering construction firms, manufacturers of motor and other pump-related equipment, libraries, universities.

Available as a 13-set document in an attractive 3-ring binder ... also in individual sets by pump types: centrifugal, vertical, rotary and reciprocating ... or as individual documents.

FOR COMPLETE INFORMATION WRITE TO:

Hydraulic Institute, Dept. 854-0107 W, P.O. Box 94020, Palatine, IL 60094-4020

OR contact us at Phone No. (708) 364-6206; or Fax No. (708) 364-1268. Credit card orders accepted.

9 Sylvan Way, Parsippany, NJ 07054

THE PULP AND PAPER MAKER'S PUMP MAKER

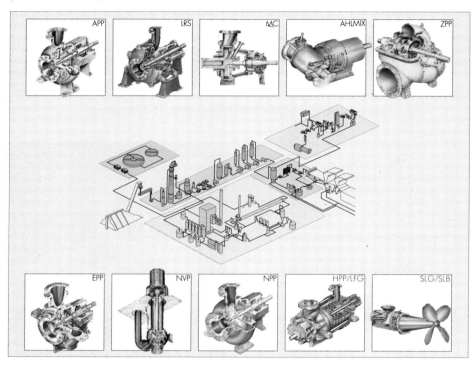

Ahlstrom has long specialized in pumps for the pulp and paper industry. This 100+ years experience, combined with our systems approach, brings you exceptional capabilities for fiber treatment and transfer.

MC technology, which started with the introduction of the MC pump, has totally changed the way that medium consistency stock is handled. Energy savings, process simplification, and reduced environmental load are all major benefits of Ahlstrom MC technology, which includes Ahlmix chemical mixers, tower discharge scrapers, flow dischargers and flow splitters in addition to MC pumps.

Ahlstar process pumps with dynamic seals have become the accepted mill standard worldwide. Low pulse headbox feed pumps and other speciality pumps and agitators have improved process reliability and pulp and paper quality in hundreds of applications.

Even our service philosophy with strong local sales and technical support expertise is designed to ensure complete customer satisfaction.

The Pulp and Paper Industry Pumps are a good example of Ahlstrom's wide range of products for pulp and paper manufacturing. Ahlstrom Machinery supplies complete fiberline and chemical recovery expertise for pulp mills, recycled fiber pulp mill systems, and machinery for fiber treatment and transfer as well as solutions for the treatment of mill effluents, sludges and flue gase.

EUROPE AND AFRICA: A. Ahlstrom Corporation, Pump Industry, Finland, Tel. +358 52 291 111, Fax +358 52 63 958
THE AMERICAS: Ahlstrom Process Equipment Inc., USA, Tel. +1 803 232 0800, Fax +1 803 232 2500
ASIA PACIFIC: Ahlstrom Machinery (Asia-Pacific) Pte Ltd, Singapore, Tel. +65 733 2522, Fax +65 732 8211
Or contact your nearest Ahlstrom representative.

PULP AND PAPER PUMPS

While other industries may have one or two critical concerns that affect pump selection, the Pulp and Paper industry has them all. The black liquor process itself has temperatures of 250°F (170°C), solids content up to 70%[1], viscosity to 1000 Cp (thixotropic), very high suction pressures, low NPSHR requirements, poor lubricity, air entrainment, (actually a three-phase solution) and aggressive, corrosive, acidic mixture of sodium or calcium bisulfite and sulfurous acid. Reliability is also a major consideration . There may be an annual maintenance shut down of the mill, meaning any failures in between are at the expense of production and the cost of the product in process at the time of the unscheduled shutdown. Pulsation is also a concern, especially for those pumps at the end of the paper process, just prior to the paper making machine itself.

The standard small process pump is the end suction, ANSI B73 or ISO 2858 chemical pump with centreline discharge open impeller and today more often than not mechanical seals. The larger pulp and low concentration transfer pumps are end suction, open impeller and adjustable wear plate, packing or mechanical seal type pumps with centerline discharge. Refer to Figs 5.43 through 5.46. Centrifugal stock pumps are used for consistencies up to approximately 5.5%, above this there are pumps with special internal features such as an inlet fluidizer, that will allow consistencies up to as high as 18% consistency to be pumped. See Fig. 23.1. There are only a few manufacturers with medium consistency pumps on the market that with the assistance of an internal vacuum rotor or external vacuum pump can pump this high a consistency. The principles of operation are shown in Fig. 23.2. In this range positive displacement pumps were once called for and many are still in use, but their failure rate was high. This together with their high initial cost has opened the door to the medium consistency pumps. From 16–20%, screw type pumps are still used.

The optimum storage consistency is 12%. On the low side of this there is excess water and on the high side there is excess air. Even at 12% there is 16% air. This is not conducive to the bleaching process where the bleaching additives are added. Hence, ways of removing this air are always under investigation. Another important function of these pumps and any stock pump is to fluidize the pulp mixture it is pumping. This is done by

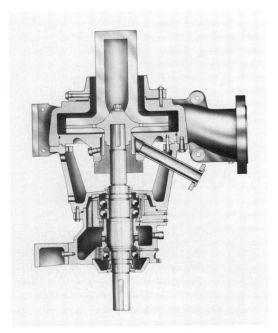

FIGURE 23.1 – Medium consistency pump.

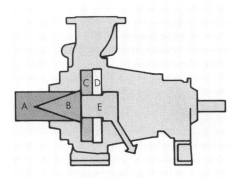

Principle of MCV pump
The patented MC Pump design (for example U.S. patents 4,780,053 and 4,854,819) has five functional zones:
A. Fluidization zone, where the fiber network in the stock is divided into single fibers
B. Gas Separation Zone, where air is separated from the stock
C. Pumping zone, where impeller vanes pump the stock toward the discharge
D. Fiber return zone, where fibers coming with air from Zone B are returned to the pump discharge
E. Degassing zone, where the vacuum rotor removes air to the degas pipe

FIGURE 23.2

PULP AND PAPER PUMPS

high shearing forces at or just prior to the pump inlet. Instead of a gradual up-sloping suction approach as is common in most other pumps (with eccentric reducers), paper pumps generally use a stepped reducer at the suction. This creates shear, turbulence and frees some of the contained air but fluidizes the stock and provides more even distribution to the impeller passages. The medium consistency pumps have these features plus the ability to mix the additives to the bleach process. Newer designs of stock pumps have more vanes and narrower passageways than their predecessors which improves the fluidization, efficiency and recirculation characteristics. Efficiency to some extent is sacrificed to function and reliability in terms of turbulence at the inlet, centreline discharge, increased clearances, etc., in stock pumps. The effect of stock consistency on its flow characteristics is shown in Fig. 23.3. One can see why fluidization is essential in the higher consistencies.

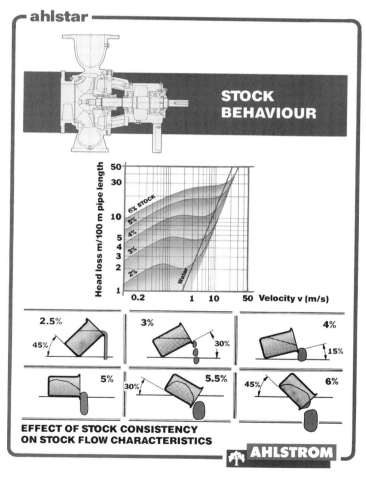

FIGURE 23.3 – Stock consistency effect on flow characteristics.

A flow diagram of a sulphite pulp and paper process is shown in Fig.23.4[2], and a block diagram of a Kraft process is shown in Fig. 23.5. Note: A section on friction loss of paper stock including tables is included in the Hydraulic Institute Data Book, 2nd Edition 1990.

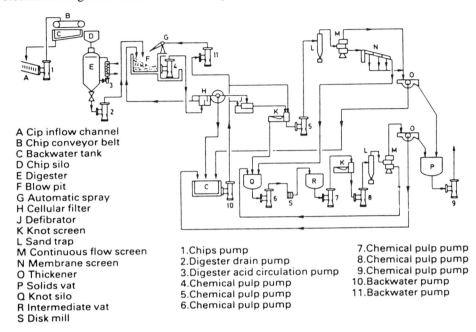

A Cip inflow channel
B Chip conveyor belt
C Backwater tank
D Chip silo
E Digester
F Blow pit
G Automatic spray
H Cellular filter
J Defibrator
K Knot screen
L Sand trap
M Continuous flow screen
N Membrane screen
O Thickener
P Solids vat
Q Knot silo
R Intermediate vat
S Disk mill

1. Chips pump
2. Digester drain pump
3. Digester acid circulation pump
4. Chemical pulp pump
5. Chemical pulp pump
6. Chemical pulp pump
7. Chemical pulp pump
8. Chemical pulp pump
9. Chemical pulp pump
10. Backwater pump
11. Backwater pump

FIGURE 23.4 – Schematic of sulphite pulp process.

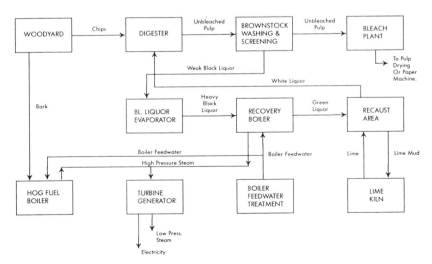

FIGURE 23.5 – Block diagram of Kraft process pulp mill.

PULP AND PAPER PUMPS

Pulp mills are pump intensive with an average of 150 pumps. The paper mills have approximately a third as many. In this process wood chips are conveyed into a digester where a cooking process takes place. In the Kraft process pumps deliver green liquor which is mostly a solution of sodium carbonate and sodium sulfide with some iron sulfide that gives it the greenish colour. These pumps are subject to erosion because of the turbulent action set up in the dissolver tank and the grit that is present. It also has a tendency to build up on the walls of the pump and piping. Specific Gravity is about 1.2. Other pumps deliver the causticizing ingredients. These are normally standard slurry pumps with good wear resistance.

As the *green liquor* is causticized it loses its green colour and becomes known as *white liquor*. This is the cooking liquor. It is corrosive, much more so in the case of the sulfite process as covered in the black liquor mentioned previously. In the batch cooking system, the pumps are generally centreline supported, see Fig. 23.6, and mounted on springs. This is an attempt to compensate for the wide swings in temperature that take place from a cold start to completion of the batch process. The make-up liquor and cold blow pumps utilize 250 psi (16-bar) pumps, 600 ft. (183 m) TDH, and 16-20 bar suction pressures that can handle the abrasion and corrosiveness of this process. In the case of continuous cooking, the transients due to temperature swings are not as great and the pump bases are grouted into the concrete.

As the cooking process continues in the digester the liquor becomes *black liquor* just before it is ejected into the blow pit where the sudden pressure reduction causes the fibres to separate from the other solids and for the first time in the process stock pumps are

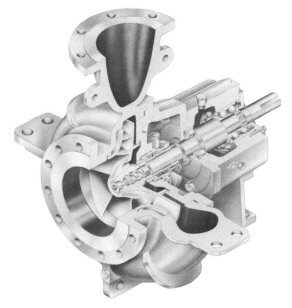

FIGURE 23.6 – Centerline supported digester pump.

required. See Figs 23.7 and 23.8. The black liquor is then put through a series of washers where the fibre is cleansed of undesirable organic residue. The black liquor is next pumped through the evaporators where it's percentage of solids is increased in stages up to 70-75%. This liquid is relatively low in corrosive aggressiveness. Evaporator pumps, however, run in cavitation so self-priming pumps are often used. *Progressive Cavity pumps* are sometimes used in this application.

The pulp separated from the black liquor is put through the *bleaching system* where

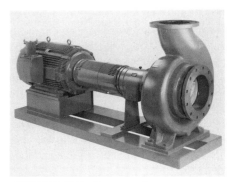

FIGURE 23.7 – Paper stock process pump.

FIGURE 23.8 – Paper stock process pump.

PULP AND PAPER PUMPS

additional delignification and/or stabilization takes place. Bleaching today is generally a chlorine process, such as hypochlorite or chlorine dioxide (ClO_2), or a peroxide process. There is a slow trend toward ozone systems. Titanium, fiberglass or high silica iron pumps are used in the bleaching process. Titanium is generally used for the reboiler recirculation, chlorate feed, salt cake filter feed, ClO2 to storage, and feed to bleach pumps. The sulfuric acid pump is generally alloy 20, and the remaining pumps have generally been 316 or 317 in the US, 329 in Europe.

Filtrate pumps see products that varies in pH to very low values, i.e. 1.5. The metallurgies that have been used here of late are titanium, SMO254, Hastelloy C, 317 and fibreglass.

Two main suppliers of pulp and paper pumps have come out with duplex steel metallurgies[3] in the last few years. One has standardized on the 22% chrome duplex 2205 (ASTM A890-Grade 4A) and the other has just standardized on a 25% chrome duplex (ASTM A-890-Grade 3A duplex), both have molybdenum and nitrogen which have beneficial effects on pitting resistance. Metallurgy 316 and 317 can be replaced by the 2205 material in most applications and alloy 20 in some whereas the 25% chrome material can be expected to provide even greater corrosion resistance and the option to utilize it instead of the other metallurgies, such as Alloy 20 in most applications, to the advantage of the user. This advantage would not only come from longer life but also from standardization.

The pumps in the paper machine area are the screen, refiner, washer, cleaner and head box units as well as the refiner reject pump. Metallurgies 316 and 329 have been used in these areas, but the new duplexes are being furnished.

The head box feed pump has to meet tight pulsation specifications to prevent degradation of the paper stock by lines caused by pulsations. Tighter restrictions in pulse levels than 0.15 psi peak to peak are now being called for. A cutaway of a fan pump is shown in Fig. 23.9. Note the skewed and staggered vanes in the impeller as well as the polished finish..

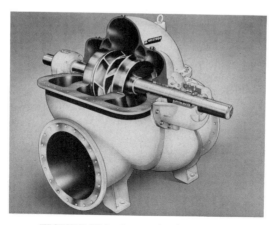

FIGURE 23.9 – Low pulse fan pump.

A flow diagram of a recycled fibre (deinking) process is shown in Fig. 23.10.
Paper coating packages containing diaphragm metering pumps and controls are shown in Figs 23.11 and 23.12.

Fig. 23.13 illustrates the effect of stock consistency on stock flow characteristics.

Fig. 23.14 shows a gear pump on a black liquor soap application.

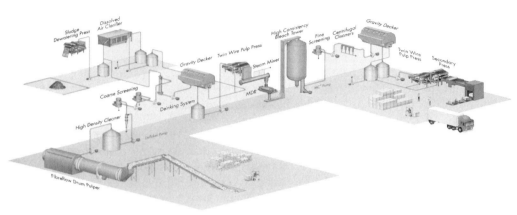

FIGURE 23.10 – Recycled fiber (Deinking) process flow diagram.

FIGURE 23.11 – Paper coating pump package for release paper production.

PULP AND PAPER PUMPS

FIGURE 23.12 – Electronically controlled resin mixing package for veneer paper coating.

FIGURE 23.13 – Closed loop color addition metering pump and control package.

FIGURE 23.14 – Rotary gear pump on black liquor soap application.

References

1. Harrison, A., "Pulping, Paper making Changes Mean More Corrosive Climate for Pumps", Pulp and Paper Magazine (Feb. 1994).
2. Dickenson, C., "The Pumping Manual" 8th Edition, Elsevier Advanced Technology (1992)
3. Rayner, R. E, "New Metallurgies for Process Pumps". World Pumps (Oct 1992).

WILLETT POSITIVE DISPLACEMENT VERTICAL RAM PUMPS

- Ideal for thick and abrasive sludges and slurries
- Hydraulic powered with versatile electronic control system
- Wide range of outputs from 155 - 3600 Ltr/min
- Pressures up to 20 bar
- Suitable for filter press feed, long distance transfer and auto-desludging of settlement tanks
- Extremely reliable with minimum maintenance requirements

EJ *Edwards & Jones*

Edwards & Jones Ltd. Whittle Road, Meir, Stoke-on-Trent ST3 7QD England. Telephone: +[44] (0)1782 599000 Fax: +[44] (0)1782 599001

SOLIDS AND SLURRY PUMPS

As with any new subject, it is best to define the terms used so that the reader's confusion is minimized even if not eliminated. *Solids handling* is a term covering a wide range of applications involving the pumping of solids in suspension. These applications may range from dredge pumping to the pumping of suspended fines or from a low percentage of solids to 100% solids, such as with powders or other fines.

Sludge

Sludge can be defined[1] as a liquid (usually water) containing large solids having a particle size of 6 mm ($^1/_4$ in.) or greater. These solids being soft rather than abrasive. Sludges may contain a high proportion of smaller solids such as sand, which can be abrasive.

Sewage

Sewage, on the other hand, contains mostly soft solids. It often contains sand or other small solids (these are often referred to as grit).

Slurries

Slurries[1] can be defined as liquids (usually water) containing abrasive solids in suspension. They can be further categorized by the size of the solids.

- *Fines* - Particle size of 75 μm (micron) or less.
- *Sands* - Particle size of 75-850 μm (micron) of less
- *Gravel* - Particle size of 850-5000 μm (micron) of less

Typical examples of slurries are drilling muds, limestone, Kaolin, phosphates, fly-ash, rock-salt, sewage sludge, crushed rock, sand, powder and pulp. Fig. 24.1 shows a classification of pumps according to solid size[2]

Suspended solids have the effect of increasing the apparent viscosity of the clear liquid. Efficiency is reduced from that of a test on the clear liquid alone. The specific gravity of the solids typically ranges from 2-3 with the solids content typically upwards of 35% on a weight basis. Slurries are either *settling* or *non-settling*. This is a function of the SG of the solids and the velocity of the flow. Heavy gravel would have a greater tendency to settle than a fine powder for instance. However, even the heavy gravel could be kept in

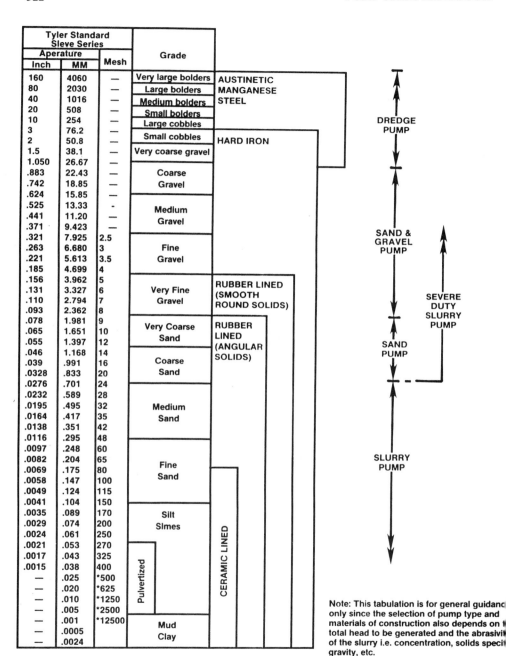

FIGURE 24.1 – Classification of pumps according to solid size.

"Sealless," self-priming & can meter as it pumps!
Specify Waukesha/Bredel Hose Pumps.

No product seals. Only hose contacts product. The rotor shoe alternately compresses and relaxes specially designed, resilient hose. Liquid lubricant in the housing minimizes friction.

Self-Priming. Boasts a 9.5 meter lift. Even runs dry without detrimental effect. Linear output. With commercially available pulse accumulators and pickup to count shaft revolutions, it's **a metering pump.**

Contact Waukesha Bredel Fluid Handling B.V.,
P.O. Box 47, 7490 AA Delden, Netherlands;
TEL: 31-7437-70000 FAX: 31-7437-61175

Waukesha Bredel Fluid Handling B.V.
® Delden, Holland

■ Spray drier ■ Transfer ■ Process Engineering ■ Mining ■ Chemical Industry

Taste it !

Pump technology à la carte
to handle
your individual problems.

In the field of fluid transfer technology, Feluwa can evaluate, design and manufacture solutions to your industrial, municipal or domestic pumping problems.

Feluwa can supply complete pumping installations from inlet to outlet with complete monitoring and control.

Our products are supplied and maintained worldwide by trained agents.
For more information, please contact:
D-54570 MÜRLENBACH
Tel. 0 65 94 / 10-0
Fax 0 65 94 / 16 40

feluwa
PUMPEN · PUMPS · POMPES
SCHLESIGER & Co. KG

Waste treatment / Pressurized drainage ■ Transfer of slurry and thickened sludge ■

SOLIDS AND SLURRY PUMPS

suspension if the flow velocities were high enough. Table 1 shows the minimum flow velocities to maintain different size particles in suspension.

TABLE 1[3] – Minimum flow velocities for solids in suspension

Particle Classification	Minimum Flow Velocity*	
	m/s	ft/s
Fines	0.9	3
Sands	1.5	5
Gravel	2.1	7
Sludge	3.35	11

* Based on a solids content of 30 to 35% by weight and Solids specific Gravity of 2.5 to 3.0.

When the flow is turbulent enough or has a high enough apparent viscosity to keep all the solids in suspension the flow is said to be *homogeneous*. When this is not the case and some of the heavier solids fall out to form a moving bed at the bottom of the pipe, the flow is said to be heterogeneous. Big differences in pressure drop and other pump application considerations exist between these two types of flow. It is not the purpose of this handbook to cover this complex subject in detail. Reference[3] covers it thoroughly.

The primary concerns in the selection of a slurry pump are the life of the pump or its major components and the time between repair. Efficiency is a concern that is subjugated to life or MTBR. The pump must also be able to pass the maximum solid sizes it will see. Since wear is approximately proportional to the cube of the pump speed there is a built in requirement of slower speeds and resulting large diameter impellers. This is offset by a countering requirement for large discharge openings and efficiency that calls for higher speeds. A compromise results in an intermediate speed. An old rule of thumb called for the impeller o.d. to be three times the eye diameter. Reference.[1] gives the following limitations on peripheral speed for various impellers on medium slurry service as follows: hard metal-7000 SFM, rubber lined-5000 SFM and ductile iron or steel-900 SFM. Another limitation that is generally followed to keep the kinetic energy down is a limit of 25 m (80 ft) of head per pump. Hence, it is common to see as many as five units in series.

Wear is a function of the abrasiveness of the slurry and the type of abrasion, the wear resistance of the pump materials and the velocity of the abrading particles. In the latter case wear is considered equal to U^n, where n = 2.5 to 3.8. There are three wear mechanisms that are generally referred to for analysis [4] (Reference[5] gives a more detailed analysis):

- *Impingement, impact or gouging abrasion* where the particles impinge on the surface of the material at large angles. Plastic deformation and fatigue wear mechanisms are present. This is the predominant mechanism at the blade inlet and casing cutwater.
- *Friction or erosion abrasion* where the particles strike the surface at small angles or slide on the surface. This mechanism acts on the impeller vane and volute surfaces.
- *Grinding abrasion* occurs when abrasive particles are ground between two surfaces such as in the wear ring and stuff box areas.

Fig. 24.2, shows the hardness values. The normal material selection philosophy is to

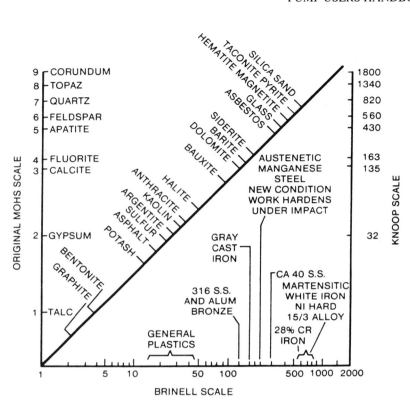

FIGURE 24.2 – The hardness values of various particulate solids and pump materials of construction. (Adapted from Wilson, 1980.)

pick pump materials that are harder than the material being pumped at least in the case of friction and grinding wear. When there is a great deal of high angle impingement or impact expected then materials with impact resistance and toughness are selected such as austenitic manganese (12-14%) steel which work hardens up from its 230 BHN under impact and resists gouging. This material is used in dredge pumps for example, where the impact loads are prevalent. Fig.24.3 shows a dredge type pump in this case the 30 x 34 x 84 is being used as a 6000 hp (4475 kW) transportable booster station for a 850 mm (33 $1/2$ in.) pipeline. Rubber linings are used for service in the ranges of particle size as shown in Fig. 24.1. Urethanes have come on strong in the last 10 years and have been the source of most developments in the lining area. Fig. 4 shows an elastomer-lined pump. Note the radial split.

The pump design itself can be a big factor in the amount of erosion that takes place. Concentric or semi-volute casings are utilized to reduce the turbulence that occurs at off design flows. Closed impellers are preferred where stringy materials are not present. The throat velocity should be made as low as practically possible, even at the expense of a small amount of efficiency. The cutwater should be as far away from the impeller discharge as feasible which is contrary to good efficient design and material costs. The characteristic

SOLIDS AND SLURRY PUMPS

FIGURE 24.3 – 6000 h.p. dredge pump.

FIGURE 24.4 – Elastomer lined slurry pump.

H/Q curve should be as steep as possible. Unfortunately this probably still means it is fairly flat. Small changes in system head will then result in big swings in capacity which could result in settling of the slurry and another increase in head. Discontinuities in the flow internal to the pump must be avoided. Three to five vanes are normal. Pumpout vanes on the back shroud should be provided. Figs. 24.6 through 24.10 show centrifugal pumps in slurry applications. Fig. 24.5 is one of eight 450 x 400 x 990 mm hard metal pumps at

FIGURE 24.5 – Hard metal slurry pump on Alumina, Australia.

FIGURE 24.6 – Elastomeric pump on copper tailings.

Nabalco's Grove Alumina Refinery in Australia, Fig. 24.6 is an elastomeric pump in a copper tailings circuit pumping 1.8 pH high density sand. Fig.24.7 is another elastomeric pump handling 32,000 gpm in a phosphate application at 30 ft. of head. Fig. 24.8 shows three elastomeric slurry pumps in series pumping ash and sludge mixture through a 10 km pipeline at the Shi Heng Power Plant in China. Fig. 24.9 shows an installation of several

SOLIDS AND SLURRY PUMPS

banks of elastomeric pumps in series, the one in the foreground is a bank of four pumps in series handling 15,000 usgpm at 500 psi. Fig. 24.10 shows two sets of six elastomeric slurry pumps in series pumping highly abrasive iron concentrate through a long pipeline to an off shore ship loading facility. To minimize wear, the kinetic energy is held down by limiting the head per stage to 23-25 m (75-80 ft).

FIGURE 24.7 – Elastomeric pump on phosphate application.

FIGURE 24.8 – 3 elastomeric slurry pumps on ash and sludge, China.

FIGURE 24.9 – Several banks of elastomeric slurry pumps in series on taconite application.

FIGURE 24.10 – 2 sets of 6 elastomeric slurry pumps on highly abrasive iron concentrate.

Submersibles

Submersibles are used for solids handling, especially solids with low percent solids such as dirty water. Fig. 24.11 shows a portable submersible being used to pump out floor surface water in a mine. Fig. 24.12 shows the same unit being used as a transfer pump in the stone industry from a settling pond to a tank truck.

SOLIDS AND SLURRY PUMPS

FIGURE 24.11 – Portable submersible used on mine floor surface water.

FIGURE 24.12 – Portable submersible on transfer duty in stone works.

Diaphragm pumps

Diaphragm pumps are used for solids handling applications also. Fig. 24.13 shows a hydraulically driven piston pump mounted on a mobile sludge loader for clearing thickened sludge from settlement sumps in coal mines. Fig. 24.14 shows a crankshaft driven piston diaphragm pump for mine dewatering. This system is a "dirty water system,

FIGURE 24.13 – Hydraulically driven piston pump on mobile sludge loader.

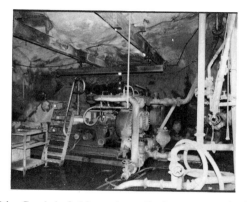

FIGURE 24.14 – Crankshaft driven piston diaphragm pump for ine dewatering.

FIGURE 24.15a – Piston diaphragm pumps in U.S. gold mine.

SOLIDS AND SLURRY PUMPS 331

FIGURE 24.15b – Piston diaphragm pumps in U.S. gold mine.

FIGURE 24.16 – Piston diaphragm pump for feeding of filter presses with kaolin slurry.

the pump transports both the water and the suspended solids from the settlement sumps to the surface. Figs 24.15 A & B show an autoclave feeding with piston diaphragm pumps in a US gold mine. These pumps are specially designed with a dropleg system for the transportation of high temperature (175°C) gold slurry. Fig. 24.16 shows a piston diaphragm pump for the feeding of filter presses with kaolin slurry.

Fig. 24.17 shows a reciprocating concrete pump and its performance characteristics. While Fig. 24.18 shows its application on a portable boom truck which can access elevated locations at construction sites and expedite the concrete pouring process. Fig. 24.19 shows high density slurry discharge for erosion control.

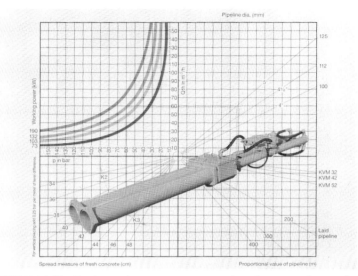

FIGURE 24.17 – Reciprocating concrete pump and performance characteristics.

FIGURE 24.18 – Application of Fig. 24.17 pump on a portable boom truck.

SOLIDS AND SLURRY PUMPS

FIGURE 24.19 – High density slurry discharge in erosion control.

FIGURE 24.20 – Axial pump on polyethylene slurry.

Axial pumps

Axial pumps are used for some slurries where high volume is required. Fig. 24.20 shows an axial pump on polyethylene slurry at 600 psig and 20,000 usgpm.

References

[1] Dickenson, C., "The Pumping Manual", 8th Edition (1992), p676.
[2] "Engineering Data Book", Hydraulic Institute, 2nd Edition (1990), p13

3 Wilson, K., G.R. Addie and R. Clift, "Slurry Transport Using Centrifugal Pumps", Elsevier Applied Science (1992).
4 Roco, M., "Wear Mechanisms in Centrifugal Slurry Pumps", Corrosion Engineering Vol. 4, No.5, pp 424-431, national Association of Corrosion Engineers (May 1991).
5 Shook, C., and Roco, M., "Slurry Flow Principles and Practice", Butterworth-Heinemann (1991).

WASTE WATER/SEWAGE PUMPS

The reader is referred to Chapter 6, 7 and 8 on Pump Types and the last Chapter (24) on Solids and Slurry Pumps for related material. It is hard to definitively distinguish between waste water and sewage because the terms are used too interchangeably. probably the best clarification is to say that sewage is one type of waste water that contains biodegradable waste. There are industrial and municipal waste waters, the former's character being tied to the products that are being manufactured and the latter being a combination of sewage and industrial wastes. Sewage is over 98% water. Pumps are used in the collection and treatment processes tied in with these waste water applications. In the sewage treatment process, the following broad steps occur: screening, during which large debris is removed; grit elimination, finer screening, during which a majority of the smaller debris is removed;

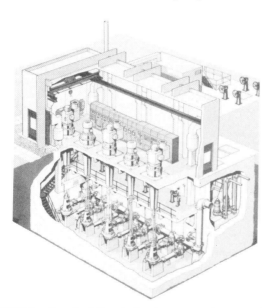

FIGURE 25.1 – 5 vertical line shaft volute pumps in a pumping station.

settling, aeration, further settling and chlorination. The sludge from the settling tanks is thickened, dewatered, incinerated and the ash disposed of. Sewage varies widely from one treatment plant to another to a degree that pumps that have thousands of successful applications will often become clogged with the ingredients of a particular treatment plant's process fluid.

Fig. 25.1 shows a cutaway view of a modern pumping station with 5 vertical line shaft type pumps in the dry pit (below ground) taking suction from the wet pit, which is not clearly shown. It is common for the dry pits in such stations to be flooded with the influent at one time or another from a variety of causes. In this case, the motors are installed in the upper floor, (above ground) to protect the motors and starters from such an occurrence. Fig. 25.2 shows 54 in. gear driven, mixed flow line shaft sewage pumps in the installation process at Sacramento, California, USA. COS are 86, 800 gpm (328, 500 l/min), 36 ft. (11 m), 1250 hp (932 kW) and 250 rpm. Horizontal waste water pumps

FIGURE 25.2 – 54 inch vertical line shaft pump.

WASTE WATER/SEWAGE PUMPS

installed down in the dry pit are shown in Fig. 25.3. The latter is a photograph of the Dokhaven sewage station in Rotterdam. Fig. 25.4 shows a horizontal centrifugal waste water pump pumping sewage water in The Netherlands. Fig. 25.24 shows 4 out of a bank of 6 vertical centrifugal sewage pumps at the Bei Xiao He treatment plant in Beijing City, China.

Piston diaphragm and progressing cavity pumps have been very successful in the

FIGURE 25.3 – Bank of horizontal waste water pumps in dry pit at Dokhaven Sewage Station, Rotterdam.

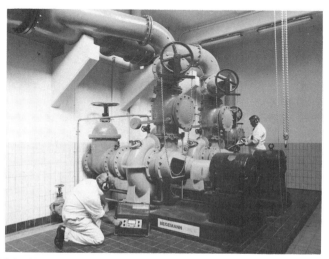

FIGURE 25.4 – Horizontal centrifugal waste water pump on sewage water, Netherlands.

transportation of sewage and industrial sludges. Fig. 25.5 shows a hydraulically driven piston pump with transfer tube for the transport of centrifuged sludge in a Dutch sewage treatment plant and Fig. 25.6 gives the accompanying internal views respectively. Fig. 25.7 shows a progressing cavity pump pumping primary abrasive sludge. Macerator pumps which reduce the solid debris to acceptable size and chopper pumps, Figs. 25.6-25.23 and 25.25-25.23 fill a need. Vortex or torque flow pumps are used to handle light gritty sludges, see Fig. 25.8, a photo of a vortex pump installation in an Ontario, Canada waste treatment plant. These pumps are used in many cases where materials in the sludge cause clogging of the conventional sewage pumps. Screw pumps are sometimes called upon to handle sludge transport with low lift and short distances. Plunger type reciprocating units are also used on occasions for sludge pumping.

FIGURE 25.5 – Hydraulically driven piston pump for transport of centrifugal sludge, Dutch sewage plant.

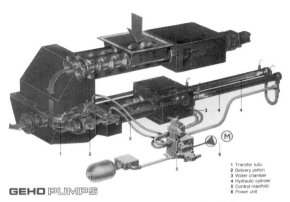

FIGURE 25.6 – Cutaway views of Fig. 25.5 pump.

WASTE WATER/SEWAGE PUMPS

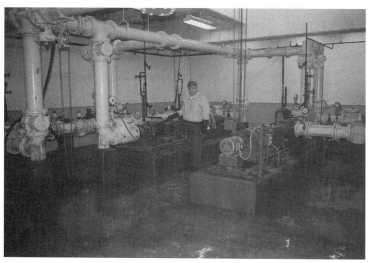

FIGURE 25.7– Progressing cavity pump on primary sludge at a wastewater treatment plant.

FIGURE 25.8 – Vortex pump in wastewater treatment plant, Ontario, Canada.

Submersibles are increasing their penetration into this market. They offer many advantages due to their hermetic nature and the only drawback appears to be a few points difference in overall efficiency. Two units in parallel operation in a wet well are shown in Fig. 25.9. A small submersible is shown in Fig. 25.10. Fig. 25.11 shows a bank of submersibles in a dry pit handling activated sludge in a Albuquerque, New Mexico, USA sewage facility. These pumps are being installed in dry wells because of their immunity

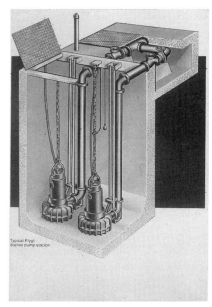

FIGURE 25.9 – Two submersibles in a wet well on sewage service.

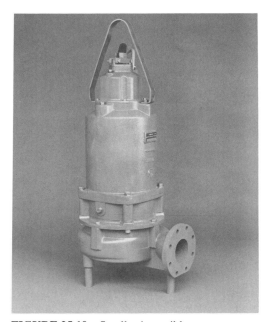

FIGURE 25.10 – Small submersible sewage pump.

FIGURE 25.11 – Bank of submersibles in a dry pit handling activated sludge. Alburquerque, New Mexico, USA.

FIGURE 25.12 – Bank of 40 MGD submersibles in a wet well. Houston, Texas, USA.

to drywell flooding mishaps. Fig. 25.12 shows a bank of 40 MGD submersibles in a 34 ft deep wet well in Oakland, California, USA. Fig 25.25 shows a pair of large submersibles being installed in a Houston, Texas, USA sewage treatment plant wet well.

FIGURE 25.13 – Bank of diaphragm pumps in municipal waste treatment plant moving sewage and sludge.

FIGURE 25.14 – Diaphragm pump in automotive plant waste treatment facility.

FIGURE 25.15 – Two pump (diaphragm) system in a licorice manufacturer's waste sludge system.

WASTE WATER/SEWAGE PUMPS

Diaphragm pumps are often used in waste water services. Fig. 25.13 shows a bank of five diaphragm pumps in a municipal waste treatment plant. Fig. 25.14 shows one in an industrial waste treatment application, and Fig. 25.15 shows a two pump system at a licourice manufacturer pumping waste sludge to a plate and frame filter press.

Two large and one smaller vertical turbine type diffuser pumps are shown in Fig. 25.16, these pumps are installed in the Metropolitan Water, Drainage and Sewerage Board service in Sydney, Australia. The two larger units when operating in parallel can achieve 5500 l/s at 34 m head. Five diesel driven axially split pumps pump reconditioned water through a 9.6 km rising main from the South Eastern Purification Plant in Melbourne, Australia to Bass Strait some 60 km away are shown in Fig. 25.17.

FIGURE 25.16 – Two large and one smaller vertical turbine type pumps. Sydney, Australia.

FIGURE 25.17 – Five diesel driven axially split pumps on reconditioned water. Melbourne, Australia.

FIGURE 25.18 – Reciprocating sewer cleaning pump.

FIGURE 25.19 – Diaphragm metering pumps on waste treatment plant additive addition, e.g. sodium hypochlorite.

A reciprocating sewer cleaning pump is shown in Fig. 25.18. This pump has a capacity of 550 l/min. with a head of 1200 bar. Diaphragm metering pumps are used for the controlled addition of additives to the waste treatment process. Fig. 25.19 shows a sodium

WASTE WATER/SEWAGE PUMPS 345

hypochlorite metering installation, Fig. 25.20 shows a potassium permanganate metering installation, Fig. 25.21 a lime slurry metering system with variable speed drive pumps in the Newport Rhode Island, USA waste treatment plant and Fig. 25.22 shows a polymer coagulant metering system at the Clinton, Massachussetts waste water treatment plant. Fig. 25.23 shows a chopper pump handling corn husks.

FIGURE 25.20 - Potassium permangenate metering installation with diaphragm pumps.

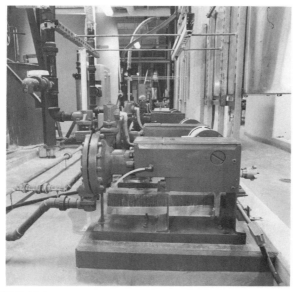

FIGURE 25.21 – Lime slurry metering system, Newport, R.I., USA.

FIGURE 25.22 – Diaphragm metering pump on polymer coagulant system. Clinton, Massachusetts, USA.

FIGURE 25.23 – Chopper pump handling corn husks.

WASTE WATER/SEWAGE PUMPS 347

FIGURE 25.24 – Four of six vertical centrifugal sewage pumps at Bei Xiao He treatment plant in Beijing City, China.

FIGURE 25.25 – Waste water submersibles, Houston, Texas, USA.

Speed control and lead /lag pump control are common in this industry. See Chapter 17.

With the reduction in cost and technological advancement of variable frequency drives one can expect the percentage of variable speed units to increase substantially in this application area.

Packages of pumps in the waste treatment area parallels that of firepumps. Many packages are furnished with typically two or more pumps in parallel or series mounted on a base complete with interconnecting piping and controls.

SECTION 8

APPENDIX

IAHR MEMBERSHIP: YOU CAN'T AFFORD TO MISS IT!

The International Association for Hydraulic Research(IAHR) founded in 1935, is a global independent organization of and for engineers and scientists engaged in hydraulics and its practical application. IAHR particularly promotes the exchange of knowledge through congresses, symposia, research committees and publications in the fields of water resources, river and coastal hydraulics, risk analysis, energy, environment, disaster prevention, industrial processes.

Take the opportunity to join the international hydraulics community now!

IAHR includes 14 Sections in 3 Technical Divisions which cater to the scientific needs of its members. The Sections regularly organize conferences and symposia in their own specific fields, provide a basis for cooperation on specific research projects, develop monographs, etc. Three regional Divisions - Latin America, Asia and Pacific, and Africa - focus on organization of activities, including Congresses, specific to their region.
In addition, broad issues of interest to the members of IAHR are tackled in task groups focused on Continuing Education and Training, and Management of Hydraulic Research.

IAHR - the international doorway to enrichment of your professional life and broadening of your insights in hydraulic research and its related fields.

Categories	1995 Fees	Most important benefits
Individual	NLG 140 US$ 80	High quality scientific Journal of Hydraulic Research and Newsletter (with extensive Conference Calendar twice a year). Other IAHR publications at reduced rates. Reduced rates at IAHR (co-sponsored) conferences and symposia.
Corporation	NLG 70 US$ 40	This option is open to engineers, and scientists whose employer is a corporate member of IAHR. Benefits are as shown above but without the Journal of Hydraulic Research. Your employer, as a corporate member, receives 2 copies for his/her library.
Corporate	NLG 1500 US$ 860 NLG 900* US$ 515*	This category is open to universities, research institutes, engineering firms, and consultants. Journal of Hydraulic Research and Newsletter in duplicate. One set of proceedings of the IAHR biennial congresses; other benefits as with 'individual'. (* = rates for small organizations)

We look forward to being able to welcome you as a member! If you have not been able to decide yet, we will be happy to provide you with more detailed information; please do not hesitate to contact us at:
telephone 31-15-569353 or fax 31-15-619674 - IAHR Secretariat - P.O.Box 177 - 2600 MH Delft, the Netherlands

MEMBERSHIP APPLICATION FORM

Name _____ Date of birth _____

Position/affiliation _____

Address _____

Telephone _____ Fax _____ Telex _____

IAHR Corporate member registration no. (See Register of Members) C _____

Year from which membership to run: 1 January 19 _____

Payment by:
☐ international postal money order
☐ postal transfer (no. 639739)
☐ bank transfer (no. 44.20.42.000 with AMRO Bank, Delft, the Netherlands)
☐ Eurocheque
☐ Credit card
My credit card may be charged with NLG _____

☐ Eurocard   ☐ MasterCard   ☐ Access   ☐ Visa   ☐ American Express

Card number ☐☐☐☐ ☐☐☐☐ ☐☐☐☐ ☐☐☐☐

Expiry date (month/year) _____ Signature _____

APPENDIX

REFERENCE DIAGRAMS
Vapour Pressure Curves for Various Liquids 351
Friction Factors for Pipes 352
Friction Factors and Relative Roughness for Commercial Pipes 353
Pipe Friction Loss Coefficients L 354
Pipe Bend, Resistance Loss Coefficiency Z 355
Nomograph for Hazon-Williams Formula 356
Kinematic Viscosity of Various Mineral Oil Distillates as a Function of
    Temperature 357
Overall Pressure Loss Across Thin-Plate Orifices 358
Overall Pressure Loss Across Flow Nozzles 359
Typical Calibration Curve of a Low β flow Nozzle with Throat Pressure Taps 360
Overall Pressure Loss Through Venturi Tubes 361

TABLES
Average Value of 'C' for Cast Iron Pipe 362
Values of 'C' for Ductile Iron Pipe 362
Representative Resistance Coefficients (K) for Valves and Fittings 363-369
Average Values of (n) for Use in Manning's Equation 370
Head Loss in Feet from Resistance Coefficient 371-373
Friction Loss in Valves and Fittings 374
Resistance Coefficient of Fittings 375
Equivalent Straight Lengths of Fittings in Feet 376
Equivalent Straight Lengths of Pipe Preceding and Following Orifices,
    Flow Nozzles and Venturi Tubes 379-380
Specific Gravity of Suspensions of Solids in Water 381

FAULT-FINDING TABLES

Packing 382
Modes, Causes and Corrections of Mechanical Face Seal Failures 383-386
Centrifugal Pumps 387
Reciprocating Pumps 388
Reciprocating (Piston) Pumps 389
Rotary Pumps 390
Lobe-Rotor Pumps 391-392
Fault-Finding and Maintenance Guide for Typical High Pressure Piston
   and Plunger Pumps 393-394

CONVERSION TABLES 395-397

APPENDIX 351

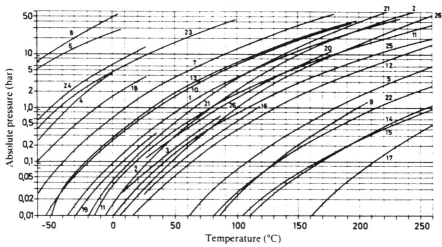

1 Acetone	10 Gasoline	19 Hexane
2 Ethyl alcohol	11 Benzene	20 Kerosene
3 Formic acid	12 Chlorobenzene	21 Methyl alcohol
4 Ammonia	13 Diethylether	22 Naphthalene
5 Aniline	14 Diphenyl	23 Propane
6 Ethane	15 Dowtherm A	24 Propylene
7 Ethyl chloride	16 Acetic acid	25 Toluene
8 Ethylene	17 Glycerine	26 Water
9 Ethylene glycol	18 Isobutane	

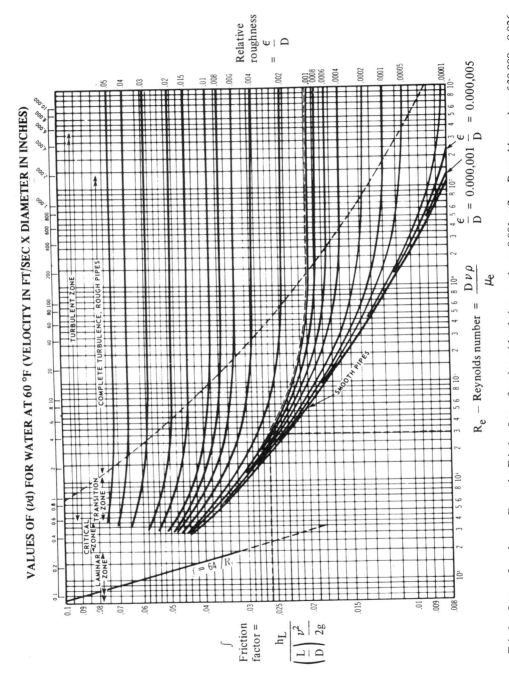

Friction factors for pipes. – Example: Friction factor for pipe with relative roughness 0.001 at flow Reynolds number of 30 000 = 0.026.

APPENDIX

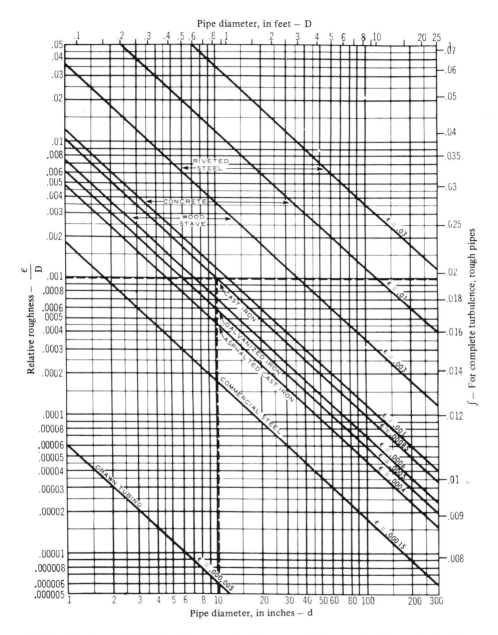

Friction Factors and Relative Roughness for Commercial Pipes with fully turbulent flow (based on ASME data originated by L. F. Moody). – Example: For 10 in diameter cast iron pipe, relative roughness (E/D) = 0.00085. Friction factor = 0.0196.

Laminar flow	Turbulent flow		
	Hydraulically smooth pipes	Hydraulically rough pipes	Pipes in transient range
	Limits: $R_e \cdot \dfrac{k}{d_i} < 65$	Limits: $R_e \cdot \dfrac{k}{d_i} > 1300$	Limits: $65 < R_e \cdot \dfrac{k}{d_i} < 1300$
Formula for λ $\lambda = \dfrac{64}{R_e}$ $\lambda = \dfrac{64 \cdot v}{c \cdot d_i}$	Formula for λ: a) Blasius formula for the range $2320 < R_e < 10^5$ $$\lambda = 0.3164 \cdot R_e^{-0.25}$$ b) Nikuradse formula for the range $10^5 < R_e < 5 \cdot 10^6$ $$\lambda = 0.0032 + 0.221 \cdot R_e^{-0.237}$$ c) Prandtl and von Kármán formula for the range $R_e > 10^6$ $$\dfrac{1}{\sqrt{\lambda}} = 2\lg(R_e \cdot \sqrt{\lambda}) - 0.8$$	Formula for λ: Nikuradse formula $$\dfrac{1}{\sqrt{\lambda}} = 2\lg\dfrac{d_i}{k} + 1.14$$	Formula for λ: Prandtl-Colebrook formula $$\dfrac{1}{\sqrt{\lambda}} = -2\lg\left[\dfrac{2.51}{R_e \sqrt{\lambda}} + \dfrac{k}{d_i} \cdot 0.269\right]$$

Pipe Friction Loss Coefficients L

APPENDIX

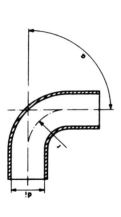

$\dfrac{L}{d_i}$ α	1	1,5	2	4	6	
15°	0.03	0.03	0.03	0.03	0.03	Interior pipe wall *smooth*
30°	0.07	0.07	0.07	0.07	0.07	
45°	0.14	0.11	0.09	0.08	0.075	
60°	0.19	0.16	0.12	0.10	0.09	
90°	0.21	0.18	0.14	0.11	0.09	
15°	0.10	0.08	0.06	0.05	0.04	Interior pipe wall *rough*
30°	0.23	0.19	0.14	0.11	0.08	
45°	0.34	0.27	0.20	0.15	0.12	
60°	0.41	0.33	0.24	0.19	0.15	
90°	0.51	0.41	0.30	0.23	0.18	

Pipe Bend, Resistance Loss Coefficient Z

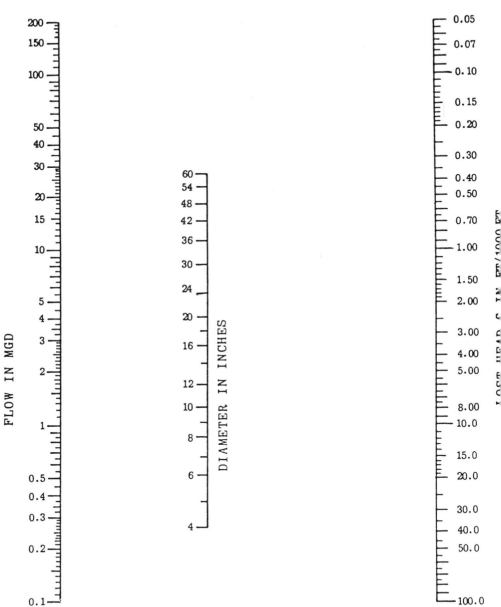

Nomograph for Hazen-Williams Formula

APPENDIX

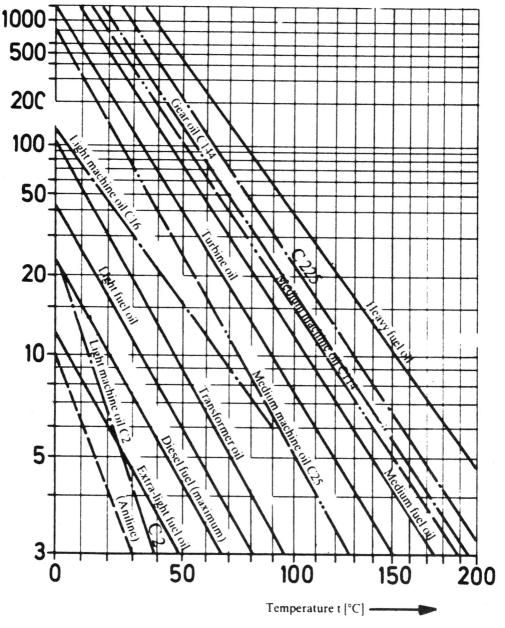

Kinematic Viscosity of Various Mineral Oil Distillates as a Function of Temperature

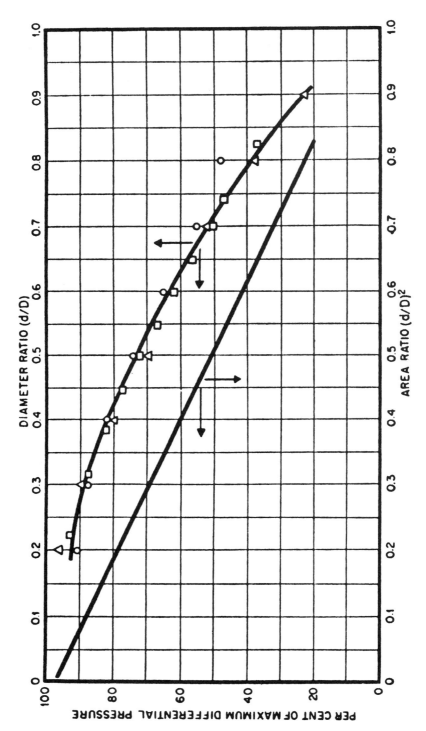

Overall Pressure Loss Across Thin-Plate Orifices

APPENDIX

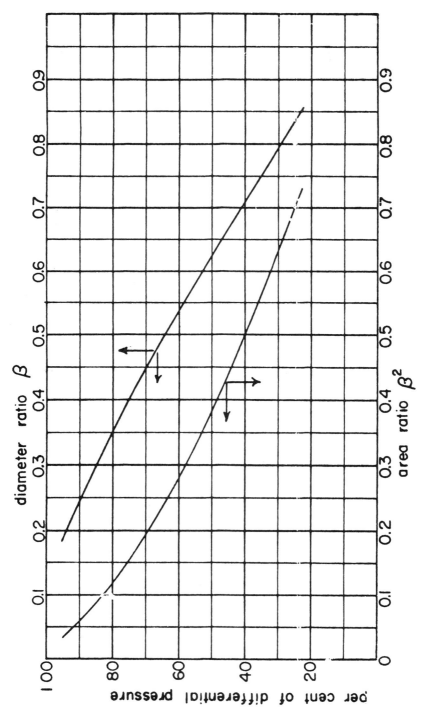

Overall Pressure Loss Across Flow Nozzles

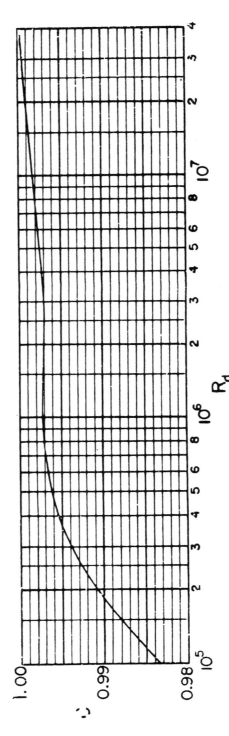

Typical Calibration Curve of a Low β Flow Nozzle with Throat Pressure Taps

APPENDIX

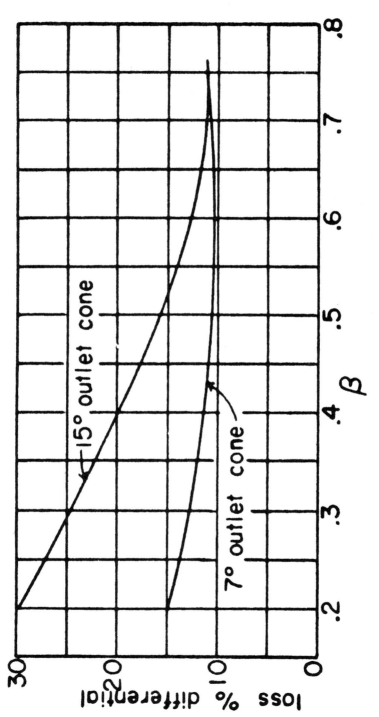

Overall Pressure Loss Through Venturi Tubes

TABLE 3A-1 – Average value of C for Cast-Iron Pipe.

Diameter of Pipe, in.	Age in Years						
	0	5	10	20	30	40	50
4	130	118	107	89	75	64	55
8	130	119	109	93	83	73	65
12	130	120	111	96	86	77	70
16	130	120	112	98	87	80	72
24	130	120	113	100	89	81	74
60	130	120	113	100	90	83	77

TABLE 3A-2 – Values of C for Ductile Iron Pipe.

Service	Type of pipe	Approximate value of 'C' for pipe of nominal size DN				
		80	150	300	600	1200
Raw and potable water pipelines	Coated spun ductile iron.	137	132	145	148	148
	Cement mortar-lined and bitumen-lined spun ductile iron.	147	149	150	152	153

APPENDIX

TABLE 5A – Representative Resistance Coefficients (K) for Valves and Fittings.

Pipe Friction Data for Clean Commercial Steel Pipe with Flow in Zone of Complete Turbulence

Metric (cm)	1.58	2.09	2.54	35.05	40.9	52.5	63, 78	102	128	154	203-255	303-381	429-57
Nominal size (in)	½	¾	1	1¼	1½	2	2½, 3	4	5	6	8-10	12-16	18-24
Friction Factor (fr)	.027	.025	.023	.022	.021	.019	.018	.017	.016	.015	.014	.013	.012

Multiply pressure drop in ft by 0.305 to get m.

Formulas for Calculating "K" Factors* for Valves and Fittings with Reduced Port

Formula 1

$$K_2 = \frac{0.8 \left(\sin \frac{\theta}{2}\right)(1 - \beta^2)}{\beta^4} = \frac{K_1}{\beta^4}$$

Formula 2

$$K_2 = \frac{0.5 (1 - \beta^2)\sqrt{\sin \frac{\theta}{2}}}{\beta^4} = \frac{K_1}{\beta^4}$$

Formula 3

$$K_2 = \frac{2.6 \left(\sin \frac{\theta}{2}\right)(1 - \beta^2)^2}{\beta^4} = \frac{K_1}{\beta^4}$$

Formula 4

$$K_2 = \frac{(1 - \beta^2)^2}{\beta^4} = \frac{K_1}{\beta^4}$$

Formula 5

$$K_2 = \frac{K_1}{\beta^4} + \text{Formula 1} + \text{Formula 3}$$

$$K_2 = \frac{K_1 + \sin \frac{\theta}{2} [0.8 (1 - \beta^2) + 2.6 (1 - \beta^2)^2]}{\beta^4}$$

*Use "K" furnished by valve or fitting supplier when available

Formula 6

$$K_2 = \frac{K_1}{\beta^4} + \text{Formula 2} + \text{Formula 4}$$

$$K_2 = \frac{K_1 + 0.5\sqrt{\sin \frac{\theta}{2}} (1 - \beta^2) + (1 - \beta^2)^2}{\beta^4}$$

Formula 7

$$K_2 = \frac{K_1}{\beta^4} + \beta (\text{Formula 2} + \text{Formula 4})$$
when $\theta = 180°$

$$K_2 = \frac{K_1 + \beta \left[0.5 (1 - \beta^2) + (1 - \beta^2)^2\right]}{\beta^4}$$

$$\beta = \frac{d_1}{d_2}$$

$$\beta^2 = \left(\frac{d_1}{d_2}\right)^2 = \frac{a_1}{a_2}$$

Subscript 1 defines dimensions and coefficients with reference to the smaller diameter.

Subscript 2 refers to the larger diameter.

TABLE 5A (continued) – Representative Resistance Coefficients (K) for Valves and Fittings.

SUDDEN AND GRADUAL CONTRACTION	SUDDEN AND GRADUAL ENLARGEMENT
If: $\theta \lesssim 45°$ .... K_2 = Formula 1 $45° < \theta \lesssim 180°$ .... K_2 = Formula 2	If: $\theta \lesssim 45°$ .... K_2 = Formula 3 $45° < \theta \lesssim 180°$ .... K_2 = Formula 4 F_T = *Friction factor in zone of complete turbulence*

GATE VALVES
Wedge Disc, Double Disc, or Plug Type

If: $\beta = 1, \theta = O$ .... $K_1 = 8 f_T$
$\beta < 1$ and $\theta \lesssim 45°$ .... K_2 = Formula 5
$\beta < 1$ and $45° < \theta \lesssim 180°$ .... K_2 = Formula 6

GLOBE AND ANGLE VALVES

If: $\beta = 1$ .... $K_1 = 340\ f_T$

If: $\beta = 1$ .. $K_1 = 150\ f_T$ If: $\beta = 1$ ... $K_1 = 55\ f_T$

All globe and angle valves,
whether reduced seat or throttled,

If: $\beta = 1$ .... $K_1 = 55\ f_T$ If: $\beta < 1$ .... K_2 = Formula 7

TABLE 5A (continued) – Representative Resistance Coefficients (K) for Valves and Fittings.

LIFT CHECK VALVES

If: $\beta = 1$ .... $K_1 = 600\, f_T$
If: $\beta < 1$ .... $K_2 =$ Formula 7
Minimum pipe velocity (fps) for full disc lift
$= 40\, \beta^2 \sqrt{\overline{V}}$

If: $\beta = 1$ .... $K_1 = 55\, f_T$
If: $\beta < 1$ .... $K_2 =$ Formula 7
Minimum pipe velocity (fps) for full disc lift
$= 140\, \beta^2 \sqrt{\overline{V}}$

SWING CHECK VALVES

$K = 100\, f_T$
Minimum pipe velocity
(fps) for full disc lift
$= 35 \sqrt{\overline{V}}$

$K = 50\, f_T$
Minimum pipe velocity (fps)
for full disc lift
$= 60 \sqrt{\overline{V}}$ except
U/L listed $= 100 \sqrt{\overline{V}}$

TILTING DISC CHECK VALVES

	$\alpha = 5°$	$\alpha = 15°$
Sizes 2 to 8" .... K =	$40\, f_T$	$120\, f_T$
Sizes 10 to 14" .... K =	$30\, f_T$	$90\, f_T$
Sizes 16 to 48" .... K =	$20\, f_T$	$60\, f_T$
Minimum pipe velocity (fps) for full disc lift =	$80 \sqrt{\overline{V}}$	$30 \sqrt{\overline{V}}$

TABLE 5A (continued) – Representative Resistance Coefficients (K) for Valves and Fittings.

STOP-CHECK VALVES
(Globe and Angle Types)

If: $\beta = 1$ .... $K_1 = 400\ f_T$
$\quad\ \ \beta < 1$ .... $K_2 =$ Formula 7

Minimum pipe velocity for full disc lift
$$= 55\ \beta^2 \sqrt{\overline{V}}$$

If: $\beta = 1$ .... $K_1 = 200\ f_T$
$\quad\ \ \beta < 1$ .... $K_2 =$ Formula 7

Minimum pipe velocity for full disc lift
$$= 75\ \beta^2 \sqrt{\overline{V}}$$

If: $\beta = 1$ .... $K_1 = 300\ f_T$
$\quad\ \ \beta < 1$ .... $K_2 =$ Formula 7

If: $\beta = 1$ .... $K_1 = 350\ f_T$
$\quad\ \ \beta < 1$ .... $K_2 =$ Formula 7

Minimum pipe velocity (fps) for full disc lift
$$= 60\ \beta^2 \sqrt{\overline{V}}$$

If: $\beta = 1$ .... $K_1 = 55\ f_T$
$\quad\ \ \beta < 1$ .... $K_2 =$ Formula 7

If: $\beta = 1$ .... $K_1 = 55\ f_T$
$\quad\ \ \beta < 1$ .... $K_2 =$ Formula 7

Minimum pipe velocity (fps) for full disc lift
$$= 140\ \beta^2 \sqrt{\overline{V}}$$

TABLE 5A (continued) – Representative Resistance Coefficients (K) for Valves and Fittings.

FOOT VALVES WITH STRAINER

Poppet Disc   **Hinged Disc**

$K_1 = 420 \, f_T$ $K_1 = 75 \, f_T$

Minimum pipe velocity (fps) for full disc lift   Minimum pipe velocity (fps) for full disc lift

$= 15 \sqrt{\overline{V}}$ $= 35 \sqrt{\overline{V}}$

BALL VALVES	BUTTERFLY VALVES
If: $\beta = 1 \;\; \theta = O$ .... $K_1 = 3 \, f_T$ $\beta < 1$ and $\theta \lesssim 45°$ .... K_2 = Formula 5 $\beta < 1$ and $45° < \theta \lesssim 180°$ .. K_2 - Formula 6	Sizes 2 to 8" .... $K = 45 \, f_T$ Sizes 10 to 14" .... $K = 35 \, f_T$ Sizes 16 to 24" .... $K = 25 \, f_T$

PLUG VALVES AND COCKS

Straight-Way   **3-Way**

If: $\beta = 1$ If: $\beta = 1$ If: $\beta = 1$
$K_1 = 18 \, f_T$ $K_1 = 30 \, f_T$ $K_1 = 90 \, f_T$

If: $\beta < 1$ .... K_2 = Formula 6

TABLE 5A (continued) – Representative Resistance Coefficients (K) for Valves and Fittings.

MITRE BENDS

α	K
0°	2 f_T
15°	4 f_T
30°	8 f_T
45°	15 f_T
60°	25 f_T
75°	40 f_T
90°	60 f_T

90° PIPE BENDS AND FLANGED OR BUTT-WELDING 90° ELBOWS

r/d	K	r/d	K
1	20 f_T	3	24 f_T
1.5	14 f_T	10	30 f_T
2	12 f_T	12	34 f_T
3	12 f_T	14	38 f_T
4	14 f_T	16	42 f_T
6	17 f_T	20	50 f_T

The resistance coefficient, K_B, for pipe bends other than 90° may be determined as follows:

$$K_B = (n-1)\left(0.25\,\pi\,f_T\,\frac{r}{d} + 0.5\,K\right) + K$$

n = number of 90° bends
K = resistance coefficient for one 90° bend (per table)

STANDARD ELBOWS

90°   45°

K = 30 f_T K = 16 f_T

APPENDIX

TABLE 5A (continued) – Representative Resistance Coefficients (K) for Valves and Fittings.

CLOSE PATTERN RETURN BENDS

$$K = 50\, f_T$$

STANDARD TEES

Flow thru run .... $K = 20\, f_T$
Flow thru branch .... $K = 60\, f_T$

PIPE ENTRANCE

Inward Projecting   **Flush**

r/d	K
0.00*	0.5
0.02	0.28
0.04	0.24
0.06	0.15
0.10	0.09
0.15 & up	0.04

*Sharp edged

$K = 0.78$ For K see table

PIPE ENTRANCE

Projecting   **Sharp-Edged**   **Rounded**

$K = 1.0$ $K = 1.0$ $K = 1.0$

TABLE 7A – Average values of (*n*) for use in Manning's Equation.

Nature of surface	
Wood	0.013
Finished concrete	0.012
Unfinished concrete	0.014
Brick	0.016
Earth, good condition	0.025
Earth, with stones or weeds	0.035
Gravel	0.028
Vitrified sewer pipe	0.013
Cast iron	0.014

APPENDIX

TABLE 8A – Head loss in feet from Resistance Coefficient

Resistance coefficient	FLOW VELOCITY – ft/s (m/s)									
	1 (0.3)	2 (0.6)	3 (0.9)	4 (1.2)	5 (1.5)	6 (1.8)	7 (2.1)	8 (2.4)	9 (2.75)	10 (3)
0.05	0.00080	0.003	0.007	0.013	0.019	0.030	0.039	0.050	0.063	0.077
0.1	0.00200	0.006	0.014	0.025	0.039	0.060	0.076	0.100	0.126	0.155
0.2	0.00300	0.012	0.028	0.050	0.078	0.112	0.152	0.199	0.252	0.310
0.3	0.00500	0.019	0.042	0.075	0.117	0.168	0.228	0.299	0.377	0.470
0.4	0.00600	0.025	0.056	0.100	0.155	0.224	0.304	0.398	0.503	0.620
0.5	0.00800	0.031	0.070	0.125	0.194	0.280	0.381	0.497	0.629	0.780
0.6	0.00900	0.037	0.084	0.149	0.233	0.335	0.457	0.597	0.755	0.930
0.7	0.01100	0.044	0.098	0.174	0.271	0.391	0.533	0.696	0.880	1.090
0.8	0.01200	0.050	0.112	0.199	0.311	0.447	0.609	0.796	1.006	1.240
0.9	0.01400	0.056	0.125	0.214	0.349	0.503	0.685	0.895	1.132	1.400
1.0	0.01600	0.062	0.140	0.249	0.388	0.559	0.761	0.995	1.258	1.550
1.1	0.01710	0.068	0.154	0.274	0.427	0.615	0.837	1.090	1.380	1.710
1.2	0.01860	0.074	0.168	0.299	0.466	0.671	0.913	1.190	1.510	1.860
1.3	0.02020	0.081	0.182	0.324	0.505	0.727	0.989	1.290	1.630	2.020
1.4	0.02170	0.087	0.196	0.349	0.544	0.783	1.065	1.390	1.760	2.170
1.5	0.02330	0.093	0.210	0.374	0.583	0.839	1.141	1.490	1.890	2.330
1.6	0.02480	0.099	0.224	0.398	0.621	0.895	1.217	1.590	2.020	2.480
1.7	0.02640	0.106	0.238	0.423	0.660	0.951	1.293	1.690	2.140	2.640
1.8	0.02800	0.112	0.252	0.448	0.699	1.007	1.369	1.790	2.270	2.800
1.9	0.02950	0.118	0.266	0.473	0.738	1.063	1.445	1.890	2.390	2.950
2.0	0.03106	0.124	0.280	0.497	0.776	1.118	1.522	1.990	2.520	3.110
2.1	0.03261	0.130	0.294	0.522	0.815	1.174	1.598	2.090	2.650	3.260
2.2	0.03416	0.136	0.308	0.547	0.854	1.230	1.674	2.190	2.780	3.420
2.3	0.03571	0.143	0.322	0.572	0.893	1.286	1.750	2.290	2.900	3.570
2.4	0.03726	0.149	0.336	0.597	0.932	1.341	1.826	2.390	3.030	3.730
2.5	0.03881	0.155	0.350	0.621	0.971	1.397	1.902	2.490	3.150	3.880
2.6	0.04036	0.161	0.364	0.646	1.009	1.453	1.978	2.590	3.280	4.040
2.7	0.04191	0.168	0.378	0.671	1.048	1.509	2.054	2.690	3.400	4.190
2.8	0.04346	0.174	0.392	0.696	1.087	1.565	2.130	2.790	3.530	4.350
2.9	0.04501	0.180	0.406	0.721	1.126	1.621	2.206	2.890	3.650	4.500
3.0	0.04659	0.186	0.419	0.746	1.165	1.677	2.283	2.990	3.770	4.660
3.1	0.04814	0.192	0.433	0.771	1.204	1.733	2.359	3.090	3.900	4.810
3.2	0.04969	0.198	0.447	0.796	1.243	1.789	2.435	3.190	4.030	4.970
3.3	0.05124	0.205	0.461	0.821	1.282	1.845	2.511	3.290	4.150	5.120
3.4	0.05279	0.211	0.475	0.845	1.321	1.901	2.587	3.390	4.280	5.280
3.5	0.05434	0.217	0.489	0.870	1.360	1.957	2.663	3.490	4.400	5.430

Footnote: For head loss in metres, factor by 0.3

contd...

TABLE 8A (continued) – Head loss in feet from Resistance Coefficient

Resistance coefficient	FLOW VELOCITY – ft/s (m/s)									
	1 (0.3)	2 (0.6)	3 (0.9)	4 (1.2)	5 (1.5)	6 (1.8)	7 (2.1)	8 (2.4)	9 (2.75)	10 (3)
3.6	0.05589	0.223	0.503	0.895	1.398	2.013	2.739	3.590	4.530	5.590
3.7	0.05744	0.230	0.517	0.920	1.437	2.069	2.815	3.680	4.650	5.740
3.8	0.05899	0.236	0.531	0.945	1.476	2.125	2.891	3.780	4.780	5.900
3.9	0.06054	0.242	0.545	0.970	1.514	2.181	2.967	3.880	4.900	6.050
4.0	0.06212	0.248	0.559	0.994	1.553	2.236	3.044	3.980	5.030	6.210
4.1	0.06367	0.254	0.573	1.019	1.592	2.292	3.120	4.080	5.160	6.370
4.2	0.06522	0.260	0.587	1.044	1.631	2.348	3.196	4.180	5.290	6.520
4.3	0.06677	0.267	0.601	1.069	1.670	2.404	3.272	4.280	5.410	6.680
4.4	0.06832	0.273	0.615	1.093	1.709	2.460	3.348	4.380	5.540	6.830
4.5	0.06987	0.279	0.629	1.118	1.748	2.516	3.424	4.480	5.660	6.990
4.6	0.07142	0.285	0.643	1.143	1.786	2.572	3.500	4.580	5.790	7.140
4.7	0.07297	0.292	0.657	1.168	1.825	2.628	3.576	4.680	5.910	7.300
4.8	0.07452	0.298	0.671	1.193	1.864	2.684	3.652	4.770	6.040	7.450
4.9	0.07607	0.304	0.685	1.218	1.903	2.740	3.728	4.870	6.160	7.600
5.0	0.07765	0.311	0.698	1.243	1.942	2.795	3.806	4.97	6.29	7.77
5.1	0.07920	0.317	0.712	1.268	1.981	2.851	3.882	5.07	6.41	7.92
5.2	0.08075	0.323	0.726	1.293	2.020	2.907	3.958	5.17	6.54	8.08
5.3	0.08230	0.330	0.740	1.318	2.058	2.963	4.034	5.27	6.66	8.23
5.4	0.08388	0.336	0.755	1.342	2.097	3.019	4.110	5.37	6.79	8.39
5.5	0.08543	0.342	0.769	1.367	2.136	3.075	4.186	5.47	6.91	8.54
5.6	0.08698	0.348	0.783	1.392	2.175	3.131	4.262	5.57	7.04	8.70
5.7	0.08853	0.355	0.797	1.417	2.213	3.187	4.338	5.67	7.17	8.85
5.8	0.09008	0.361	0.811	1.442	2.252	3.243	4.414	5.77	7.30	9.01
5.9	0.09163	0.367	0.825	1.467	2.290	3.299	4.490	5.87	7.42	9.16
6.0	0.09318	0.373	0.839	1.491	2.329	3.354	4.567	5.97	7.55	9.32
6.1	0.09473	0.379	0.853	1.516	2.368	3.410	4.643	6.07	7.68	9.47
6.2	0.09628	0.385	0.867	1.541	2.407	3.466	4.719	6.17	7.81	9.63
6.3	0.09783	0.392	0.881	1.566	2.446	3.522	4.795	6.27	7.93	9.78
6.4	0.09938	0.398	0.895	1.590	2.485	3.572	4.871	6.37	8.06	9.94
6.5	0.10093	0.404	0.909	1.615	2.524	3.634	4.947	6.47	8.18	10.09
6.6	0.10248	0.410	0.923	1.640	2.562	3.690	5.023	6.57	8.31	10.25
6.7	0.10403	0.417	0.937	1.665	2.601	3.746	5.099	6.67	8.43	10.40
6.8	0.10558	0.423	0.951	1.690	2.640	3.802	5.175	6.76	8.56	10.56
6.9	0.10713	0.429	0.965	1.715	2.679	3.858	5.251	6.86	8.68	10.71
7.0	0.10871	0.435	0.979	1.740	2.717	3.913	5.328	6.96	8.80	10.87
7.1	0.11026	0.441	0.993	1.765	2.756	3.969	5.404	7.06	8.93	11.03
7.2	0.11181	0.447	1.007	1.790	2.795	4.025	5.480	7.16	9.06	11.18
7.3	0.11336	0.454	1.021	1.814	2.834	4.081	5.556	7.26	9.18	11.34
7.4	0.11491	0.460	1.035	1.839	2.873	4.137	5.632	7.36	9.31	11.49

Footnote: For head loss in metres, factor by 0.3

contd..

APPENDIX

TABLE 8A (continued) – Head loss in feet from Resistance Coefficient

Resistance coefficient	\multicolumn{10}{c}{FLOW VELOCITY – ft/s (m/s)}									
	1 (0.3)	2 (0.6)	3 (0.9)	4 (1.2)	5 (1.5)	6 (1.8)	7 (2.1)	8 (2.4)	9 (2.75)	10 (3)
7.5	0.11646	0.466	1.049	1.864	2.912	4.193	5.708	7.46	9.43	11.65
7.6	0.11801	0.472	1.063	1.888	2.950	4.249	5.784	7.56	9.56	11.80
7.7	0.11956	0.479	1.077	1.913	2.989	4.305	5.860	7.66	9.68	11.96
7.8	0.12111	0.485	1.091	1.938	3.028	4.361	5.936	7.76	9.81	12.11
7.9	0.12266	0.491	1.105	1.963	3.066	4.417	6.012	7.86	9.93	12.27
8.0	0.12424	0.497	1.118	1.988	3.105	4.472	6.089	7.96	10.06	12.42
8.1	0.12579	0.503	1.132	2.013	3.144	4.528	6.165	8.06	10.19	12.58
8.2	0.12734	0.509	1.146	2.038	3.183	4.584	6.241	8.16	10.32	12.73
8.3	0.12889	0.516	1.160	2.063	3.221	4.640	6.317	8.26	10.44	12.89
8.4	0.13044	0.522	1.174	2.087	3.260	4.696	6.393	8.36	10.57	13.04
8.5	0.13199	0.528	1.188	2.112	3.299	4.752	6.469	8.46	10.69	13.20
8.6	0.13354	0.534	1.202	2.137	3.338	4.808	6.545	8.55	10.81	13.35
8.7	0.13509	0.541	1.216	2.162	3.377	4.864	6.621	8.65	10.94	13.51
8.8	0.13664	0.547	1.230	2.187	3.416	4.920	6.697	8.75	11.07	13.66
8.9	0.13819	0.553	1.244	2.212	3.455	4.976	6.773	8.85	11.19	13.82
9.0	0.13977	0.559	1.258	2.237	3.493	5.031	6.850	8.95	11.32	13.98
9.1	0.14132	0.565	1.272	2.262	3.532	5.087	6.926	9.05	11.45	14.13
9.2	0.14287	0.571	1.286	2.287	3.571	5.143	7.002	9.15	11.58	14.29
9.3	0.14442	0.578	1.300	2.312	3.609	5.199	7.078	9.25	11.70	14.44
9.4	0.14597	0.584	1.314	2.336	3.648	5.255	7.154	9.35	11.83	14.60
9.5	0.14752	0.590	1.328	2.361	3.687	5.311	7.230	9.45	11.95	14.75
9.6	0.14907	0.596	1.342	2.386	3.726	5.367	7.306	8.55	12.08	14.91
9.7	0.15062	0.603	1.356	2.411	3.765	5.423	7.382	9.65	12.20	15.06
9.8	0.15217	0.609	1.370	2.435	3.804	5.479	7.458	9.75	12.33	15.22
9.9	0.15372	0.615	1.384	2.460	3.843	5.534	7.534	9.85	12.45	15.37
10.0	0.15530	0.621	1.398	2.485	3.882	5.589	7.612	9.94	12.58	15.53
11	0.17083	0.68	1.54	2.74	4.27	6.15	8.37	10.9	13.8	17.08
12	0.18636	0.74	1.68	2.99	4.66	6.71	9.13	11.9	15.1	18.64
13	0.20189	0.81	1.82	3.24	5.05	7.27	9.89	12.9	16.3	20.19
14	0.21742	0.87	1.96	3.49	5.44	7.83	10.65	13.9	17.6	21.74
15	0.23295	0.93	2.10	3.74	5.83	8.39	11.41	14.9	18.9	23.30
16	0.24848	0.99	2.24	3.98	6.21	8.95	12.17	15.9	20.2	24.85
17	0.26401	1.06	2.38	4.23	6.60	9.51	12.93	16.9	21.4	26.40
18	0.27954	1.12	2.52	4.48	6.99	10.07	13.69	17.9	22.7	27.95
19	0.29507	1.18	2.66	4.73	7.38	10.63	14.45	18.9	23.9	29.51
20	0.31060	1.24	2.80	4.97	7.76	11.18	15.22	19.9	25.2	31.06

Footnote: For head loss in metres, factor by 0.3

TABLE 8A-3 – Friction Loss in Valves and Fittings.

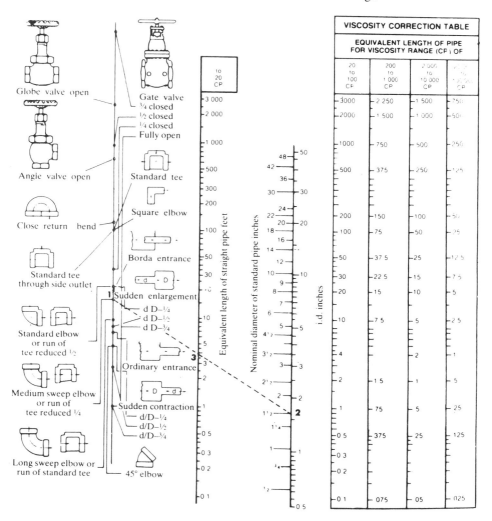

APPENDIX 375

TABLE 8A-2 – Resistance Coefficient of Fittings

Fitting	Pipe diameter – in (mm)									
	⅜ (10)	½ (12.5)	¾ (20)	1 (25)	1¼ (35)	1½ (40)	2 (50)	3 (75)	4 (100)	5 (125)
Integral pipe bend (turbulent flow) bend 3 x D	Approximately 0.04 (all sizes)									
Integral pipe bend (turbulent flow) bend 4 x D	Approximately 0.025 (all sizes)									
Integral pipe bend (laminar flow) bend 3 x D	Approximately 0.1 (all sizes)									
Integral pipe bend (laminar flow) bend 5 x D	Approximately 0.06 (all sizes)									
Integral pipe bend (laminar flow) bend 10 x D	Approximately 0.04 (all sizes)									
Standard 90° elbow (screwed)	2.40	2.10	1.70	1.50	1.25	1.15	1.00	0.80	0.70	0.55
Standard 90° elbow (flanged)	–	–	–	0.45	–	0.40	0.38	0.34	0.32	0.30
Large radius 90° elbow (screwed)	–	1.00	0.90	0.75	0.60	0.50	0.40	0.30	0.25	0.20
Large radius 90° elbow (flanged)	–	–	–	0.40	–	0.34	0.30	0.25	0.21	0.18
Standard 45° elbow (screwed)	0.39	0.37	0.36	0.35	0.34	0.32	0.30	0.29	0.28	0.27
Standard 45° elbow (flanged)	0.37	0.35	0.34	0.33	0.32	0.31	0.29	0.28	0.27	0.26
Large radius 45° elbow (screwed)	–	–	–	0.24	0.23	0.22	0.21	0.20	0.19	0.18
Large radius 45° elbow (flanged)	–	–	–	0.22	0.21	0.21	0.20	0.18	0.18	0.17
Return bend 180° (screwed)	2.40	2.10	1.70	1.50	1.38	1.25	0.96	0.78	0.68	0.58
Return bend 180° (flanged)	–	–	–	0.43	0.40	0.37	0.34	0.32	0.30	0.28
Large radius return bend 180° (screwed)	–	–	–	0.80	0.70	0.60	0.50	0.40	0.35	0.30
Large radius return bend 180° (flanged)	–	–	–	0.42	0.39	0.36	0.30	0.25	0.22	0.20
T – line flow (screwed)	Approximately 0.9 (all sizes)									
T – line flow (flanged)	–	–	–	0.26	0.24	0.22	0.18	0.16	0.14	0.13
T – branch flow (screwed)	2.50	2.40	2.00	1.80	1.60	1.50	1.40	1.20	1.10	1.00
T – branch flow (flanged)	–	–	–	1.00	–	0.90	0.80	0.72	0.64	0.62
Screw-down valve (straight)	–	12.00	–	10.00	–	–	8.00	7.00	6.00	5.00
Screw-down valve (right angle flow)		6.00	–	5.00	–	–	4.00	3.50	3.00	2.50
Gate valve (typical) (screwed)	–	–	–	–	0.23	0.20	0.17	0.15	0.13	0.11
Gate valve (typical) (flanged)	–	–	–	–	–	–	0.30	0.23	0.15	0.13
Gate valve – ¼ closed	0.8 to 0.2 this range									
Gate valve – ½ closed	4.0 to 0.8 this range									
Gate valve – ¾ closed	16.0 to 2.0 this range									
Globe valve (screwed)	–	15.00	–	12.50	–	–	8.50	7.50	6.50	6.00
Globe valve (flanged)	–	–	–	12.50	–	–	8.50	7.50	6.50	6.00
Swing check valve (screwed)	–	5.00	–	3.00	–	–	2.00	2.00	2.00	2.00
Swing check valve (flanged)	Typically 2.0 (all sizes)									
Foot valve	Typically 0.8 (all sizes)									
Basket strainer	Typically 1.5 to 1.0 this range									

TABLE 8A-4 – Equivalent Straight Lengths of Pipe in Metres

Fitting	Pipe diameter – in (mm)									
	6 (150)	8 (200)	10 (250)	12 (300)	14 (350)	16 (400)	18 (450)	20 (500)	24 (600)	36 (900)
Welded 90° L-bends R/D, where R = bend radius D = pipe diameter										
0.5	19	25	32	28	44	50	56	–	–	–
1.0	8	11	14	17	20	23	26	–	–	–
1.5	6	8	10	12	14	16	18	–	–	–
2.0	4	6	8	10	12	14	16	–	–	–
3.0	4	6	7	9	11	13	15	–	–	–
Standard 90° elbow (screwed)	12	16	–	–	–	–	–	–	–	–
Standard 90° elbow (flanged)	12	16	20	24	28	32	36	40	48	72
Large radius 90° elbow (screwed)	–	–	–	–	–	–	–	–	–	–
Large radius 90° elbow (flanged)	9	12	15	18	21	24	27	30	36	55
Standard 45° elbow (screwed)	–	–	–	–	–	–	–	–	–	–
Standard 45° elbow (flanged)	3	4	5	6	7	8	9	10	12	18
Large radius 45° elbow (screwed)	–	–	–	–	–	–	–	–	–	–
Large radius 45° elbow (flanged)	2	2.5	3.3	4	4.5	5.5	6	6.5	8	12
Return bend 180°	36	47	62	73	–	–	–	–	–	–
Large radius return bend 180°	19	26	33	39	–	–	–	–	–	–
T – line flow	30	40	50	60	70	80	90	100	120	180
T – branch flow	150	200	250	300	350	400	440	480	570	850
Cast elbow (90°) standard	16	20	26	32	36	42	46	52	63	94
Cast elbow (90°) large radius	11	14	17	20	23	27	30	34	40	60
Screw down valve (straight)	156	208	260	310	363	415	467	519	622	934
Screw down valve (right-angled run)	75	100	125	150	175	200	225	250	300	450
Gate valve (typical) (flanged)	3.5	4.5	5.7	6.7	8.0	9.0	10.2	12.0	14.0	20.0
Gate valve – ¼ closed	19	26	33	39	–	–	–	–	–	–
Gate valve – ½ closed	100	130	160	190	–	–	–	–	–	–
Gate valve – ¾ closed	400	540	700	800	–	–	–	–	–	–
Globe valve (flanged)	160	214	267	320	373	427	480	534	640	960
Swing check valve (screwed)	–	–	–	–	–	–	–	–	–	–
Swing check valve (flanged)	40	53	67	80	93	107	120	134	160	240
Foot valve (typical)	12	16	20	24	28	32	36	40	48	72
Basket strainer – (typical)	11	14	16	18	21	24	30	35	-	-
Sudden enlargement										
d/D = ¼	16	20	26	31	36	42	–	–	–	–
d/D = ½	11	15	18	22	25	29	–	–	–	–
d/D = ¾	3	4	6	7	8	10	–	–	–	–
Sudden contraction										
d/D = ½	6	8	10	12	14	16	18	20	24	36
Entrance (typical)	9	12	15	18	21	24	27	30	–	–

d = smaller pipe diameter
D = larger pipe diameter

TABLE 8A-4 – Equivalent Straight Lengths of Pipe in Metres.

Fitting		1000	800	600	500	400	300	200	150	100	80	65	50	40	32	25	15
Gate valve		6	5	4	3	2.7	2.2	1.5	1	0.7	0.5	0.43	0.35	0.27	0.2	0.18	0.1
Non-return flap valve		110 to 200	90 to 170	70 to 130	60 to 110	50 to 90	35 to 70	25 to 50	20 to 35	13 to 25	10 to 20	8 to 15	6 to 12	5 to 9	4.7	3.6	2.4
Screw down valve		300	250	200	160	130	100	70	50	35	28	22	17	13	10	8	5
Screw down valve, right-angled		150	130	90	80	60	50	32	25	16	13	10	8	7	5	4	2.5
Bends and elbows		18	15	12	10	8	6	4	3	2	1.7	1.4	1	0.8	0.6	0.5	0.3
		25	20	15	13	10	8	5	4	2.8	2	1.8	1.5	1	0.8	0.7	0.4
		12	10	7	6	5	4	2.5	2	1.5	1	0.9	0.7	0.5	0.4	0.3	0.2
		30	25	18	15	13	10	6.5	5	3.2	2.5	2	1.8	1.4	1	0.8	0.5
		75	60	50	40	33	25	17	13	8	6.5	5.5	4	3.2	2.6	2	1.4

TABLE 8A-4 (Continued) – Equivalent Straight Lengths of Pipe in Metres.

	100	80	60	50	40	30	20	15	10	8	7	5	4	3.2	2.7	1.7
T's	70	58	45	35	30	24	15	12	8	6	5	4	3	2.5	2	1.2
	18	15	12	10	8	6	4	3	2	1.7	1.4	1	0.8	0.6	0.5	0.3
Taper connectors d/D = 3/4	25	20	16	14	11	8	5.5	4	3	2.5	2	1.5	1.3	0.9	0.7	0.4
Taper connectors d/D = 1/2	30	25	20	16	13	10	7	5	3.5	2.8	2.2	1.7	1.4	1	0.8	0.5
Abrupt 90° bend	60	50	40	35	28	20	14	10	7	5.5	4.4	5	3	2	1.8	1
Abrupt changes of section d/D = 3/4	6	5	4	3.5	3	2	1.5	1	0.7	0.6	0.5	0.37	0.3	0.24	0.18	0.11
d/b = 1/4	13	10	8	7	5	4	3	2	1.5	1.2	0.9	0.7	0.55	0.45	0.35	0.2
d/b = 1/2	10	8	6	5	4	3.2	2.2	1.6	1.1	0.9	0.7	0.55	0.45	0.35	0.27	0.17
d/b = 1/4	30	25	20	16	13	10	7	5	3.5	2.8	2.2	1.7	1.4	1	0.8	0.5
d/b = 1/2	15	13	10	9	7	5.5	3.5	3	2	1.5	1.3	0.9	0.7	0.6	0.5	0.3
	15	13	10	8	7	5	3.5	2.5	1.8	1.5	1.2	0.9	0.7	0.5	0.4	0.25

APPENDIX

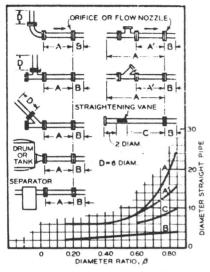

(A) for Orifices and Flow Nozzles.
All fittings in same plane.

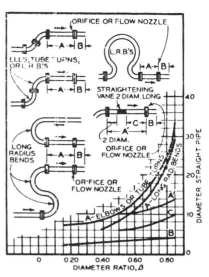

(B) for Orifices and Flow Nozzles.
All fittings in same plane.

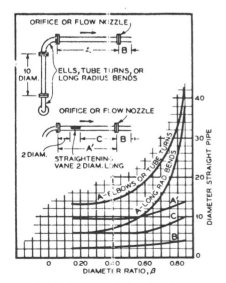

(C) for Orifices and Flow Nozzles.
Fittings in Different Planes.

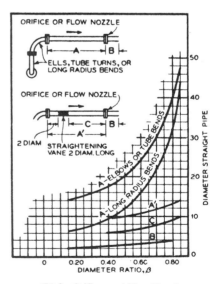

(D) for Orifices and Flow Nozzles.
Fittings in Different Planes.

TABLE 10A – Recommended Minimum Lengths of Pipe Preceding and Following Orifices, Flow Nozzles and Venturi Tubes

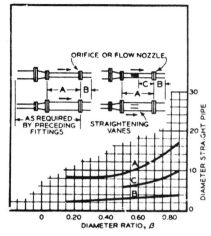

(E) for Orifices and Flow Nozzles.
With Reducers and Expanders.

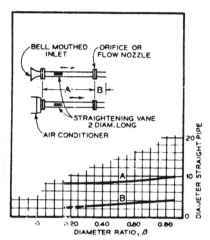

(F) for Orifices and Flow Nozzles
in Atmospheric Intake.

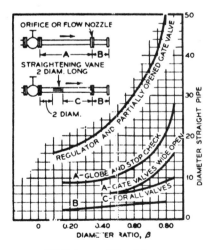

(G) Valves and Regulators.

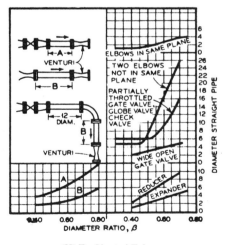

(H) For Venturi Tubes.

TABLE 10A – Recommended Minimum Lengths of Pipe Preceding and Following Orifices, Flow Nozzles and Venturi Tubes

APPENDIX

TABLE 8 – Specific Gravity of Suspensions of Solids in Water.

Percentage by weight of solids	Ratio water to solids	Specific gravity of dry solids															
		2.0	2.1	2.2	2.3	2.4	2.5	2.6	2.7	2.8	2.9	3.0	3.1	3.2	3.3	3.5	4.0	4.5	5.0
		Specific gravity of solution																	
10	9:1	1.05	1.05	1.06	1.06	1.06	1.06	1.06	1.07	1.07	1.07	1.07	1.07	1.07	1.07	1.08	1.08	1.09	1.09
15	5.66:1	1.08	1.08	1.09	1.09	1.09	1.10	1.10	1.10	1.11	1.11	1.11	1.11	1.11	1.12	1.12	1.13	1.13	1.14
20	4:1	1.11	1.11	1.12	1.12	1.13	1.14	1.14	1.14	1.15	1.15	1.15	1.16	1.16	1.16	1.17	1.18	1.19	1.19
25	3:1	1.14	1.15	1.15	1.16	1.17	1.18	1.18	1.19	1.19	1.19	1.20	1.20	1.21	1.21	1.22	1.23	1.24	1.25
30	2.33:1	1.17	1.18	1.19	1.20	1.21	1.22	1.23	1.23	1.24	1.24	1.25	1.25	1.26	1.26	1.27	1.29	1.31	1.31
35	1.87:1	1.21	1.22	1.23	1.25	1.26	1.27	1.28	1.28	1.29	1.30	1.30	1.31	1.32	1.32	1.33	1.35	1.37	1.39
40	1.5:1	1.25	1.26	1.28	1.29	1.30	1.32	1.33	1.34	1.35	1.36	1.36	1.37	1.38	1.39	1.40	1.43	1.45	1.47
45	1.22:1	1.29	1.30	1.32	1.34	1.36	1.37	1.38	1.40	1.41	1.42	1.43	1.44	1.45	1.46	1.47	1.51	1.54	1.56
50	1:1	1.33	1.35	1.37	1.39	1.41	1.43	1.44	1.46	1.47	1.49	1.50	1.51	1.52	1.53	1.55	1.60	1.63	1.67
55	0.91:1	1.37	1.38	1.41	1.43	1.44	1.49	1.51	1.53	1.55	1.56	1.58	1.59	1.61	1.62	1.65	1.70	1.75	1.79
60	0.67:1	1.43	1.46	1.48	1.51	1.54	1.56	1.58	1.61	1.63	1.65	1.67	1.68	1.70	1.72	1.75	1.82	1.87	1.92
65	0.54:1	1.48	1.51	1.55	1.58	1.61	1.64	1.67	1.69	1.72	1.74	1.76	1.79	1.81	1.83	1.87	1.95	2.03	2.08
70	0.43:1	1.54	1.57	1.62	1.65	1.69	1.72	1.75	1.79	1.82	1.85	1.88	1.90	1.93	1.95	2.00	2.10	2.20	2.27
75	0.33:1	1.60	1.65	1.69	1.73	1.78	1.82	1.86	1.90	1.93	1.97	2.00	2.03	2.06	2.09	2.15	2.29	2.40	2.50

TABLE 11A-1 – Packing Fault-Finding Chart

	Observations	Possible causes/actions
At start-up	No leakage	1. Incorrect installation — negative stuffing box pressure requiring lantern ring. 2. Slight overtightening — follow stop/start procedure.
	Excessive leakage	1. Wrong section packing. 2. Incorrect installation procedure. 3. Shaft run-out.
Packing set removal	Packing section reduced under horizontal shaft	1. Shaft misaligned with stuffing box. 2. Worn bearings — packing acting as bearing.
	Packing section reduced above horizontal shaft	1. Shaft misaligned with stuffing box. 2. Worn bearings.
	Whole or part of an end ring missing	Excessive clearance between shaft and stuffing box neck or gland follower allowing extrusion.
	Wear on the outside of one or more rings	Rings rotating with shaft, packing section too small.
	Rings next to gland follower worn, other all right.	Incorrect installation procedure, gland follower overtightened.
	Ring i.d.'s burnt, dried or charred, remaining material all right.	1. Incorrect packing selection re : temperature limit/shaft speed. 2. Inadequate lubrication.
	Innermost ring deteriorated	Incorrect packing selection re : compatibility with product.
	Packing seizure on shaft after equipment shutdown	Product crystallizing or solidifying in packing bore through: 1. Inadequate lubrication. 2. Lack of heating or cooling.

APPENDIX

TABLE 11A-2 – Modes, Causes and Corrections of Mechanical Face Seal Failures

Failure mode	Cause	Correction
PRIMARY SEAL FACES		
Overall corrosion	Improper materials.	Upgrade materials.
Intergranular corrosion	Improper materials.	Upgrade materials.
	Improper processing.	Correct processing, such as welding and heat treating.
Stress-corrosion cracking	Improper material.	Upgrade material.
	Improper processing.	Correct processing, such as stress relief.
Leaching corrosion	Improper material.	Upgrade material.
Seal-face distortion (concavity, convexity, waviness or non-uniform contact)	Excessive fluid pressure on seal.	Lower fluid pressure; upgrade material, upgrade design.
	Excessive swell of confined secondary seal.	Change secondary-seal materials or design.
	Improper seal assembly — excessive or non-uniform clamping or bolting stresses, or weight or misalignment of suction piping on end-suction over-hung-type pumps.	Correct assembly.
	Internal stress.	Stress relieve and refinish.
	Improper finishing.	Refinish.
	Foreign material trapped between faces.	Remove foreign material.
	Excessive PV value of seal operation.	Reduce PV value or upgrade design or materials.
	Improper equipment operation resulting in adverse seal environment (for example, insufficient seal lubrication).	Correct environment, such as cooling of fluid flow rate; upgrade materials or design.
Fracture	Improper handling.	Handle parts with care.
	Improper assembly (jamming).	Correct assembly.
	Stress-corrosion cracking.	See above, at Stress Corrosion Cracking.
	Excessive thermal stress by: Improper operation such as insufficient seal lubrication.	Improve equipment operation or environment; upgrade design or materials.
	Excessive PV value.	Reduce PV value.
	Excessive fluid pressure on seal.	Reduce fluid pressure or upgrade design or materials.
	Excessive swell of secondary seal.	Upgrade materials; change fluid or lower temperature.

cont...

TABLE 11A-2 (continued) – Modes, Causes and Corrections of Mechanical Face Seal Failures

Failure mode	Cause	Correction
PRIMARY SEAL FACES contd...		
Edge chipping	Excessive fluid pressure on seal.	Reduce fluid pressure; upgrade design or materials.
	Excessive shaft run-out.	Reduce run-out or upgrade seal design.
	Excessive shaft deflection.	Reduce deflection or upgrade seal design.
	Excessive shaft whip.	Reduce shaft whip or upgrade seal design.
	Seal faces out-of-square.	Square faces to shaft axis.
	Seal face vibration.	Reduce fluid temperature to 28 °C below boiling point.
Severe, uniform adhesive wear	PV value too high.	Reduce PV value; upgrade materials; improve seal balance or lubrication.
	Deposition of dissolved solids.	Reduce fluid temperature to 28 °C below boiling point; reduce dissolved solids; upgrade face materials.
	Incorrect assembly (jamming).	Correct assembly.
	Fine abrasives in sealed fluid.	Remove abrasives from fluid; upgrade materials.
	Failure of axial holding hardware (jamming).	Correct for axial slippage of set screws or collar.
	Excessive shaft end play.	Correct end play.
	Poor environment, such as insufficient seal lubrication or loss of cooling.	Correct equipment operation; upgrade face materials or design.
Heavy, non-uniform wear, galling and grooving	Abrasive contaminants.	Remove contaminants or upgrade materials.
	PV value too high with one or both faces of metal.	Reduce PV value; upgrade materials or design.
Erosion	Abrasive flow into seal face.	Remove abrasive from sealed fluid; shroud seal faces or direct flow away from seal.
Blistering of carbon-graphite faces	Improper materials.	Upgrade seal-face material.
	PV value excessive or cyclic.	Reduce PV value and cycling.
	Inadequate seal cooling.	Improve seal cooling.

cont...

APPENDIX

TABLE 11A-2 (continued) – Modes, Causes and Corrections of Mechanical Face Seal Failures

Failure mode	Cause	Correction
SECONDARY SEALS		
Extrusion	Excessive pressure.	Reduce pressure or upgrade design or material.
	Excessive temperature.	Reduce temperature or upgrade design or materials.
	Excessive swell.	Upgrade materials or change seal fluid.
Chemical attack	Improper material.	Select proper material.
Cracking	Excessive temperature.	Reduce temperature (by cooling); upgrade materials; reduce PV value.
	Chemical attack.	Select proper material.
	Ozone attack.	Upgrade materials; reduce ozone concentration; reduce stress.
Cuts, tears and splits	Improper handling.	Correct handling.
	Poor dispersion.	Correct manufacturing.
	Poor knit.	Correct manufacturing.
	Inclusion.	Correct manufacturing.
	Material overstressed.	Lower stress or upgrade materials.
Corrosion of interface	Crevice corrosion.	Eliminate crevice; apply corrosion-resistant coating.
	Fretting corrosion.	Eliminate vibration caused by out-of-square faces or excessive deflection; end-play or run-out of the shaft.; apply corrosion-resistant coatings. Lubricate interface. Upgrade resistance to corrosion and wear.
HARDWARE, GENERAL		
Overall corrosion	Improper material.	Upgrade material.
Stress-corrosion cracking	Improper material.	Upgrade material.
	Improper processing.	Correct processing, such as stress relief.
Intergranular corrosion	Improper material.	Upgrade material.
	Improper processing.	Correct processing, such as heat treating and welding.
Hydrogen embrittlement	Improper processing.	Correct processing.
Fatigue	Excessive stress or vibration.	Reduce stress or vibration.
	Seal-face vibration.	Reduce fluid temperature to

TABLE 11A-2 (continued) — Modes, Causes and Corrections of Mechanical Face Seal Failures

Failure mode	Cause	Correction
MECHANICAL DRIVE		
Torsional shear	Excessive torque due to: Improper lubrication.	Correct for stoppage of lubricant flow to seal faces; clean out or install (if needed) bypass flushing line.
	Failure of axial holding device.	Check set screws or other means of securing collar for slippage or jamming.
	Excessive fluid pressure.	Reduce fluid pressure or re-design (for example, seal balance).
Axial shear	Excessive shaft end play.	Correct end play.
	Improper assembly (jammed).	Correct assembly.
	Excessive pressure loading.	Reduce pressure or upgrade design.
Wear	Excessive torque.	See Torsional shear.
	Out-of-square seal faces.	Square faces to shaft axis.
	Excessive shaft run-out.	Reduce run-out.
	Excessive shaft deflection.	Reduce deflection.
	Excessive shaft end play.	Reduce end play.
	Seal face vibration.	Reduce fluid temperature to 28 °C below boiling point.
Seal hang-up	Deposition of dissolved solids, or products of corrosion, seal-fluid oxidation or decomposition.	Change to non-pusher seal; upgrade corrosion resistance of material; lower temperature of fluid; quench seal.

APPENDIX

TABLE 11A-3 – Centrifugal Pumps

Fault	Cause	Remedy or Action
Pump not turning	Driver not running	Check fuses, circuit breakers.
	Keys sheared	Replace.
	Drive belt slip	Check and adjust.
	Coupling fault	Check if slipping or broken; replace if necessary.
	Shaft or gears sheared	Check; replace if necessary.
Pump not priming	Inlet valve closed	Open valve.
	Inlet clogged or restricted	Check and clear.
	Air leaks on suction side	Replace seals; check line(s) for leaks.
	Liquid drained or syphoned from system	Fit check or foot valve to prevent draining.
	Worn pump impeller	Inspect; increasing pump speed might help; also fitting foot valve.
No discharge	Lack of prime (see also above)	Open all vent cocks to release trapped air and fill pump and suction pipe completely with fluid.
	Excessive suction lift	Check pump inlet for clogging, *etc*, causing excessive friction head. Check suction head.
	Excessive discharge head	Check that valves are open. Check piping for obstructions or blockage. Check total head.
	Speed too low	Check that pump rev/min is consistent with manufacturer's recommendations.
	Pump clogged	Check that impeller is not clogged.
	Wrong direction of rotation	Check that pump is running in the correct direction.
	Vapour lock	Bleed suction pipe to clear air lock. Check that suction pipe is properly submerged.
	Relief valve not properly adjusted	Check adjustment; check for dirt on valve seat.
	Air leak	Check seals; check line(s) for air leaks.
Low delivery	Air leaks	Check suction piping and pump for air leaks. Check pump gaskets.
	Vapour lock	Check NPSH and fluid temperature to ensure that liquid in suction line is not 'flashing'.
	Low NPSH or damage	As above, and also check suction pipe, foot valve, *etc*.
	Clogged strainer(s)	Check and clean if necessary.
	Excessive inlet friction	Suction line too small; or too many fittings adding fluid friction.
	Relief valve incorrectly set, or jammed.	Check and adjust as necessary.
	Excessive system back-pressure	Reduce system friction by re-design.
	Worn impeller	Inspect and replace if necessary.
	Worn wear rings	As above.
	Wrong direction of rotation	Check against manufacturer's specification.
	Constriction in suction line	Check that foot valve size is adequate (if fitted). Check for other possible obstruction.
	Wrong pump size	Check that pump is adequate for the job.
	Poor suction	Check that suction pipe is properly submerged and in best position.

TABLE 11A-4 – Reciprocating Pumps

Fault	Cause	Remedy or Action
Low delivery (cont)	High fluid viscosity	Check that fluid viscosity is consistent with anticipated performance.
	Excessive fluid temperature	Reduce speed and/or delivery; decrease suction head.
	Speed too low	Check operating rev/min against specification for pump performance.
Over-heating	Stuffing box over-heats	Check that packing is not too tightly or badly fitted.
		Check packing lubricant (where applicable).
		Check that packing is consistent with manufacturer's specification.
		Check cooling flow (where applicable).
	Bearings over-heating	Check oil level or lubricant condition.
		Check if correct lubricant is being used.
		Check bearing for misalignment or excessive tightness.
		Check fitting and condition of oil seals.
		Check that operating speed is not excessive.
	Fluid too viscous	Reduce fluid viscosity (eg by heating).
	Excessive pressure	Reduce pump speed; increase delivery line size(s).
Vibration and noise	Cavitation	Check pump operating conditions.
	Excessive fluid viscosity	Check product suitability.
	Entrained air	Check for air leaks.
	High vapour pressure fluid	Check product/pump suitability
	Improper pump assembly	Check and rectify.
	Unbalanced impeller	Check that impeller is not damaged or clogged.
	Misalignment	Check alignment with driver.
	Non-rigid mount	Check mounting for rigidity.
	Bent shaft; faulty bearings	Check and replace if necessary.
	Pump wear	Strip down and check for wear.
	Relief valve chattering	Re-adjust, repair or replace as necessary.
Excessive wear	Misalignment	As above.
	Out-of-balance	As above.
	Non-rigid mount	As above.
	Bent shaft	As above.
	Lack of lubrication	Check quantity and quality of lubricant.
	Dirt in pump	Use filter to remove.
	Corrosion	Check that pump materials are compatible with fluid being handled.
	Too high operating speed	Check against manufacturer's recommendations for fluid viscosity.
	Operating pressure too high	Reduce speed or pressure, eg change in system.
	Abrasives present in fluid	Check product/pump suitability.
Pump requires excessive power	Speed too high	Check against recommended rev/min.
	Misalignment	Check alignment of pump and driver, also foundations.
	Internal friction	Check for rubbing contact, clogging, etc.
	Tight bearings	Check bearings and packings (bearing temperature will be a clue).
	Lack of lubrication	Check quantity and quality of lubricants.
	High fluid viscosity	Check that fluid viscosity is not too high for economic handling.

APPENDIX

TABLE 11A-4 – Reciprocating (Piston) Pumps

Fault	Cause	Remedy or Action
No discharge	Not primed	Prime. Open vents on discharge side to release trapped air and leave open until all air is discharged.
	Excessive suction lift	Reduce suction lift or reduce suction friction with larger diameter pipe or eliminate bends, *etc*.
	Air leaks	Check system and eliminate air leaks by sealing, *etc*.
	Vapour bound	Check fluid temperature and vapour pressure – fluid temperature may be too high, or suction lift excessive for fluid temperature.
	Blockage	Check for blockage in suction pipe, foot valve or strainer. Check pump suction valves.
	Deterioration	Check suction valves, piston packing, piston rod packing, worn valves or badly scored cylinder.
Low discharge pressure	Low steam pressure	Check for obstruction, leak or partially closed valve in steam system.
	Tight packing	Loosen gland until leakage is apparent.
	Excessive backpressure	Check that pump is not operating against excessive system head.
	Deterioration	Check cylinder bore for wear, also condition of piston packings and valves.
Pump stops or hesitates	Intermittent steam supply	Check for blockage, *etc*.
	Valve trouble	Check for excessive valve wear and leakage. Check that valve timing is correct.
	Excessive backpressure	Check that system head is not excessive.
Variable delivery	Air leaks	Check system for air leaks.
	Misalignment	Check alignment of pump and possible distortion due to unsupported piping connected to pump cylinder.
	Excessive suction lift	As above.
	Vapour bound	Reduce fluid temperature, or reduce suction lift.
	Tight packing	Loosen gland.
	Excessive speed	Check that operating speed is consistent with specification.
Pump short-strokes	Excessive cushioning	Adjust cushioning valves to obtain correct stroke.
	Worn valve	Check valve for leakage, re-face as necessary.
	Entrained gas	Modify suction intake as necessary.
	Incorrect valve timing	Check against specification and adjust as necessary.
	Worn bore	Replace liner, or re-bore and fit oversize piston.
Piston over-strokes (hits head)	Cushion valves	Close down as necessary to reduce stroke.
	Piston leakage	Replace piston packing; replace worn liner or re-bore cylinder.
	Valve leakage	Check valves on liquid head for leakage – re-grind or re-set if necessary.
Excessive wear	Misalignment	Check alignment.
	Bent piston rod	Check for straightness.
	Worn bore	Replace liner or re-bore.
	Fluid	Check that pump is compatible with fluid being handled.

TABLE 11A-5 – Rotary Pumps

Fault	Cause	Remedy or Action
No discharge	Not primed	Prime to fill pump.
	Excessive suction lift	Reduce suction lift or reduce friction in suction side with large pipe.
	Air leaks	Check and rectify; check gaskets.
	Blockage	Check adjustment and setting.
	Excessive wear	Check components for wear against manufacturer's permitted tolerances.
	Wrong rotation	Check that pump is being rotated in the correct direction.
	Insufficient speed	Check that pump is running at rated speed.
Low discharge pressure or reduced capacity	Insufficient speed	As above.
	Wrong rotation	As above.
	Excessive suction lift	As above.
	Air leaks	As above, check gaskets particularly.
	Air entrainment	Re-position suction inlet.
	Relief valve or bypass valve	Setting may be too low. Check and re-set.
	Excessive wear	Check as above.
Excessive noise	Misalignment (where applicable)	Check alignment of driver and pump and drive connection.
	Internal damage	Bent or broken rotor.
	Unbalance	If suspected, check rotor for static and dynamic balance.
	Air entrainment	Re-position suction inlet.
	Air leaks	Check and rectify.
	Cavitation	Check against causes of cavitation.
	Excessive pressure	Relief valve set too high, adjust to correct setting consistent with pump rating.
	Deterioration	Check for excessive wear or clearances on components.
Excessive discharge pressure	System pressure	If system pressure is too high for pump rating a larger pump will have to be used, when some relief may be possible.
	Relief valve or bypass valve	Check and re-set relief valve for correct pressure.
	System throttled	Discharge valve may be partially closed or system partially blocked.
Excessive wear	Abrasive liquid	Check that pump is suitable for handling liquid if abrasive solids are present; or check that filter or strainer used is adequate.
	Distortion	Pipework loads transmitted directly to the casing may cause distortion.
	Excessive pressure developed	See above.
	Excessive speed	Check that speed is consistent with pump specification for viscosity of liquid handled.
Excessive input power required	Damage	Check for bent or damaged shaft, etc.
	Excessive pressure	As above.
	Excessive fluid viscosity	Check speed rating against actual viscosity of fluid; reduce speed for higher viscosities.
	Excessive speed	Check against pump rating for fluid viscosity handled.
Pump overheats	Relief or bypass valves	Check that settings are correct.
	Excessive speed for fluid handled	Check that speed is consistent with rating for fluid viscosity.
	Excessive pressure	As above.
	Discharge throttle	Looped flow through relief valve will cause heating; may be relieved by separate relief valve discharging to tank.

APPENDIX

TABLE 11A – Lobe-Rotor Pumps

SYMPTOMS													CAUSES	REMEDIES
No discharge	Under capacity	Irregular discharge	Prime lost after starting	Pump stalls when starting	Pump over-heats	Motor over-heats	Excessive power absorbed	Noise and vibration	Pump element wear	Excessive gland/seal wear	Product loss through gland	Seizure		
X	X												Incorret direction of rotation.	Reverse motor.
X													Pump unprimed.	Expel gas from supply line and pumping chamber and introduce liquid.
	X	X	X					X					Insufficient NPSH available.	Increase supply line diameter increase static suction head. Simplify supply line configuration and reduce length.
	X	X	X					X					Product vaporising in supply line.	Reduce speed. Decrease product temperature – check effect of increased viscosity on available and permitted power inputs.
			X					X					Air entering supply line.	Remake pipework joints. Adjust or repack gland.
X		X	X					X					Gas in supply line.	Expel gas from supply line and pumping chamber and introduce liquid.
	X	X	X					X					Insufficient head above supply vessel outlet	Raise produce level. Lower outlet position.
X	X	X	X					X					Foot valve/strainer obstructed or blocked.	Increase submergence of supply pipe. Service fittings.
				X	X	X	X	X					Product viscosity above rated figure.	Decrease pump speed. Increase product temperature.
	X												Product viscosity below rated figure.	Increase pump speed. Decrease product temperature.
	X				X	X		X	X			X	Product temperature above rated figure.	Cool the product/pumping chamber.
				X		X	X						Product temperature below rated figure.	Heat the product/pumping chamber. (Check with pump maker).
								X	X	X	X	X	Unexpected solids in product.	Clean the system. Fit strainer to supply line.
	X				X	X	X	X	X	X		X	Delivery pressure above rated figure.	Check for obstructions. Service system and revise to prevent problem recurring. Simplify delivery time.

TABLE 11A-6 (continued) – Lobe-Rotor Pumps

Possible cause	Remedy
Gland over-tightened.	Slacken and readjust gland.
Gland under-tightened.	Adjust gland.
Gland flushing inadequate.	Check that fluid flows freely into gland. Increase flow rate.
Pump speed above rated figure.	Decrease pump speed.
Pump speed below rated figure.	Increase pump speed.
Rotorcase strained by pipework.	Check alignment of pipes. Fit flexible pipes or expansion fittings. Support pipework. Re-tension to maker's recommendations.
Belt drive slipping.	
Flexible coupling misaligned.	Check flange alignment and adjust mountings accordingly. Fit lock-washers to slack fasteners and retighten.
Insecure pump/driver mountings.	Refer to pump maker for advice and replacement parts.
Shaft bearing wear or failure.	Refer to pump maker for advice and replacement parts.
Worn unsynchronized timing gears.	Refer to pump maker's instructions.
Gearcase oil quantity/quality incorrect.	
Metal-to-metal contact of pumping element.	Check rated and duty pressure. Refer to pump maker. Fit new components.
Worn pumping element.	
Front cover relief valve leakage.	Check pressure setting and readjust if necessary. Examine and clean seating surfaces. Replace worn parts.
Relief valve chatter.	Check for wear of sealing surfaces, guides, etc. – replace as necessary.
Relief valve incorrectly set.	Readjust spring compression. Valve should lift about 10% above duty pressure.

DIAGNOSIS WILL BE GREATLY ASSISTED BY TAKING ON-STREAM PRESSURE READINGS AT THE PUMP'S INLET AND OUTLET PORTS

APPENDIX

TABLE 11A-7 – Fault-finding and Maintenance Guide for Typical High Pressure Piston and Plunger Pumps

Problem	Probable Cause	Solution
Pulsation	Faulty Pulsation Dampener.	Check precharge, if low recharge it or install a new one.
Low pressure	Worn nozzle.	Replace nozzle, of proper size.
	Belt slippage.	Tighten or replace; use correct belt.
	Air leak in inlet plumbing.	Disassemble, reseal, and reassemble.
	Pressure gauge inoperative or not registering accurately.	Check with new gauge; replace worn or damaged gauge.
	Relief valve stuck, partially plugged or improperly adjusted; valve seat worn.	Clean, and adjust relief valve; check for worn and dirty valve seats.
	Inlet suction strainer clogged or improper size.	Clean. Use adequate size. Check more frequently.
	Worn piston assembly or packing. Abrasives in pumped fluid or severe cavitation. Inadequate water supply.	Install proper filter. Suction at inlet manifold must be limited to lifting less than 20 feet of water.
	Fouled or dirty inlet or discharge valves.	Clean inlet and discharge valve assemblies.
	Worn inlet or discharge valves.	Replace worn valves, valve seats and/or discharge hose.
	Leaky discharge hose.	
Pumps run extremely rough, pressure very low	Restricted inlet or air entering the inlet plumbing.	Proper size inlet plumbing; check for air tight seal.
	Inlet restrictions and/or air leaks. Damaged cup or stuck inlet or discharge valve.	Replace worn cup or cups, clean out foreign material, replace worn valves.
	Worn inlet manifold seals.	Replace worn seals.
	Worn inlet seals allow pump to draw air.	Install new inlet manifold seals.
Cylinder O-rings blown next to discharge manifold.	Pressures in excess of rated.	Check for plugged nozzle, closed valves or improperly adjusted bypass valve.
	Warped manifold.	Replace manifold.
Leakage at the cylinder O-rings at the discharge manifold and black, powdery substance in the area of the O-rings.	Loose cylinders. Cylinder motion caused by improper shimming of the discharge manifold.	Remove spacer shims on manifold studs. Do not remove too many shims or the ears of the manifold will be bowed when the manifold is retightened, causing looseness in the centre cylinder.
Water leakage from under the inlet manifold.	Worn inlet manifold seals. Leaking sleeve O-ring.	Install seals. If piston rod sleeves are scored, replace sleeves and sleeve O-rings.
Oil leak between crankcase and pumping section.	Worn crankcase piston rod seals. Excess oil from wicks.	Replace crankcase piston rod seals. Reduce quantity of oil per oiling.
Oil leaking in the area of crankshaft.	Worn crankshaft seal or improperly installed oil seal retainer packing, O-ring.	Remove oil seal retainer and replace damaged gasket and/or seals.
	Bad bearing.	Replace bearing.

cont...

TABLE 11A-7 (continued) – Fault-finding and Maintenance Guide for Typical High Pressure Piston and Plunger Pumps

Problem	Probable Cause	Solution
Excessive play in the end of the crankshaft pulley.	Worn main ball bearing from excessive tension on drive belt.	Replace ball bearing. Properly tension belt.
Water in crankcase.	May be caused by humid air condensing into water inside the crankcase.	Change oil.
	Leakage of manifold inlet seals and/or piston rod sleeve O-ring.	Replace seals, sleeve and O-rings.
Oil leaking from underside of crankcase.	Worn crankcase piston rod seals.	Replace seals, sleeve and O-rings.
Oil leaking at the rear portion of the crankcase.	Damaged or improperly installed oil gauge or crankcase rear cover O-ring, and drain plug O-ring.	Replace oil gauge or cover O-ring, and drain plug O-ring.
Oil leakage from drain plug.	Loose drain plug or worn drain plug seal.	Tighten drain plug or replace seal.
Loud knocking noise in pump.	Pulley loose on crankshaft.	Check key and tighten set screw.
	Broken or worn bearing.	Replace bearings.
Frequent or premature failure of the inlet manifold seals or the packing.	Scored rods or sleeves or plungers.	Replace rods and sleeves.
	Over-pressure to inlet manifold.	Reduce inlet pressure per instructions.
Short cup or packing life.	Damaged or worn chrome plating of the cylinders.	Replace cylinders.
	Damaged or worn plungers.	Replace plungers.
	Abrasive material in the fluid being pumped.	Install proper filtration on pump inlet plumbing.
	Excessive pressure and/or temperature of fluid being pumped.	Check pressures and fluid inlet temperature; be sure they are within specified range.
	Over-pressure of pumps.	Reduce pressure.
	Running pump dry.	Do not run pump without water.
	Front edge of piston sharp.	Replace with new piston.
	Chrome plating of cylinders damaged causing excessive wear of cups. May be caused by pumping acid solution.	Install new cups and cylinders. Pump only fluid compatible with chrome.
Strong surging at the inlet and low pressure on the discharge side.	Foreign particles in the inlet or discharge valve or worn inlet and/or discharge valves.	Check for smooth lap surfaces on inlet and discharge valve seats. Discharge valve seats and inlet valve seats may be lapped on a very fine oil stone; damaged cups and discharge valves cannot be lapped but must be replaced.

APPENDIX

Conversion Factors

Multiply	By	To obtain
Atmospheres	33.93	Feet of water
Atmospheres	29.92	Inches of mercury
Atmospheres	1.0332	Kilograms / Sq. centimetre
Atmospheres	760	Millimetres of mercury
Atmospheres	14.7	Pounds / Sq. inch
Barrels (oil)	42.0	U.S. gallons
Barrels (U.S. liquid)	31.5	U.S. gallons
Bars	0.9869	Atmospheres
Bars	1×10^6	Dynes / Sq. centimetre
Bars	750.1	Millimetres of mercury
Bars	14.5	Pounds / Sq. inch
B.T.U.	778.2	Foot pounds
B.T.U. / Second	1055	Watts
Centimetres	0.3937	Inches
Centipoises	0.001	Pascal seconds
Centipoises	0.01	Poises
Centistokes	Sp.Gr.	Centipoises
Centistokes	0.01	Stokes
Centistokes	0.000001	Sq. metres / Second
Cubic centimetres	0.06102	Cubic inches
Cubic centimetres	1.0	Millilitres
Cubic Centimetres	0.000264	U.S. gallons
Cubic feet	1728	Cubic inches
Cubic feet	28.32	Litres
Cubic feet	7.481	U.S. gallons
Cubic feet of water	62.4	Pounds of water
Cubic feet / Minute	0.1247	U.S. Gallons / Minute
Cubic inches	16.39	Cubic centimetres
Cubic inches	5.787×10^{-4}	Cubic feet
Cubic inches	1.639×10^{-5}	Cubic metres

Conversion Factors *(continued)*

Cubic inches	0.01639	Litres
Cubic inches	0.004329	U.S. Gallons
Cubic metres	35.31	Cubic feet
Cubic metres	264	U.S. Gallons
Degrees (Celsius)	$(C° \times 9/5) + 32$	Degrees (Fahrenheit)
Degrees (Fahrenheit)	$(F° - 32)\, 5/9$	Degrees (Celsius)
Dynes	2.248×10^{-6}	Pounds
Dynes / Sq. centimetre	1.45×10^{-5}	Pounds / Sq. inch
Feet of water	0.8819	Inches of mercury
Feet of water	304.5	Kilograms / Sq. metre
Feet of water	0.4331	Pounds / Sq. inch
Foot pounds	1.356×10^7	Dyne centimetres
Food pounds	1.383×10^4	Gram Centimetres
Gallons (Imperial)	277.4	Cubic inches
Gallons (Imperial)	1.2	Gallons (U.S.)
Gallons (U.S.)	231	Cubic inches
Gallons (U.S.)	0.003785	Cubic metres
Gallons (U.S.)	0.833	Gallons (Imperial)
Gallons (U.S.)	3.785	Litres
Gallons (U.S.)	128	Ounces (fluid)
Gallons (U.S.) of water	8.337	Pounds of water
Horsepower	42.4	B.T.U. / minute
Horsepower	33000	Foot pounds / Minute
Horsepower	550	Foot pounds / Second
Horsepower	1.014	Horsepower (metric)
Horsepower	10.7	Kilogram calories / Minute
Horsepower (metric)	75	Kilogram metres / Second
Horsepower	745.7	Watts
Inches	2.54	Centimetres
Inches of mercury	0.03342	Atmospheres
Inches of mercury	1.134	Feet of water
Inches of mercury	345.3	Kilograms / Sq. metre
Inches of mercury	25.4	Millimetres of mercury
Inches of mercury	0.4912	Pounds / Sq. inch
Kilograms	2.205	Pounds
Kilograms / Sq. centimetre	14.22	Pounds / Sq. inch
Kilowatts	1.341	Horsepower
Litres	1000	Cubic centimetres
Litres	0.03531	Cubic feet
Litres	61.02	Cubic inches
Litres	0.001	Cubic metres
Litres	0.220	Gallons (Imperial)
Litres	0.2642	Gallons (U.S.)
Litres	33.81	Ounces (fluid)
Metres	3.281	Feet
Metres	39.37	Inches
Metres	10^6	Microns

Conversion Factors *(continued)*

Metres / Second	3.281	Feet / Second
Microns	3.937 x 10−5	Inches
Microns	0.001	Millimetres
Millilitres	0.06102	Cubic inches
Millimetres of mercury	1333.22	Dynes / Sq. centimetre
Ounces (fluid)	1.805	Cubic inches
Ounces (fluid)	0.02957	Litres
Pounds	453.6	Grams
Pounds / Foot	1.488	Kilograms / Metre
Pounds / Inch	178.6	Grams / Centimetre
Pounds / Sq. foot	4.882	Kilograms / Sq. Centimetre
Pounds / Sq. inch	2.309	Feet of water
Pounds / Sq. inch	2.036	Inches of mercury
Pounds / Sq. inch	0.07031	Kilograms / Sq. centimetre
Pounds / Cubic foot	16.02	Kilograms / Cubic metre
Pounds / Cubic inch	27680	Kilograms / Cubic metre
Radians	57.3	Degrees
Radians / Second	0.1592	Revolutions / Second
Square centimetres	0.155	Square inches
Square feet	929	Square centimetres
Square inches	6.452	Square centimetres
Slugs	14.594	Kilograms
Slugs	32.174	Pounds
Tons (metric)	2205	Pounds
Watts	0.0569	B.T.U. / Minute

SECTION 9

BUYERS GUIDE

TRADE NAMES INDEX

ADVERTISERS' NAMES, ADDRESSES AND
CONTACT NUMBERS

EDITORIAL INDEX

ADVERTISERS INDEX

50 Years Serving Design

THE INSTITUTION OF ENGINEERING DESIGNERS

The Institution is a professional body for designers who operate in widely diverse fields of design practice, including product, domestic appliance, jig and tool, special purpose machinery, electro-mechanical, electronic, piping, etc. in industry, consultative practice, management and education. It embraces all forms of technology relevant to design, particularly computer-aided design and drafting, computer-aided engineering and rapid prototyping. Although primarily a design body the Institution has very active and close associations with engineering in all its forms and disciplines.

Membership of the Institution is open to persons working in the fields of Engineering and/or Product Design as designers, managers or educators.

Fellows (FIED), Members (MIED) and Associate members (AMIED) are corporate members.

Associates (AIED) and Competent Design Associations (CDAIED) are non-corporate members.

Student and Diplomates may join as Affiliate members.

The basic educational qualification for corporate membership is an HNC in engineering or product design or their equivalent, however in the foreseeable future this will rise to an Hon Degree or equivalent, in line with the latest Engineering Council registration requirements. There are also training and experience requirements. The full assessment procedure culminates in a professional review for Corporate Membership. The Institution also specifically welcomes applications from other mature professional-experienced designers.

The Institution is a nominated and Institution-affiliated body of the Engineering Council. Engineering Registration is available to qualifying members, after election, in the Engineering Technician (EngTech), Incorporated Engineering (IEng), and Chartered Engineering (CEng) sections of the Engineering Council's Register of Engineers.

Qualification to all three registration grades is also possible via a Mature Candidate route and to CEng and IEng by an appropriate Open University degree, providing the degree profile meets the requirements of the Institution or by sitting the appropriate level of Engineering Council set examinations.

Among its services to Members and Industry the Institution provides

- A Register of Designers
- Consultant & Contract Engineering Designer List
- Courses in the Management of Engineering Design
- A Professional Indemnity Insurance Scheme
- Conferences, Seminars & an Information Service
- An Annual Reference Book
- A Bi-Monthly Journal - "Engineering Designer"
- A professional qualification
- A Continuing Professional Development Scheme

Please remember the Institution is here to serve all designers so your problem is our problem. Just ring 01373 822801 if you require any further help about membership of if you have any queries.

FOR THOSE WHO CREATE OUR FUTURE

For full details write to or phone: Margaret Jackson or Alison Parker, Institution of Engineering Designers, Courtleigh, Westbury Leigh, Westbury, Wiltshire BA13 3TA. Fax: (01373) 858085.

TEL: (01373) 822801

CLASSIFIED INDEX OF MANUFACTURERS AND SUPPLIERS BY PRODUCT CATEGORY

PUMP TYPES
Centrifugal Pumps
1. Close Coupled
Albany Engineering Co. Ltd
Ansimag Inc.
Hyundai Heavy Industries Co. Ltd
ITT Richter Chemie-Technik GmbH
KSB Aktiengesellschaft
Pumpenfabrik Ernst Vogel
Stork Pumps Ltd
Sunstrand Fluid Handling
Waukesha Bredel Fluid Handling

2. End Suction Back Pullout
Ahlstrom Corporation
Albany Engineering Co. Ltd
Ansimag, Inc.
ITT Richter Chemie-Technik GmbH
KSB Aktiengesellschaft
Pumpenfabrik Ernst Vogel
Stork Pumps Ltd
Sunstrand Fluid Handling

3. Single Stage Process
Ahlstrom Corporation
Ansimag, Inc.
Hyundai Heavy Industries Co. Ltd
ITT Richter Chemie-Technik GmbH
KSB Aktiengesellschaft
Mackley Pumps
Pumpenfabrik Ernst Vogel
Stork Pumps Ltd
Sunstrand Fluid Handling
Wanner International
Waukesha Bredel Fluid Handling

4. Hygienic Pump
KSB Aktiengesellschaft
Stork Pumps Ltd
Waukesha Bredel Fluid Handling

5. Single Stage Double Entry
Ahlstrom Corporation
Hyundai Heavy Industries Co. Ltd
KSB Aktiengesellschaft
Mackley Pumps
Stork Pumps Ltd

6. Multistage Single Entry
Ahlstrom Corporation
Hyundai Heavy Industries
KSB Aktiengesellschaft

Mackley Pumps
Pumpenfabrik Ernst Vogel
Stork Pumps Ltd

7. Multistage Double Entry
Hyundai Heavy Industries
KSB Aktiengesellschaft
Stork Pumps Ltd

8. Multistage Caisson
Stork Pumps Ltd

9. Multistage Barrel Insert
Hyundai Heavy Industries
KSB Aktiengesellschaft
Stork Pumps Ltd

11. Sump Pump
Albany Engineering Co. Ltd
Feluwa Schlesiger & Co KG
Grundfos A/S
Mackley Pumps
Pumpenfabrik Ernst Vogel
Stork Pumps Ltd
Wilo GmbH

12. In-Line Pump
Grundfos A/S
KSB Aktiengesellschaft
Pumpenfabrik Ernst Vogel
Stork Pumps Ltd
Sunstrand Fluid Handling

13. Solids Handling Pump
Ahlstrom Corporation
Feluwa Schlesiger & Co KG
KSB Aktiengesellschaft
Mackley Pumps
Pumpenfabrik Ernst Vogel
Wilo GmbH

14. Sludge Pump
Ahlstrom Corporation

Feluwa Schlesiger & Co KG
Grindex AB
Pumpenfabrik Ernst Vogel
Wilo GmbH

16. Macerator
Feluwa Schlesiger & Co KG

17. Abrasive Handling
Ahlstrom Corporation
Feluwa Schlesiger & Co KG
KSB Aktiengesellschaft
Mackley Pumps
Pumpenfabrik Ernst Vogel
Stork Pumps Ltd
Wanner International
Wilo GmbH

18. Glandless Canned Motor
KSB Aktiengesellschaft
Sunstrand Fluid Handling

19. Glandless Magnetic Drive
Ansimag, Inc.
ITT Richter Chemie-Technik GmbH
KSB Aktiengesellschaft
Pumpenfabrik Ernst Vogel
Stork Pumps Ltd
Sunstrand Fluid Handling

20. Glandless Hot Water Circulator
KSB Aktiengesellschaft
Grundfos A/S
Pumpenfabrik Ernst Vogel

21. Reactor
Feluwa Schlesiger & Co KG
KSB Aktiengesellschaft

22. Recessed Impeller (Vortex) Pump
Ahlstrom Corporation
Grindex AB
ITT Richter Chemie-Technik GmbH
Pumpenfabrik Ernst Vogel

23. Regenerative
ITT Richter Chemie-Technik GmbH

24. Channel Impeller
Ahsltrom Corporation
Feluwa Schlesiger & Co KG
KSB Aktiengesellschaft
Pumpenfabrik Ernst Vogel

25. Side Channel
ITT Richter Chemie-Technik GmbH
KSB Aktiengesellschaft
Pumpenfabrik Ernst Vogel
Stork Pumps Ltd

26. Swimming Pool
KSB Aktiengesellschaft
Pumpenfabrik Ernst Vogel

27. Axial Flow
Ahlstrom Corporation
Hyundai Heavy Industries Co. Ltd
KSB Aktiengesellschaft
Pumpenfabrik Ernst Vogel
Stork Pumps Ltd

28. Mixed Flow
Hyundai Heavy Industries Co. Ltd
KSB Aktiengesellschaft
Stork Pumps Ltd

29. Mixed Flow Volute
Ahlstrom Corporation
Hyundai Heavy Industries Co. Ltd
KSB Aktiengesellschaft
Pumpenfabrik Ernst Vogel
Stork Pumps Ltd

30. Borehole shaft-driven
Hyundai Heavy Industries Co. Ltd
KSB Aktiengesellschaft
Mackley Pumps

31. Sumbersible Axial/Mixed Flow
KSB Aktiengesellschaft
Pumpenfabrik Ernst Vogel

33. Submersible Sewage
KSB Aktiengesellschaft
Feluwa Schlesiger & Co KG
Pumpenfabrik Ernst Vogel

34. Submersible Contractor
Grindex AB
Grundfos A/S
Pumpenfabrik Ernst Vogel
Wilo GmbH

35. Submersible Solids Handling
Feluwa Schlesiger & Co KG
Grindex AB
Mackley Pumps
Wilo GmbH

Rotary Positive Displacement Pumps

38. Internal Gear
Alfa Laval Pumps Ltd
Pumpenfabrik Ernst Vogel
Stork Pumps Ltd

39. External Gear
Albany Engineering Co. Ltd

40. Helical Rotor/Progressing Cavity/ Eccentric Screw
Alfa Laval Pumps Ltd
Netzsch Mohnopumpen GmbH

41. Peristaltic
Alfa Laval Pumps Ltd
Autoclude
Blue White Industries
Stork Pumps Ltd
Waukesha Bredel Fluid Handling

43. Lobe
Alfa Laval Pumps Ltd
Stork Pumps Ltd
Waukesha Bredel Fluid Handling

44. Orbital Lobe
Waukesha Bredel Fluid Handling

47. Semi Rotary
Sunstrand Fluid Handling

49. Roller Vane
Reciprocating Displacement Pumps

Reciprocating Displacement Pumps

50. Diaphragm, Oil-free
Alfa Laval Pumps Ltd
Stork Pumps Ltd
Wallace & Tiernan Ltd
Wanner International
Warren Rupp Inc.
Wilo GmbH

51. Diaphragm, Double or Single Acting
Alfa Laval Pumps Ltd
Feluwa Schlesiger & Co KG
Stork Pumps Ltd
Warren Rupp Inc.
Wilo GmbH
Wirth Maschinen

52. Double Disk
Alfa Laval Pumps Ltd
Warren Rupp Inc.

53. Piston
Donkin/Charlie Chapman Marinse
Feluwa Schlesiger & Co KG
Wallace & Tiernan Ltd
Wirth Maschinen
WOMA Apparatebau GmbH

54. Plunger/Ram
Stork Pumps Ltd

Wallace & Tiernan Ltd
WOMA Apparatebau GmbH

55. Jet
Stork Pumps Ltd

56. Proportioning and dosing/meter pumps
Albany Engineering Co. Ltd
Alfa Laval Pumps Ltd
Blue White Industries
Feluwa Schlesiger & Co KG
Wallace & Tiernan Ltd
Wanner International
WOMA Apparatebau GmbH

57. Solenoid Pumps
Wallace & Tiernan Ltd

OTHER PUMPS

59. Hand Pumps
Albany Engineering Co. Ltd
Feluwa Schlesiger & Co KG

60. Seal-Less Pumps
Alfa Laval Pumps Ltd
Wanner International
Warren Rupp Inc.
Wilo GmbH

62. Non Metallic/Plastic Pumps
Alfa Laval Pumps Ltd
Wanner International
Warren Rupp Inc.
Wilo GmbH

PUMPS ANCILLARIES & SERVICES

63. Engines
Alfa Laval Pumps Ltd
Hyundai Heavy Industries Co. Ltd

64. Electric Motors
Alfa Laval Pumps Ltd

Brook Hansen
Hyundai Heavy Industries Co. Ltd
SEW Eurodrive

66. Mechanical Seals
Alfa Laval Pumps Ltd
Flexibox Ltd
Hyundai Heavy Industries Co. Ltd
James Walker Britco Ltd
Pac Seal Inc. International

67. Bellows Seals
Alfa Laval Pumps Ltd
Flexibox Ltd
Hyundai Heavy Industries Co. Ltd

68. Gland Packings
Alfa Laval Pumps Ltd
Hyundai Heavy Industries Co. Ltd
IHC Lagersmit

69. Couplings
Alfa Laval Pumps Ltd
Brook Hansen
Flexibox Ltd
Hyundai Heavy Industries Co. Ltd

70. Filters
Hyundai Heavy Industries Co. Ltd
Wilo GmbH

71. Inverters
Brook Hansen
Hyundai Heavy Industries Co. Ltd
SEW Eurodrive

73. Overload Protection Devices
Wanner International

74. Coatings and Linings
Hyundai Heavy Industries Co. Ltd

75. Speed Controllers
Brook Hansen
Hyundai Heavy Industries Co. Ltd
SEW Eurodrive
Warren Rupp Inc.
Wilo GmbH

76. Shaft Alignment
Alpha Laval Pumps Ltd
Flexibox Ltd
Hyundai Heavy Industries Co. Ltd

78. Control and Measurement
Hyundai Heavy Industries Co. Ltd
Wallace & Tiernan Ltd
Warren Rupp Inc.

79. Condition Monitoring
Flexibox Ltd
Hyundai Heavy Industries Co. Ltd
IHC Lagersmit
KSB Aktiengesellschaft

80. Computer Aided Pump Selection Software (CAPS)
Alfa Laval Pumps Ltd
KSB Aktiengesellschaft

PUMP APPLICATIONS

81. Ash Handling
Feluwa Schlesiger & Co KG
KSB Aktiengesellschaft
Wirth Maschinen

83. Boiler Circulating
Hyundai Heavy Industries Co. Ltd
KSB Aktiengesellschaft
Mackley Pumps
Pumpenfabrik Ernst Vogel

84. Boiler Feed (Industrial)
Feluwa Schlesiger & Co KG
Hyundai Heavy Industries Co. Ltd
KSB Aktiengesellschaft
Mackley Pumps
Pumpenfabrik Ernst Vogel

Stork Pumps Ltd
Sunstrand Fluid Handling
Wallace & Tiernan Ltd
Wanner International
WOMA Apparatebau GmbH

85. Boiler Feed (Power Station)
Feluwa Schlesiger & Co KG
Hyundai Heavy Industries Co. Ltd
IHC Lagersmit
KSB Aktiengesellschaft
Mackley Pumps
Stork Pumps Ltd

86. Brewery Stuff
Alfa Laval Pumps Ltd
Autoclude
SEW Eurodrive
Stork Pumps Ltd
Wallace & Tiernan Ltd
Warren Rupp Inc.
Waukesha Bredel Fluid Handling
Wilo GmbH

87. Cargo Oil
Hyundai Heavy Industries Co. Ltd
Mackley Pumps
Pumpenfabrik Ernst Vogel
Stork Pumps Ltd
Warren Rupp Inc.
Wilo GmbH

88. Cargo Stripping
Stork Pumps Ltd
Warren Rupp Inc.
Wilo GmbH

89. Cement Slurry
Alfa Laval Pumps Ltd
Feluwa Schlesiger & Co KG
IHC Lagersmit
Warren Rupp Inc.
Waukesha Bredel Fluid Handling
Wilo GmbH

90. Chemical Process
Alfa Laval Pumps Ltd
Ansimag, Inc.
Autoclude
Blue White Industries
Feluwa Schlesiger & Co KG
Hyundai Heavy Industries Co. Ltd
IHC Lagersmit
ITT Richter Chemie-Technik GmbH
Mackley Pumps
Netzsch Mohnopumpen GmbH
Pumpenfabrik Ernst Vogel
Stork Pumps Ltd
Sunstrand Fluid Handling
Wallace & Tiernan Ltd
Wanner International
Warren Rupp Inc.
Waukesha Bredel Fluid Handling
Wilo GmbH
WOMA Apparatebau GmbH

91. Chemical Abrasive
Alfa Laval Pumps Ltd
Feluwa Schlesiger & Co KG
IHC Lagersmit
ITT Richter Chemie-Technik GmbH
KSB Aktiengesellschaft
Netzsch Mohnopumpen GmbH
Pumpenfabrik Ernst Vogel
Stork Pumps Ltd
Wanner International
Warren Rupp Inc.
Waukesha Bredel Fluid Handling
Wilo GmbH
Wirth Maschinen
WOMA Apparatebau GmbH

92. Coal Washing
Feluwa Schlesiger & Co KG
IHC Lagersmit
Mackley Pumps
Waukesha Bredel Fluid Handling
Wirth Maschinen

BUYER'S GUIDE

93. Concrete Handling
Warren Rupp Inc.
Wilo GmbH

94. Condensate Extraction
Autoclude
Hyundai Heavy Industries Co. Ltd
KSB Aktiengesellschaft
Mackley Pumps
Pumpenfabrik Ernst Vogel
Stork Pumps Ltd
WOMA Apparatebau GmbH

95. Cooling Water
Ahlstrom Corporation
Alfa laval Pumps Ltd
Hyundai Heavy Industries Co. Ltd
IHC Lagersmit
KSB Aktiengesellschaft
Mackley Pumps
Pumpenfabrik Ernst Vogel
Stork Pumps Ltd
Warren Rupp Inc.
Wilo GmbH
WOMA Apparatebau GmbH

97. Descaling
Hyundai Heavy Industries Co. Ltd
Mackley Pumps
Stork Pumps Ltd
Warren Rupp Inc.
WOMA Apparatebau GmbH
Wilo GmbH

98. Dredging
Hyundai Heavy Industries Co. Ltd
IHC Lagersmit
KSB Aktiengesellschaft
Warren Rupp Inc.

99. Dry and Floating Docks
Warren Rupp Inc.

100. Fire (Certified) Portable
KSB Aktiengesellschaft

101. Fire (Certified) Stationary
Hyundai Heavy Industries Co. Ltd
IHC Lagersmit
KSB Aktiengesellschaft
Pumpenfabrik Ernst Vogel

102. Fish Pulping
Warren Rupp Inc.

103. Flue Gas Desulphurisation
KSB Aktiengesellschaft
Wanner International
Waukesha Bredel Fluid Handling

104. Flue Gas Desulphurisation Pumps (FGD)
Feluwa Schlesiger & Co KG
ITT Richter Chemie-Technik GmbH
KSB Aktiengesellschaft

105. Foodstuffs and Drink
Alfa Laval Pumps Ltd
KSB Aktiengesellschaft
Stork Pumps Ltd
Sunstrand Fluid Handling
Wanner International
Warren Rupp Inc.
Waukesha Bredel Fluid Handling
Wilo GmbH

106. Garden Fountain
KSB Aktiengesellschaft
Pumpenfabrik Ernst Vogel

107. Glue
Albany Engineering Co. Ltd
Alfa Laval Pumps Ltd
Feluwa Schlesiger & Co KG
Stork Pumps Ltd
Wanner International
Warren Rupp Inc.
Waukesha Bredel Fluid Handling
Wilo GmbH

108. Gravel/Sand
Feluwa Schlesiger & Co KG
IHC Lagersmit
KSB Aktiengesellschaft
Warren Rupp Inc.
Waukesha Bredel Fluid Handling

109. Fuel Oil (Heavy)
Albany Engineering Co. Ltd
Alpha Laval Pumps Ltd
KSB Aktiengesellschaft
Netzsch Mohnopumpen GmbH
Pumpenfabrik Ernst Vogel
Stork Pumps Ltd
Wanner International
Warren Rupp Inc.
Wilo GmbH
WOMA Apparatebau GmbH

110. Fuel Oil (Light)
Albany Engineering Co. Ltd
Alpha Laval Pumps Ltd
KSB Aktiengesellschaft
Mackley Pumps
Netzsch Mohnopumpen GmbH
Pumpenfabrik Ernst Vogel
Stork Pumps Ltd
Wanner International
Warren Rupp Inc.
Wilo GmbH
WOMA Apparatebau GmbH

111. Grease/Lubricating Oil
Albany Engineering Co. Ltd
Alpha Laval Pumps Ltd
Pumpenfabrik Ernst Vogel
Stork Pumps Ltd
Warren Rupp Inc.
Wilo GmbH

112. Heating
KSB Aktiengesellschaft
Pumpenfabrik Ernst Vogel

113. High Pressure Pumps
Brook Hansen
Feluwa Schlesiger & Co KG
Hyundai Heavy Industries Co. Ltd
IHC Lagersmit
KSB Akteingesellschaft
Pumpenfabrik Ernst Vogel
Sunstrand Fluid Handling
Wanner International
Warren Rupp Inc.
WOMA Apparatebau GmbH

114. Hydraulic System
Hyundai heavy Industries Co. Ltd
Pumpenfabrik Ernst Vogel
Warren Rupp Inc.

115. Hydro-Pneumatic Water Booster
Hyundai Heavey Industries Co. Ltd
Pumpenfabrik Ernst Vogel

116. Irrigation (Intake)
Ahlstrom Corporation
Hyundai Heavy Industries Co. Ltd
IHC Lagersmit
KSB Aktiengesellschaft
Pumpenfabrik Ernst Vogel
Warren Rupp Inc.
Wilo GmbH

117. Irrigation (Spray)
KSB Aktiengesellschaft
Mackley Pumps
Pumpenfabrik Ernst Vogel
Wanner International
Warren Rupp Inc.

118. Laboratory
Alpha Laval Pumps ltd
Ansimag, Inc.
Autoclude
Netzsch Mohnopumpen GmbH
Wallace & Tiernan Ltd

BUYER'S GUIDE 407

Warren Rupp Inc.
Waukesha Bredel Fluid Handling
Wilo GmbH

119. Land Drainage
Hyundai Heavy Industries Co. Ltd
IHC Lagersmit
KSB Aktiengesellschaft
Pumpenfabrik Ernst Vogel
Warren Rupp Inc.
Wilo GmbH

121. Liquor Pumps
Alpha Laval Pumps Ltd
Autoclude
Mackley Pumps
Netzsch Mohnopumpen GmbH
Pumpenfabrik Ernst Vogel
Warren Rupp Inc.
Wilo GmbH

122. Machine Tool Lubricating (Suds)
Albany Engineering Co. Ltd
Wanner International
Warren Rupp Inc.
Wilo GmbH

123. Marine (Bilge)
Hyundai Heavy Industries Co. Ltd
IHC Lagersmit
Mackley Pumps
Netzsch Mohnopumpen GmbH
Wanner International
Warren Rupp Inc.
Wilo GmbH

124. Marine (Sewage)
Alfa Laval Pumps Ltd
Feluwa Schlesiger & Co KG
IHC Lagersmit
Netzsch Mohnopumpen GmbH
Warren Rupp Inc.
Wilo GmbH

125. Marine (Ballast)
Hyundai Heavy Industries Co. Ltd
IHC Lagersmit
Mackley Pumps
Warren Rupp Inc.

126. Marine (Boiler Feed)
Feluwa Schlesiger & Co KG
IHC Lagersmit
Mackley Pumps
Wanner International

127. Marine (Fire)
IHC Lagersmit
Mackley Pumps

128. Mine Draining and Dewatering
Feluwa Schlesiger & Co KG
Hyundai Heavy Industries Co. Ltd
IHC Lagersmit
KSB Aktiengesellschaft
Mackley Pumps
Netzsch Mohnopumpen GmbH
Pumpenfabrik Ernst Vogel
Warren Rupp Inc.
Waukesha Bredel Fluid Handling
Wilo GmbH
Wirth Maschinen

129. Mine Tailings
Ahlstrom Corporation
Feluwa Schlesiger & Co KG
KSB Aktiengesellschaft

130. Molasses
Ahlstrom Corporation
Albany Engineering Co. Ltd
Alfa Laval Pumps Ltd
Autoclude
Netzsch Mohnopumpen GmbH
Pumpenfabrik Ernst Vogel
Stork Pumps Ltd
Wanner International

Warren Rupp Inc.
Waukesha Bredel Fluid Handling
Wilo GmbH
Wirth Maschinen

131. Mud
Ahlstrom Corporation
Feluwa Schlesiger & Co KG
IHC Lagersmit
Netzsch Mohnopumpen GmbH
Wanner International
Warren Rupp Inc.
Waukesha Bredel Fluid Handling
Wilo GmbH
Wirth Maschinen

132. Oil Burner/Fuel Injection
Albany Engineering Co. Ltd
Stork Pumps Ltd
Wanner International

133. Oil Extraction
KSB Aktiengesellschaft
Netzsch Mohnopumpen GmbH

134. Oil Pipeline
KSB Aktiengesellschaft
Netzsch Mohnopumpen GmbH
Stork Pumps Ltd
Wilo GmbH
WOMA Apparatebau GmbH

135. Oil Transfer
Albany Engineering Co. Ltd
Hyundai Heavy Industries Co. Ltd
KSB Aktiengesellschaft
Netzsch Mohnopumpen GmbH
Sunstrand Fluid Handling
Warren Rupp Inc.
Wilo GmbH
WOMA Apparatebau GmbH

136. Paper Stock
Ahlstrom Corporation

Alfa Laval Pumps Ltd
Netzsch Mohnopumpen GmbH
Pumpenfabrik Ernst Vogel
Wilo GmbH

137. Petrol/Light Fuel/Solvents
Albany Engineering Co. Ltd
Alfa Laval Pumps Ltd
Hyundai Heavy Industries Co. Ltd
ITT Richter Chemie-Technik GmbH
Mackley Pumps
Netzsch Mohnopumpen GmbH
Pumpenfabrik Ernst Vogel
Stork Pumps Ltd
Sunstrand Fluid Handling
Wanner International
Warren Rupp Inc.
Waukesha Bredel Fluid Handling
Wilo GmbH
WOMA Apparatebau GmbH

138. Portable Pumps
Albany Engineering Co. Ltd
Alfa Laval Pumps Ltd
KSB Aktiengesellschaft
Pumpenfabrik Ernst Vogel
Warren Rupp Inc.
Grundfos A/S
Wilo GmbH

139. Pulp & Paper
Ahlstrom Corporation
Alpha Laval Pumps Ltd
Feluwa Schlesiger & Co KG
IHC Lagersmit
ITT Richter Chemie-Technik GmbH
KSB Aktiengesellschaft
Netzsch Mohnopumpen GmbH
Sunstrand Fluid Handling
Wanner International
Warren Rupp Inc.
Waukesha Bredel Fluid Handling
Wilo GmbH

BUYER'S GUIDE

140. Printer's Ink
Albany Engineering Co. Ltd
Albany Engineering Co. Ltd
Alpha Laval Pumps Ltd
Autoclude
Netzsch Mohnopumpen GmbH
Stork Pumps Ltd
Wanner International
Warren Rupp Inc.
Waukesha Bredel Fluid Handling
Wilo GmbH

141. Radioactive Liquid
KSB Aktiengesellschaft
Wanner International
Warren Rupp Inc.
Waukesha Bredel Fluid Handling
Wilo GmbH

142. Sewage (Raw)
Albany Engineering Co. Ltd
Alpha Laval Pumps Ltd
Feluwa Schlesiger & Co KG
IHC Lagersmit
KSB Aktiengesellschaft
Netzsch Mohnopumpen GmbH
Pumpenfabrik Ernst Vogel
SEW Eurodrive
Warren Rupp Inc.
Waukesha Bredel Fluid Handling
Wilo GmbH

143. Sewage (Sludge)
Alpha Laval Pumps Ltd
Feluwa Schlesiger & Co KG
IHC Lagersmit
Netzsch Mohnopumpen GmbH
Pumpenfabrik Ernst Vogel
SEW Eurodrive
Warren Rupp Inc.
Waukesha Bredel Fluid Handling
Wilo GmbH
Wirth Maschinen

144. Sewage (Treated)
Ahlstrom Corporation
Albany Engineering Co. Ltd
Alpha Laval Pumps Ltd
Autoclude
Feluwa Schlesiger & Co KG
IHC Lagersmit
KSB Aktiengesellschaft
Mackley Pumps
Netzsch Mohnopumpen GmbH
Pumpenfabrik Ernst Vogel
SEW Eurodrive
Warren Rupp Inc.
Waukesha Bredel Fluid Handling
Wilo GmbH

145. Sewage (Effluent)
Ahlstrom Corporation
Albany Engineering Co. Ltd
Alpha Laval Pumps Ltd
ITT Richter Chemie-Technik GmbH
Feluwa Schlesiger & Co KG
KSB Aktiengesellschaft
Mackley Pumps
Pumpenfabrik Ernst Vogel
Warren Rupp Inc.
Waukesha Bredel Fluid Handling
Wilo GmbH

146. Sewage (Packaged Sets)
Alpha Laval Pumps Ltd
KSB Aktiengesellschaft
Netzsch Mohnopumpen GmbH
Pumpenfabrik Ernst Vogel

147. Storm Water/Flood Control
Hyundai Heavy Industries Co. Ltd
KSB Aktiengesellschaft
Mackley Pumps

148. Shower Booster
Pumpenfabrik Ernst Vogel

149. Sinking/Dewatering Control
Wanner International

150. Sludge
Ahlstrom Corporation
Alpha Laval Pumps Ltd
Feluwa Schlesiger & Co KG
IHC Lagersmit
Netzsch Mohnopumpen GmbH
Pumpenfabrik Ernst Vogel
Warren Rupp Inc.
Waukesha Bredel Fluid Handling
Wilo GmbH
Wirth Maschinen

151. Slurry
Ahlstrom Corporation
Alpha Laval Pumps Ltd
Autoclude
Feluwa Schlesiger & Co KG
IHC Lagersmit
KSB Aktiengesellschaft
Mackley Pumps
Netzsch Mohnopumpen GmbH
Pumpenfabrik Ernst Vogel
Wallace & Tiernan Ltd
Wanner International
Warren Rupp Inc.
Waukesha Bredel Fluid Handling
Wilo GmbH
Wirth Maschinen
WOMA Apparatebau GmbH

152. Solids Handling
Alpha Laval Pumps Ltd
Autoclude
Feluwa Schlesiger & Co KG
IHC Lagersmit
KSB Aktiengesellschaft
Mackley Pumps
Netzsch Mohnopumpen GmbH
SEW Eurodrive
Warren Rupp Inc.

Waukesha Bredel Fluid Handling
Wilo GmbH
Wirth Maschinen

153. Sugar Beet
Ahlstrom Corporation
Alpha Laval Pumps Ltd
IHC Lagersmit
KSB Aktiengesellschaft
Netzsch Mohnopumpen GmbH
Pumpenfabrik Ernst Vogel
Wallace & Tiernan Ltd
Wanner International
Warren Rupp Inc.
Wilo GmbH

155. Tar & Liquor
Ahlstrom Corporation
Albany Engineering Co. Ltd
Alpha Laval Pumps Ltd
Feluwa Schlesiger & Co KG
Pumpenfabrik Ernst Vogel
Stork Pumps Ltd
Warren Rupp Inc.
Wilo GmbH
WOMA Apparatebau GmbH

156. Utilities (Paper Generation etc)
Alpha Laval Pumps Ltd
KSB Aktiengesellschaft
Warren Rupp Inc.
Wilo GmbH

157. Viscous
Albany Engineering Co. Ltd
Alpha Laval Pumps Ltd
Feluwa Schlesiger & Co KG
Netzsch Mohnopumpen GmbH
Pumpenfabrik Ernst Vogel
Stork Pumps Ltd
Wanner International
Warren Rupp Inc.
Waukesha Bredel Fluid Handling
Wilo GmbH

BUYER'S GUIDE 411

158. Water (Jetting Pumps)
IHC Lagersmit
Wanner International
WOMA Apparatebau GmbH

159. Water (Raw)
Ahlstrom Corporation
Hyundai Heavy Industries Co. Ltd
IHC Lagersmit
KSB Aktiengesellschaft
Mackley Pumps
Pumpenfabrik Ernst Vogel
Stork Pumps Ltd
Warren Rupp Inc.
Waukesha Bredel Fluid Handling
Wilo GmbH
WOMA Apparatebau GmbH

160. Water (Potable)
Grundfos A/S
Hyundai Heavy Industries Co. Ltd
IHC Lagersmit
KSB Aktiengesellschaft
Mackley Pumps
Pumpenfabrik Ernst Vogel
Stork Pumps Ltd
Warren Rupp Inc.
Wilo GmbH
WOMA Apparatebau GmbH

161. Water (Automatic)
Grundfos S/A
KSB Aktiengesellschaft
Mackley Pumps
Pumpenfabrik Ernst Vogel
Stork Pumps Ltd

162. Water (Domestic)
Grundfos A/S
IHC Lagersmit
KSB Aktiengesellschaft
Mackley Pumps

Pumpenfabrik Ernst Vogel
Stork Pumps Ltd
Warren Rupp Inc.
Wilo GmbH

163. Water (Packaged Sets)
KSB Aktiengesellschaft
Mackley Pumps
Pumpenfabrik Ernst Vogel
Wallace & Tiernan Ltd

164. Wine & Distillers
Alpha Laval Pumps Ltd
Autoclude
Netzsch Mohnopumpen GmbH
Stork Pumps Ltd
Sunstrand Fluid Handling
Wanner International
Warren Rupp Inc.
Waukesha Bredel Fluid Handling
Wilo GmbH

165. Variable Speed Drives
Alfa Laval Pumps Ltd
Brook Hansen

166. Aseptic
Alfa Laval Pumps Ltd

167. Pharmaceuticals
Alfa Laval Pumps Ltd

168. High Purity Media
ITT Richter Chemie-Technik GmbH

169. Self-Priming
ITT Richter Chemie-Technik GmbH

170. Bushings & Wear Rings
Graphite Metallizing Corporation

TRADE NAMES INDEX

AHLSTAR – Stock/Process pump – Ahlstrom Corporation
ALBANY – Rotary gear and centrifugal pumps – The Albany Engineering Co. Ltd
ALPHA LAVAL – Gear pumps – Alfa Laval Pumps Ltd
ANSIMAG – Heavy-duty, Mag-drive, Seal-less, Non-metallic, Centrifugal Pumps (ANSI & ISO) – Ansimag, Inc.
AP range – High capacity rotary lobe pumps s/s – Alfa Laval Pumps Ltd
ARGUS 55 – High specification hostile environment motor – Brook Hansen
BREDEL – Peristaltic hose pump – Waukesha Bredel Fluid Handling B.V.
BRITCO – Mechanical Seals – James Walker Britco Ltd
CHEM-FEED – Diaphragm type metering injector – Blue-White Industries
COMBISYSTEM – Modular DIN standard centrifugal pump – Stork Pumps Ltd
CROWN – Rotary gear and centrifugal pumps – The Albany Engineering Co Ltd
DEPA – Air operated diaphragm pumps – Alfa Laval Pumps Ltd
DIGI-FLO – Electronic flowmeters – Blue-White Industries

DRM range – Rotary lobe sewage sludge pumps – Alfa Laval Pumps Ltd
ENCORE – Mechanical diaphragm pumps – Wallace & Tiernan Ltd
ESADS+Plus – Externally serviceable Air distribution system – Warren Rupp, Inc.
FLEXFLO – Peristaltic metering injector – Blue-White Industries
FRE-FLOW – Self priming centrifugal pump – Stork Pumps Ltd
FUMEX – High temperature smoke extraction motors – Brook Hansen
GRAPHALLOY – Self lubricating non galling bushings – Graphite Metallizing Corporation
GRAPHILON – Corrosion resistant bushings and seals – Graphite Metallizing Corporation
GRAPHALLAST Abrasion resistant bushings – Graphite Metallizing Corporation
GP range – High capacity rotary lobe pumps – sludge – Alfa Laval Pumps Ltd
HYCLEAN LOBE – Hygienic rotary lobe pump – Stork Pumps Ltd
HYDUTY LOBE – High pressure rotary lobe pump – Stork Pumps Ltd
HYDRA-CELL – Positive displacement seal-less pump – Wanner International Ltd

IBEX Pumps – Rotary lobe pumps – stainless steel – Alfa Laval Pumps Ltd
IP range – Peristaltic pumps – Alfa Laval Pumps Ltd
KGE – Self priming centrifugal pump – Stork Pumps Ltd
KONTRO – Magnet drive centrifugal pumps – Sunstrand Fluid Handling
LIQUIDYNE – Suitable for shaft sealing of all 'water bases', centrifugal pumping pumps and replaces shaft seals as mechanical seals – IHC Lagersmit
MCH – High pressure rotary lobe pump – Stork Pumps Ltd
MC Pump – Medium Consistency (8-18% pump – Ahlstrom Corporation
MCV – Vertical multistage centrifugal pump – Stork Pumps Ltd
METASTREAM – Non lubricated shaft couplings – Flexibox Ltd
MULTI-MOUNT – Motor giving adaptable mounting positions – Brook Hansen
NEMO-Pumps – Progressing Cavity Pumps – NETZSCH Mohnopumpen GmbH
PHL – Horizontal API 610 refinery pump – Stork Pumps Ltd
PVML – Vertical API 610 refinery pump – Stork Pumps Ltd
Phoenix – High pressure diaphragm pumps – Warren Rupp, Inc.
PortaPump – Electric Submersible Utility Pump – Warren Rupp, Inc.
PoweRupp – Motor-driven diaphragm pumps – Warren Rupp, Inc.
PREMIER 75 – Solenoid operated diaphragm pumps – Wallace & Tiernan Ltd
RuppTech – Pump controls – Warren Rupp, Inc.
SAFEGLIDE – Dry-run optimized SiC-plain bearings – UTT Richter Chemie-Technik GmbH

SandPIPER – Air-operated double diaphragm pumps – Warren Rupp, Inc.
SCMP – Centrifugal canned motor pumps – Sunstrand Fluid Handling
SHP – High volume peristaltic pump – Stork Pumps Ltd
SINE – Sinusoidal rotor food grade pumps – Sunstrand Fluid Handling
SK range – Ultra Hygienic Rotary lobe pumps – Alfa Laval Pumps Ltd
SRT – Gear pump for viscous liquids – Stork Pumps Ltd
SUNDYNE – High speed API centrifugal pumps – Sunstrand Fluid Handling
SUNFLO – High speed centrifugal industrial pumps – Sunstrand Fluid Handling
S3 – Geared motors – strong, silent, standard – Brook Hansen
SAFEGLIDE – Dry-run optimized SIC-plain bearings – ITT Richter Chemie-Technik GmbH
SludgeMaster – Air-driven Submersible Utility Pump – Warren Rupp, Inc.
SSP Pumps – Rotary lobe pumps – stainless steel/ductile iron – Alfa Laval Pumps Ltd
STANCOR – Chemically resistant centrifugal process pump – Wanner International Ltd
STANHOPE – Rotary gear and centrifugal pumps – The Albany Engineering Co Ltd
Tranquilizer – Flow surge supressors – Warren Rupp, Inc.
'W' Range – High efficiency motors – Brook Hansen
WAUKESHA – Cemtrifugal pump/Corrosion resistant rotary pumps – Waukesha Bredel Fluid Handling B.V.
WILO – Heating Circulator – Wilo GmBH

ADVERTISERS' NAMES, ADDRESSES AND CONTACT NUMBERS

Ahlstrom Corporation, P.O. Box 18, FIN 48601 Karhula, Finland
Tel: +35 8 52 2291111 Fax: +35 8 52 63958

Albany Engineering Limited,
Church Road, Lydney, Gloucester, GL15 5ER, United Kingdom
Tel: 01594 842275 Fax: 01594 842574

Alfa Laval Pumps Ltd, Birch Road, Eastbourne, BN23 6PQ, United Kingdom
Tel: 01323 412555 Fax: 01323 412515

Ansimag, Inc., 1090 Pratt Boulevard, Elk Grove Village, Illinois 60007, USA
Tel: +1 708 290 0482 Fax: +1 708 290 0481

Autoclude, Victor Pyrate Works, Arisdale Avenue, South Ockendon, Essex, RM15 5DP, United Kingdom
Tel: 01708 856125 Fax: 01708 857366

Blue White Industries, 14931 Chestnut St., Westminster, CA 92683, USA
Tel: +1 714 893 8529 Fax: +1 714 894 9492

Brook Hansen, St Thomas Road, Huddersfield, HD1 3LJ, United Kingdom
Tel: 01484 422150 Fax: 01484 519661

Chemical Equipment, PO Box 650, Morris Plains, NJ 07950-0650, USA
Fax: +1 201 898 9281

Ernst Vogel, A-2000 Stockerau, Ernst-Vogel-Straße 2, Austria
Tel: 02266 604 Fax: 02266 6531

Feluwa Schlesiger & Co KG, Beulertweg, Muerlenbach/Eifel 54570, Germany
Tel: +49 6594/10-0 Fax: +49 6594/16 40

Flexibox Ltd, Nash Road, Trafford Park, Manchester, M17 1SS, United Kingdom
Tel: +44 (0)161 827 2484 Fax: +44 (0)161 827 1654

Gorman-Rupp Co, 305 Bowman Street, P.O. Box 1217, Mansfield, OH 44901-1217, USA
Tel: +1 419 755 1338 Fax: +1 419 755 1404

Graphite Metallizing Corporation, PO Box 110, Yonkers, NY 10702, USA
Tel: +1 914 968 8400 Fax: +1 914 968 8468

Grindex AB, Box 538, S-13625 Hanige, Sweden
Tel: +46 8 6066 29 Fax: +46 8 7455 328

Grundfos A/S, Poul de Jensens Vej, 8850 Bjerringbro, Denmark
Tel: 45 86681400 Fax: 45 86684245

Hyundai Heavy Industries Co., (Engine & Machinery Division), 1 Cheona-Dong, Dong-Ku, Ulsan 682 060, South Korea
Tel: 82 5223 02114 Fax: 82 5223 07694

Intereco (CVB), 10042 Nickelino (70), Via X1 Febbraio 11, Italy
Tel: +39 11 6054 297 Fax: +39 11 6054 244

ITT Richter Chemie-Technik GmbH, Otto-Schott-Strasse 2, D-47906 Kempen, D-47906, Germany
Tel: +49 02152 146-0 Fax: +49 02152 146-190

James Walker Britco Ltd, Dore House Industrial Estate, Orgreave Crescent, Handsworth, Sheffield, S13 9NQ, UK
Tel: 01142 690776 Fax: 01142 540455

KSB a.g., Johann-Klein-Strabe, Frankenthal 67227, Germany
Tel: +49 6233 860 Fax: +49 6233 863401

Netzsch Mohnopumpen GmbH, Liebigstrabe 28, Postfach 1120, D-84464 Waldkraiburg 84464, Germany
Tel: +49 8638 63-0 Fax: +49 67 981 + 67 999

Pac-Seal International, 211 Frontage Road, Burr Ridge, Illinois 60521, USA
Tel: +1 708 986 0430 Fax: +1 708 986 1033

SEW Eurodrive, Beckbridge Industrial Estate, Normanton, Yorkshire WF6 1QR, UK
Tel: 01924 893855 Fax: 01924 893702

Stork Pumps, Meadow Brook Industrial Centre, Maxwell Way, Crawley, East Sussex RH10 2SA, UK
Tel: 01293 553495 Fax: 01293 524635

Sunstrand Fluid Handling, 14845 W, 69th Avenue, Arvada, Colorado 80007, USA
Tel: +1 303 425 0800 Fax: +1 303 425 0896

Wallace & Tiernan Ltd, Priory Works, Tonbridge, Kent, TN11 OQL, UK
Tel: 01732 771717 Fax: 01732 771800

Wanner International, Grange Court, Grange Road, Tongham, Surrey GU10 1DW, UK
Tel: 01252 781234 Fax: 01252 781235

Warren Rupp, Inc., A unit of IDEX Corporation, P.O. Box 1568, 800 North Main Street, Mansfield, OH 44901-1568, USA
Tel: +1 419 524 8388 Fax: +1 419 522 7867

Waukesha Bredel BV, P.O. Box 47, 7490 AA Delden, Holland
Tel: +31 5407 70000 Fax: +31 5407 61175

Wilo GmbH, Nortkirchenstr. 100, 44263 Dortmund-Horde, Germany
Tel: +49 0231 41020 Fax: +49 0231 4102363

Wirth, Maschinen-un Bohregerate-Fabrick GmbH, PO Box 1327 D-41803 Erkelenz, Germanny
Tel: 49 2431 830 Fax: 49 2431 83267

Woma Apparatebau GmbH, Werthauser Str. 77-79, D-47226 Duisberg, Germany
Tel: +49 2065 304-0 Fax: +49 2065 304-200

EDITORIAL INDEX

A

Abrasive wear resistance, 177, 241
Absolute pressure, definition, 2
Absolute viscosity, 8
AC drives, adjustable voltage, 234
AC motors, 223-6
Acceleration head, 163-5
Actual vs reference curve speed, 58
Adjustable voltage a.c. drives, 234
Affinity laws, 54-57
Ahlstrom paper pumps, 75-6
Air lift pump, 32
Airport jet fuel transfer, 298
Alignment, 196-200
Angular alignment, 196-200
Anticorrosion agent injection, 307
API (American Petroleum Institute Standards), 295
Aseptic operation, 151
Aseptic operations, 127-8
Atmospheric pressure
    definition, 1
    effects, 27
Austenitic steels, 173
Automobile water pumps, 239
Axial flow propeller pumps, 272
Axial flow pumps, 61, 250
    slurries, 333
Axial piston pumps, 135, 139-40

B

Backing-off, 58
Back-vanes, 79
Balance holes, 79
Balancing, 213-22
    effects of unbalance, 214
    quality classification (G grades), 215-20
    requirements, 214-15
    rotor assembly, 222
    single plane, 220
    two plane, 222
    types of balance, 220-2
Barrel pumps, 243, 244, 269-70
Barrier pump design, 126
Bearing frame, 84
Bearing housing, 84
Bearing isolator, 92
Bearing protectors, 91
Bearings, 84-8
    lubrication of, 91-4
    water lubricated rubber, 91
Bellows seals, 186
Bends, secondary flow in, 19
BEP (best efficiency point), 46,95
Bernoulli equation, 33
Bernoulli theorem, 5-8
Beverages
    centrifugal pump, 293
    metering system, 288
Boiler feed pumps, 243-6
Booster pumps, 128
Breakdown torque, 225
Brewery industry, 293
Bronze, 172

C

Canned motor pumps, 41, 120-6, 265
Capacity, definition, 137, 163
Carbon steel, 172
Cartridge seals, 186-90
Casing wear ring, 75-6
Casings, 71-4
Cast-iron, 171
Cavitation, 97
   definition, 3
   incipient, definition, 4
Cavitation erosion
   definition, 4
   resistance, 177
Cavitation inception, 95-6
Cavitation resistance, 178
Central heating circulating pump, 249
Centrifugal pumps, 23, 290
   aseptic operations, 127-8
   beverages, 293
   characteristic curves, 26, 39-69
   combined flow diagram for residential well water supply, 33
   configurations and types, 111-31
   diffuser type, 25
   end suction configuration, 111
   in-line configurations, 111
   intake design, 102
   nomenclature, 39
   operation at off-design conditions, 94-100
   parallel - series operation, 100-102
   pulp and paper industry, 311
   relative pressures at entrance, 93
   reversible, 32
   rotation, 39
   rotor critical speed in, 210
   slurry applications, 325
   specific speeds, 37
   submersible type configuration, 111
   turbine-type vertical, 111
   volute type, 25
   waste water/sewage pumps, 337, 347
   see also under special designations
Ceramic membrane filtration package system, 292
Certified tests, 192
Charge pumps for hydrocarbon tower, 308
Chemical environments, 176
Chemical pumps, 125, 178, 263-81
   12 stage 20,000 ft head, 281
   centerline supported, 264
   cooling, 125-6
   materials, 264
   standards, 264-5
Chezy-Manning formula, 11
Chilled water circulation, 246
Chipping, 58
Chopper pumps, 128, 129, 273, 276
   handling corn husks, 345
Chromium steels, 172
Circular (or concentric) casing, 74
Circulating pumps, 242
Circumferential piston pumps, 136, 145
   aseptic, 151
Clean in place applications, 241, 292
Closed impeller, 66
Concrete pump, 331, 332
Condensate injection into natural gas, 307
Condensate pump, 247
Condensate services, 246
Condition monitoring, 211
Containment shells, 126
Continuity equation, definition, 4
Continuity of momentum equation, 33-4
Controlled volume bellows pumps, definition, 163
Controlled volume diaphragm pumps, definition, 163
Controlled volume (metering) pumps, 169
Controlled volume piston (or plunger) pumps, definition, 156
Controlled volume pumps, definition, 156
Cooling systems, gear pumps, 147
Cooling tower water circulation, 246
Corrosion, 171-81
Corrosion rate, 176
Corrosion resistance, 174
Corrosive applications, 171
Couplings, 228-32
   comparison chart, 231
Cover wear rings, 76
Cream pasteurization system, 291
Crevice corrosion, 174
Crevice corrosion temperature, 176
Critical corrosion temperature, 176
Cutter pumps, 128

EDITORIAL INDEX

D

Darcy-Weisbach formula, 10, 11
Data acquisition systems, 193
Datum, definition, 4
D.C. excited motor, 223
D.C. motors with SCR's, 234
Density, definition, 1
Detergent blending, 286
Dial indicators, 198
Diameter modifications, 54-5
Diaphragm pumps, 161, 162, 168, 169, 270, 271, 285
    characteristic curves, 167
    metering, 168, 169, 275
    multihead, 287
    slurry, 331
    solids handling, 329
    speed characteristics, 167
    synthetic rubber process, 274
    waste water/sewage pumps, 337-8, 342-3
Diesel driven axially split pumps, waste water/sewage, 343
Diffuser casing, 74-5
Diffusers, resistance coefficients for, 13
Dirty water system, 329-31
Disc coupling, 230
Discharge head, 34
Discharge piping, 165
Discharge recirculation, 96-7
Displacement, definition, 137
Dosing pumps, 167, 169, 272, 277
    definition, 156
Double acting pumps, 155
Double suction horizontal split-case pump, 82
Double suction impeller, 66
Double suction pumps, 64
Double volute casing, 71
Dredge pumps, 325
Drum pumps, 269-70
Dry pit, definition, 102
Dry pit pumps, 250
Ductile iron, 172
Duriron's Pump Engineering Manual, 103
Dynamic seal pump, 80
Dynamic seals, 79

E

Eddy current drives, 234

Eductor, 31
Elastomeric pumps, 327
Elbows, secondary flow in, 19
Electric motors, 223-6
    enclosures, 226
    insulation systems, 226
    NEMA designs A, B, C and D, 224
    starting, 226
    torque and system curves, 225
Elevation head, definition, 3
End suction pumps, 283, 290, 292
Energy, definition, 4
Energy conservation, 20-1
Erosion-corrosion, 171, 241
Erosion resistance, 176
Erosive applications, 171
Euler head curve, 64

F

Fault finding, 201
Fibreglass, 171
Field tests, 193
Fire protection systems, 252-3
Fire pumps, 257-61
    diesel engine driven, 259
    main requirements, 257-8
    motor driven, 258
    mounting, 258
    vertical in-line, 261
Flexible coupling, 228
Flexible line pumps, 141
Flexible liner pump, 135
Flexible tube pumps, 135, 140
Flexible vane pumps, 135, 140-1
Flood control pumps, 253
Flow diagram, 33
Flow-work, definition, 4
Fluid, definition, 1
Fluid drives, 234
Fluid mechanics, principles of, 1-22
Food processing pumps, 127-8
Forced vortex flow, 20
Foundation, 195-6
Free vortex flow, 20
Friction loss
    on instruments, 12
    on pipe fittings, 12

G

Gas turbine driven pumps, 300
Gauge head, definition, 3
Gauge pressure, definition, 2
Gear coupling, 229
Gear pumps
    characteristic curve, 144
    cooling systems, 147
    external, 136, 141-4, 148
    internal, 136, 144-5, 147, 148
Gears, 228
Gradient diagram, 10
Grease lubrication, 92
Grinder pumps, 128
Grouting, 200

H

Hardness and corrosion resistance, 176
Hardness values of ores and materials, 323-4
Hazen-Williams formula, 11
Head, definition, 2
Head/capacity curves, 45
Head/capacity range, 25
Head losses, 13-17, 64-5
Head ratio, 34
Head wear rings, 76
Heating pumps, 246-8
Hermetic chemical pump, 263
Hermetic magnetic drive pump, 267
Hermetic pumps, 118-26
    canned motor, 120-6, 265
    definition, 163
    magnetic drive, 120-6
    monitoring of, 126
Hermetic rotary pumps, 147
High pressure multistage pumps, 130
High pressure plunger pumps, 273-80
Homogeneous flow, 323
Homogenizer, 290
Hose pumps, 168, 169
H/Q characteistic curve, rotary pumps, 138
H/Q characteristic, lobe pumps, 141
H/Q characteristic curves, 49-54, 57
Hydraulic Institute, 98
    Data Book, 314
    Standards, 39, 103, 134
Hydraulic radius, 11
Hydraulic ram pump, 31
Hydraulic resonance, 211
Hydrostatic testing, 191, 193
Hysteresis units, 223

I

IC engines, 227-8
Ideal head curve, 64
Impeller
    configurations, 66-71
    modifications, 57
    performance curve shapes for various specific speed design, 117
    wear rings, 75, 76
Incipient cavitation, definition, 4
Increasers, resistance coefficients for, 13
Inducers, 128, 130
Induction motors, 223, 234
Industrial type circulating pump, 248
Installation, 195-200
Instruments, friction loss on, 12
Intake, definition, 102
Internal energy, definition, 4
Irrigation ram pump, 240
Irrigation service, 250, 251
Iso-efficiency curve plots, 44

J

Jet pumps, 23, 31, 32

K

Kinematic viscosity, 8
Kinetic energy, definition, 4
Kinetic pumps, 23
    selection criteria, 26
Kw (bhp) curves, 44

L

Laminar flow, 10
Laser alignment, 198-200
Leveling, 196
Liquid, definition, 1
Lobe pumps, 141
    H/Q characteristic, 141
Losses, 64-66
Lubrication

bearings, 91-4
pumped product, 94

M

Main utility supply pumps, 251
Manning formula, 11
Mass, definition, 1
Mass ratio, 34
Materials, 171-81
    ASTM Standards designations, 173
    chemical pumps, 264
    commonly used, 171
    liquid-end combinations, 171
    petroleum production, 297
    physical properties, 171, 178
    selection criteria, 171
    water pumps, 239-40
Maximum service speed of rotation, 216
Mean time between failure (MTBF), 264
Mechanical seal pump, 80
Mechanical seals, 183-90, 261
Metering effectiveness, definition, 137
Metering pumps, 272, 286, 307, 308
    beverages, 288
    definition, 156
    waste water/sewage pumps, 344-6
Metering system, 287-8
Midpoint, 19
Minimum allowable flow due to temperature rise, 97-98
Mixed-flow pumps, 76
Model testing, 191
Multi-stage pumps, 303

N

Natural frequency, 208-10
Net inlet pressure, definition, 137
Net inlet pressure required, definition, 137
Newtonian fluids, 8-9
Non-clogging wear resistant process pump, 81
Non-metallic pumps, 127
Non-Newtonian fluids, 9, 127
Non-residential well water pumps, 241
NPSHA, 4, 25, 27-9, 64, 94, 95, 99, 103, 117, 165
NPSHR, 4, 25, 44, 64, 94, 95, 165, 192, 311
Nuclear reactors, 266

O

Oak Tree curves, 44
Oil line shipping pump, 304
Oil lubrication, 91-2
Oil pumps, 302
Open channel flow, 19
Open impeller, 66, 71
Operating head, 33
Overall efficiency, definition, 137
Overfiling, 57-58

P

Packed stuffing box, 80
Packing, 183
    failures, 190
Paper pumps. *See* Pulp and paper industry
Parallel alignment, 196-200
Performance correction chart for viscous liquids, 304, 305
Performance tests, 191, 192
Peristaltic pumps, 140, 142, 143, 268, 269
    characteristic curve, 144
Permanent magnet units, 223
Petroleum production, materials, 297
Petroleum products, 295-309
Pharmaceutical service, rotary pumps, 291
Pipe fittings, friction loss on, 12
Pipe flow, 9-21
Pipeline pumps, 301-3, 309
Pipes
    aged, 11
    non-circular, 11
    roughness, 10
Piping plans for seals, 190
Piston pumps, 168-70, 289
    definition, 156
    hydraulically driven, 329, 338
    waste water/sewage pumps, 337-8
Pitting index, 174, 175
Pitting resistance, 174
Plastics, 171, 181
    properties of, 179
    pump body blocks compared with elastomeric liners, 180
Plunger pumps, 161, 166, 170, 289
    definition, 156
Portable submersible pump, 129

Portable submersible vortex pump, 129
Positive displacement pumps, 23, 153, 169, 305
  capacity range, 27
  chemical plant, 267-9
  selection criteria, 25
Potential energy, definition, 4
Power characteristic curves, 100
Power generating plant, 244
Power input, definition, 137, 163
Power losses, 64
Power output, definition, 137, 163
Power pumps, 157-60, 165-9, 166
  definition, 156
Pressurized reactor primary circulating pump, 255
Pre-start actions, 200
Production test rigs, 193
Progressive cavity pumps, 145, 254, 273, 276, 284, 285
  pulp and paper industry, 316
  waste water/sewage pumps, 337-8
Proportioning pumps, definition, 156
Pseudo-plastic fluid, 9
Pulp and paper industry, 311-20
  black liquor, 315-16, 320
  bleaching system, 316-17
  centerline supported digester pump, 315
  centrifugal pumps, 311
  closed loop colour addition metering pump and control package, 319
  duplex steel metallurgies, 317
  green liquor, 315
  head box feed pump, 317
  Kraft process pulp mill, 314-15
  low pulse fan pump, 317
  medium consistency pumps, 311-13
  paper coating pump package, 318
  paper stock process pump, 316
  progressive cavity pumps, 316
  recycled fibre (deinking) process flow diagram, 318
  standard small process pump, 311
  stock consistency effect on flow characteristics, 313
  sulphite pulp process, 314
  veneer paper coating, 319
  white liquor, 315

Pulsometer, 31
Pump classification chart, 23
Pump displacement, definition, 163
Pump efficiency
  calculation, 56
  definition, 138
Pump User's Handbook, 103
Pump location, 195
Pump manufacturers minimum capacity limitation, 97
Pump selection
  criteria, 23-9
  rough selection criteria, 25-7
Pump testing, 191-3
Pumped product lubrication, 92
Pumping Manual, 103
Pumping station, 335-6
Pumpout vanes, 79

R

Radial flow pumps, 61
Radial piston pumps, 140, 142
Radial plunger pumps, 140
Radial thrust vs volute type, 74
Reciprocating pumps, 23, 153-70
  characteristic curve, 26, 156
  duplex, 155
  flow characteristics, 154
  number of cylinders, 155
  simplex, 155
  triplex, 155
  types, 153-70
Recirculation, 96-7
Refined hydrocarbons, 297
Regenerative turbine pumps, 35-8
  nomenclature, 37
  performance map, 36-7
  self-priming, 37
Relative roughness, 11
Reluctance units, 223
Residential centrifugal hot water circulating pump, 248
Residential submersible sump pump, 253
Residential submersible well water pump, 240
Residential well pumping systems, 239
Resistance coefficients
  for diffusers, 13

for increasers, 13
Resonance, 208, 211
Reverse dial indicator method, 200
Reversible centrifugal pump, 32
Reynolds number, 10-11, 305
Rheopectic fluid, 9
Rinsing, 241
Rotary piston pumps, 267
    chemical plant, 269
Rotary pumps, 23, 133-52, 283, 290
    capacity, 133
    characteristic curves, 26, 138
    classification chart, 138
    H/Q and power curve, 138
    nomenclature, 134
    operation, 133
    pharmaceutical service, 291
    rotation, 138
    types, 134, 138-51
    see also under specific types
Rotary vane pumps, 305
Rotating casing (Pitot) pump, 31-2
Rotating stall, 97
Rotor assembly balance, 222
Rotor critical speed in centrifugal pumps, 210

S

Sampling pumps, 241
Sanitary pumps, 127, 241, 284
Sanitary rotary gear motor drive pump, 150
Screw and wheel pump, 136
Screw pumps, 145-6
    mag drive, 151
    performance curve, 142
Seal chambers, 80-84
Sealless pumps, 118-26
Seals
    failures, 190
    mechanical. *See* Mechanical seals
    piping plans for, 190
Secondary flow
    in bends, 19
    in elbows, 19
Self-excited units, 223
Self-priming pumps, 113-14, 133, 139
Semi-open (or semi-closed) impeller, 66
Sewage, definition, 321

Sewage pumps. *See* Waste water/sewage pumps
Sewer cleaning pump, 344
Shape of H/Q characteristic curves, 49, 53
Shockless flow point, 95
Sine pumps, 146-7
Sine wave pumps
    performance curve, 150
    rotor, 149
Single acting pumps, 155
Single lobe pump, 135
Single screw pump, 136
Single suction impeller, 66
Sliding vane pump, 135
Slip, definition, 163
Sludge, definition, 321
Slurries
    definition, 321
    settling or non-settling, 321
Slurry pumps, 321-34
    classification according to solid size, 321
    elastomer lined, 325-6
    elastomeric, 327
    life, 323
    MTBR, 323
    wear, 323
Solid coupling, 228
Solids
    specific gravity of, 321
    suspended, 321-2
Solids handling, 321-34
    classification according to solid size, 321
    diaphragm pumps for, 329
    submersible pumps for, 328
Spacer couplings, 228, 230
Special effect pumps, 31-4
Specific gravity, definition, 1
Specific gravity of solids, 321
Specific speed, 61
Specific volume, definition, 1
Specific weight, definition, 1
Speed control, 233-4
Sprinkling pumps, 249
Squirrel cage induction motors, 225
Squirrel cage winding, 223-4
Stainless steels, 128, 171
Standard atmospheric pressure, definition, 2

Start up, 200-1
Steady flow energy equation, 5, 32
Steam pumps, 161, 165
    definition, 156
Steelflex coupling, 229
Step-up gear, 130
Storage, 195
Stress cracking resistance, 177
Stuffing boxes, 80
Submersible dewatering pump, 120
Submersible pumps, 111, 117-19, 129, 240, 253
    for solids handling, 328
    waste water/sewage pumps, 339-41, 347
Submersible slurry pump with liner, 119
Submersible waste water pump, 119
Suction piping, 108
Suction recirculation, 96, 97
Suction specific speed (S), 64, 97
Suction specific speed available (SA), 64
Sump pumps, 272, 274
Sumps, 102
    definition, 103
    model tests, 103
Synchronous motors, 223
System-head curve, 20, 27
System/pump interaction, 27

T

Tandem seal, 185
Tanks, 102, 103
Temperature rise, 125-6
    vs. capacity, 99
Testing, 191-3
Thixotropic fluid, 9
Three-lobe pump, 135
Three-screw pump, 136
Thrusts, 72-73
Torque pumps, 114
Torsional vibration, 210
Torus coupling, 229
Total discharge head, definition, 3
Total head, definition, 3
Total suction head, definition, 3
Trouble shooting, 201
Turbine diffuser pump, 247
Turbine pumps, 23, 259
    regenerative, 244
    see also under specific types
Turbines, 227
Two-screw pump, 136

U

Undercutting, 58
Underfiling, 58

V

Vacuum, definition, 2
Vacuum pressure, definition, 2
Vane pumps, 149
    characteristic curves, 139
    external, 135, 138-9
    internal, 138-9
Vane shapes, 69, 70
Vapour pressure, definition, 2
Variable frequency drives, 234-7
Variable speed drives, 234-7
    selection criteria, 236
V-belt drives, 228
Velocity head, definition, 3
Vertical radially split bowl turbine pumps, 114-17
Vertical turbine diffuser pumps, 116
    waste water/sewage pumps, 343
Vertical turbine pumps, 117
Vibration, 195, 203-11
    definition, 203
    frequency spectrum, 208
    limits, 205-7
    monitoring, 211
    problems, 203, 208
    severity chart, 204-5
    tolerance, 205
    torsional, 210
Viscosity
    conversion table, 306
    definition, 7-8
    effects, 305
    shear diagram, 7
Viscous liquids, 12
    performance correction chart for, 304, 305
Volumetric efficiency, definition, 138, 163
Volute casing, 71
Volute pumps, 44

EDITORIAL INDEX 425

Vortex, definition, 102
Vortex breakers, 105
Vortex classification system, 104
Vortex impeller, 66
Vortex impeller pump, 81
Vortex pumps, 114
    wastewater, 339

W

Waste water/sewage pumps, 118, 335-47
    centrifugal pumps, 337, 347
    diaphragm pumps, 337-8, 342-3
    diesel driven axially split pumps, 343
    gear-driven mixed-flow line-shaft, 336
    horizontal pumps in dry pit, 337
    metering pumps, 344-6
    piston pumps, 337-8
    progressing cavity pumps, 337-8
    submersible pumps, 339-41, 347
    vertical turbine type diffuser pumps, 343
Water eductor, 32
Water injection pumps, 299, 300
Water pumps, 239-55
    circulation duty, 241
    large utility circulating, 239
    materials, 239-40
Waterflood pumps, 299, 300
Wear mechanisms, 323
Wear plates, 75, 76, 79
Wear rings, 75
Weight, definition, 1
Wet pit, definition, 102
Wet pit pumps, 252
Wet well, definition, 102
Wobble plate rotary piston pump, 280
Wound rotor motors, 225, 234

INDEX TO ADVERTISERS

Ahlstrom Corp.	Facing page 311
Albany Engineering	Page 59 and Facing page 135
Alfa Laval	Facing page 133
Ansimag	Facing page 262
Autoclude	Page 170
Beresford Pumps	Page 77
Blue White Industries	Facing page 155
Brook Hansen	Facing page 223
CAT	Page 160
Centrifugal Pumps & Allied Machinery	Page 78
Chemical Equipment	Page 50
Durametallic Eurtope	Page 182
Durco Europe	Bookmark
Edwards & Jones	Page 320
Elsevier Advanced Technology	Page 90
Engineering Designer	Facing page 399
Europump Terminology	Page viii and Page 42
Ernst Vogel	Facing page 2
ESK	Facing page viii
Feluwa	Facing page 323
Flex-a-Seal	Page 52
Flexibox	Page 182
Graphite Metallizing	Page 43
Grindex	Facing page 113
Grundfos	Facing page 239
Hydraulic Institute	Facing page 310
Hyundai Heavy Industries	Page 51
IAHR	Facing page 349
IHC Lagersmit	Facing page 185
Intereco (CVB)	Page 159

ITT Richter	Facing page 263
Kestner	Page 60
Kinder Janes	Page 59
KSB	Facing page 2
Leakfree Pumps	Page 110
Mackley Pumps	Facing page 112
Netzsch Mohnopumpen	Facing page 134
Pac-Seal	Facing page 184
Pumping Manual	Page 67 and Facing Page 241
Pumps Profile	Page 68
Sealco	Facing page 184
SEW Eurodrive	Facing page 224
Speck Pumps	Facing page 154
Stork Pumps	Facing page 112 and Facing page 134
Sunstrand Fluid Handling	Facing page ix
Tapflo	Page 159
Twaki	Page 60
Valves, Pipes & Pipelines	Page 89
Vaughan	Facing page 3
James Walker	Facing page 185
Wallace & Tiernan	Facing page 155
Warren Rupp	Between pages 154 & 155
Wanner International	Page 43
Waukesha Bredel	Facing page 323
Wilo GmbH	Facing page 240
Wirth	Facing page 322
Woma	Facing page 154
World Pumps	Facing page 225